Solutions to Equipment Failures

P.F. Timmins

ASM INTERNATIONAL®
The Materials Information Society

First printing, February 1999

Library of Congress Cataloging-in-Publication Data
Timmins, P.F.

Solutions to equipment failures / P.F. Timmins.
Includes bibliographical references and index.
1. System failures (Engineering). 2. Fracture mechanics.
3. Corrosion and anti-corrosives. 4. Welded joints.
I. Title
TA169.5.T56 1998 620'.00452-dc21 98-29456

ISBN: 0-87170-662-8
SAN: 204-7586
ASM International®
Materials Park, OH 44073-0002

Printed in the United States of America

About the Author

Peter Timmins, P.Eng., is a Metallurgical Engineer, a member of APEGBC, PEO, APEGGA, and APEGS (all in Canada) and is a graduate of the Universities of Surrey, Bradford, and Sheffield in Britain. He is a registered Chartered Engineer and a Chartered Physicist in Britain and is registered in Paris as a European Engineer. In the United States, he is active in ASM International, NACE International, ASME, and API. He has authored more than 60 publications, including three books on fracture and process control, along with two software programs, additional papers on composite materials, and three patents on brazing alloys. Timmins has 25 years graduate experience in steelmaking, sour gas, petrochemical, nuclear, and refinery engineering. His experience internationally in nuclear and refinery engineering includes work in Canada, the United States, the United Kingdom, the Middle East, Japan, Korea, South Africa, the Philippines, and Australia. Since January 1995, Timmins has been president of Risk Based Inspection Inc. and Associated Companies.

Antipov had been pestering the repair shops about the quality of the replacements for the tracks. The steel was not sufficiently tensile, the rails failed the test for strains and Antipov thought that they would crack in frosty weather. The management paid no attention to his complaints.

Doctor Zhivago - Boris Pasternak

ASM International Technical Books Committee (1998–1999)

Contents

Preface

Twenty-six years ago I published my first paper on fracture, "The Relative Importance of Crack Propagation and Crack Nucleation in the Ductile Fracture of FeCo-2%V," in *Nature_Physical Science* (Vol 235, p113–114, 7 Feb 1972). Shortly after, I wrote my first public work on fracture mechanics, "The Application of the Theory of Fracture Mechanics to the Failure Behavior of Springs and Spring Materials" (currently available as Report #254 from the Institute of Spring Technology in Sheffield, England, or at *http://www.istec.demon.co.uk/res_index.htm*). Subsequent to this and other research-oriented work, I have written several hundred service failure reports for various companies in different sectors of industry. Up until quite recently, one of my objectives was to write a book on service failure analysis. However, as a consequence of the background reading I did in this field, it became abundantly clear to me that yet another book on service failure analysis would be redundant in light of the excellent texts currently available.

I turned my attention to "analyzing the analyses," in the hope that this approach would yield a direction for providing some means to solve the problems which give rise to equipment failures, to create a proactive approach, rather than simply performing reactive service failure analysis. For example, I discovered that failure rates in the aviation sector were decreasing, whereas in the hydrocarbon processing sector, failure rates were increasing. I examined mechanisms and causes and broadened the direction into "What if?" type thinking, with some thought of risk assessment, to develop what is now Chapter 1 of this work. That path led to studying what precisely was being done in the aviation sector to give rise to the observed decrease in failure rates. From this, the methodology described in Chapter 2 evolved, which appears to work for at least the cases cited.

The use of fracture mechanics is inevitable if some consideration is given to what is taking place in the aviation sector. Chapter 3 enlarges on fracture mechanics as a whole in an attempt to cover as much ground as possible within a reasonable space.

The notion of the prediction of corrosion and establishing growth rates for the stress=corrosion cracking (SCC) and corrosion=fatigue (CF) mechanisms, for example, has much appeal for several sectors of industry, as outlined in Chapter 4. Further, I have attempted to use observations on other aspects of corrosion as a guide word approach to design and changes in design, such that the reader/user can use this information in the "What if?" and HAZOP approaches described in Chapter 1.

The behavior at weldments is of concern to almost every sector of industry, but I have attempted to extend the thought from Chapter 3 to Chapter 5, which considers the fatigue behavior primarily of weldments and expands upon the use of BSI PD 6493.

The "game plan" which evolved from the preceding chapters is brought together in Chapter 6, which applies this overall approach to the hydrocarbon processing sector specifically to the refining of oil.

I have attempted to weave a thread through all the chapters that allows the reader/user easy reference to key facets of this work. I think the summation and direction given in Chapter 7 broaden the approach sufficiently to allow managers and engineers to apply this thinking to solving their problems with equipment failures in many other sectors of industry.

P.F. Timmins

Acknowledgments

I am grateful to my wife, Ann Timmins A.M.I.M., for her help in compiling the manuscript, to my son Chris, for typing some of the References and to my daughter Rachel, who prepared Table and Figure captions.

Introduction

This work is concerned primarily with the reduction of equipment failures in sectors of industry.

For this work, the focus is on the development and application of methodologies which have led to the observed reduced failure rates in the aviation sector and how these can be employed in reducing failure rates in other sectors of industry, where there is a rising trend in equipment failures. Principal among these sectors is hydrocarbon processing.

There are seven chapters to this work.

Chapter 1 is on trends in equipment failure rates, costs, mechanisms, and causes.

Chapter 2 is on solutions to equipment failures in aviation and Chapter 3 offers solutions to equipment failures based on fracture mechanics.

Chapter 4 focuses on solutions based on corrosion prediction and Chapter 5 on welded joints.

Chapter 6 considers how the previous factors can be drawn together to provide a "game plan" for reducing failure rates in hydrocarbon processing.

Chapter 7 is a summary of the work which includes evidence of the system working at one specific location with a direction for the future, in applications to gas processing facilities, power generating facilities and pulp and paper production, in addition to the petrochemical and chemical sectors.

CHAPTER 1

Trends in Equipment Failure Rates, Costs, Mechanisms, and Causes

It is important to determine trends in failure rates and costs in various industrial sectors in order to direct effort into those sectors in which failure rates are highest. Further, it is necessary to determine a priority of dominant failure mechanisms and causes, in order to direct effort into improving the resistance of materials to failure by these mechanisms and to evolve methods of identifying causes. By reviewing trends in equipment failure rates over time for a given industry, an idea of how that industry as a whole is performing is revealed. The sectors of industry reviewed include aviation and hydrocarbon processing, where massive capital investment equipment failures can result in the loss of human life as well as substantial financial loss.

Aviation

The annual Boeing Corporation survey (Ref 1) indicates a decrease in jet aircraft failure rates between 1964 and 1992. This is illustrated in Fig. 1-1, in which the annual percentage loss of the world fleet of jet aircraft decreased over time.

The cause of loss of jet aircraft, according to the Boeing survey (Ref 1), is shown in Fig. 1-2, and the type of mechanical failure leading to total loss of jet aircraft as a percentage of total failure is given in Fig. 1-3 (Ref 1). The mechanisms of mechanical failure, including both those that were revealed in an accident and those revealed during routine inspection, are shown in Fig. 1-4 (Ref 2). Fatigue cracking is indicated as the most frequent failure mechanism. Corrosion is also a major problem in aircraft maintenance, although it rarely causes accidents. Stress corrosion cracking and corrosion fatigue cracking are also indicated as mechanisms which may result in mechanical failure.

It is clear from these data that the overall failure rate of jet aircraft equipment is decreasing, despite the infrequent occurrence of failures involving the mechanisms outlined. Evidently, the annual decrement in loss rate is matched by the annual increase in size of the world fleet of jets, such that the average annual number of passenger jet aircraft hull losses is about 16; see Fig. 1-5 (Ref 1).

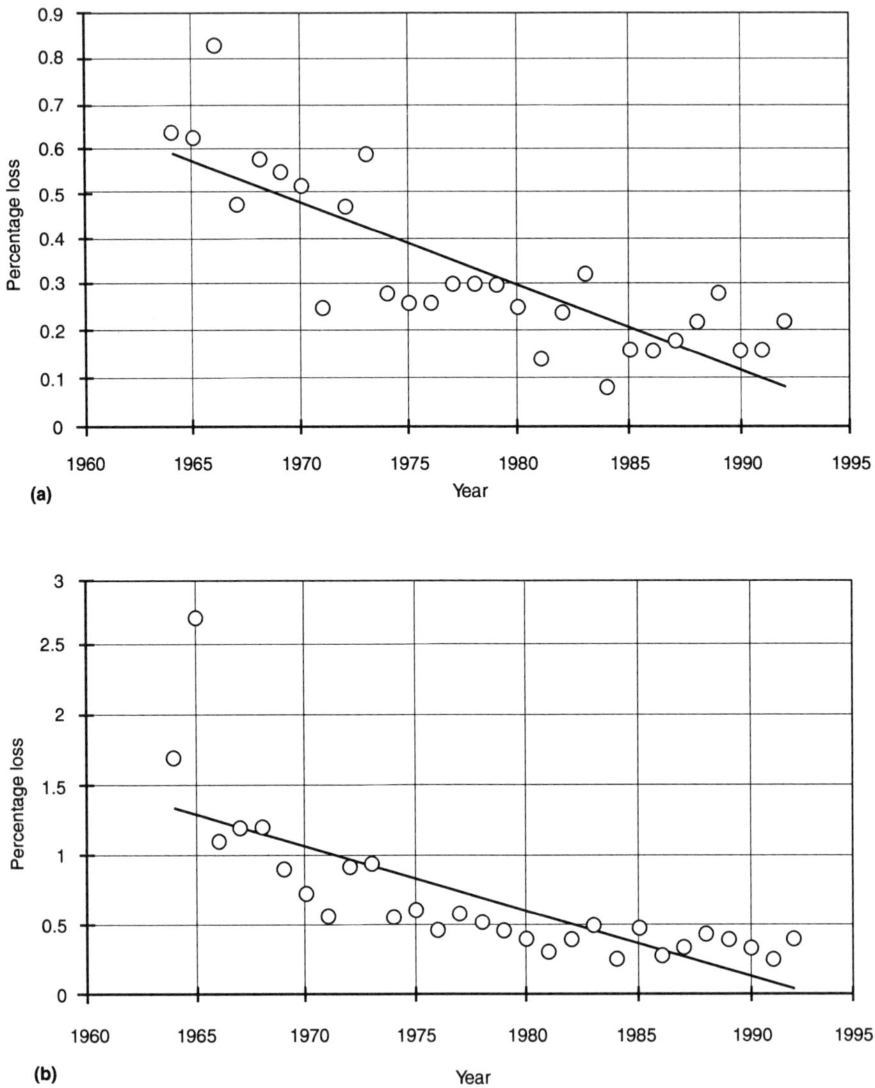

Fig. 1-1 (a) Jet aircraft, total loss per year as a percentage of those in service. Annual decrement = 5.2%. (b) Jet aircraft, all accidents, per year as a percentage of those in service. Correlation coefficient of semi-ln plot = – 0.8689; annual decrement = 5.9%. Source: Ref 1

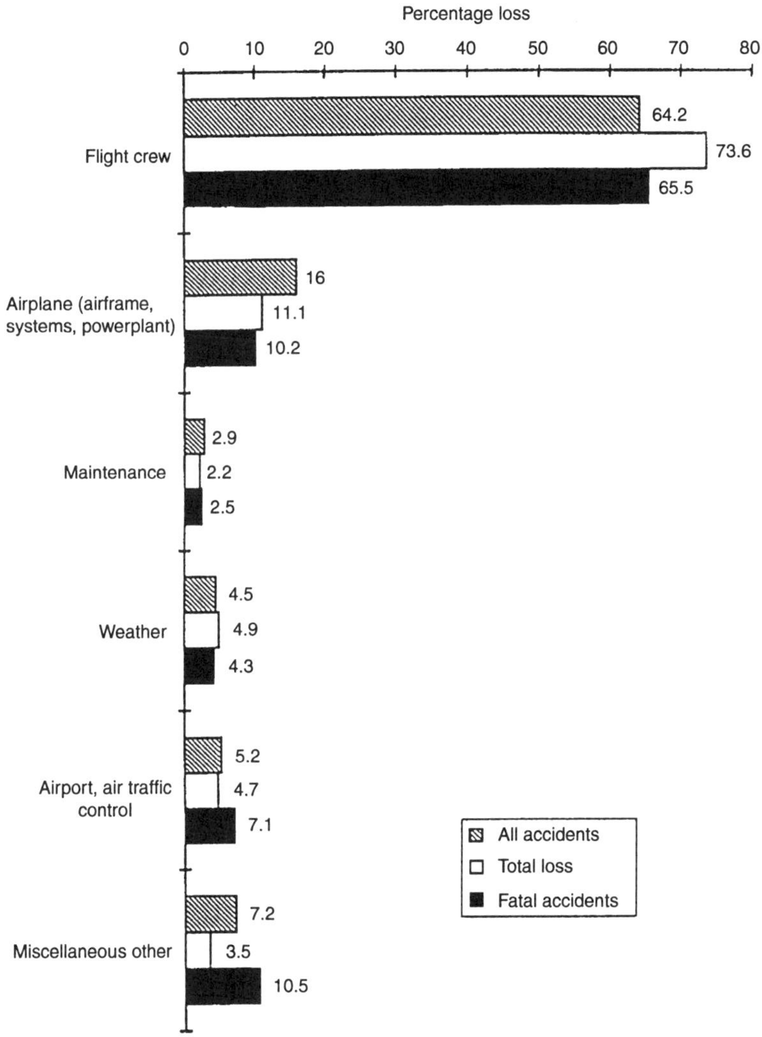

Fig. 1-2 Cause of loss of jet aircraft 1959-1992. Percentage of losses by category: (a) all accidents: (b) total loss (hull loss); (c) accidents in which fatalities occurred. Source: Ref 1

To further elucidate the failure mechanisms and the equipment in which they occur, typical examples of these failures are given in the following section.

Aviation Service Failures and Analysis

The following are a series of case histories from *Handbook of Case Histories in Failure Analysis,* Vol 1 and 2.

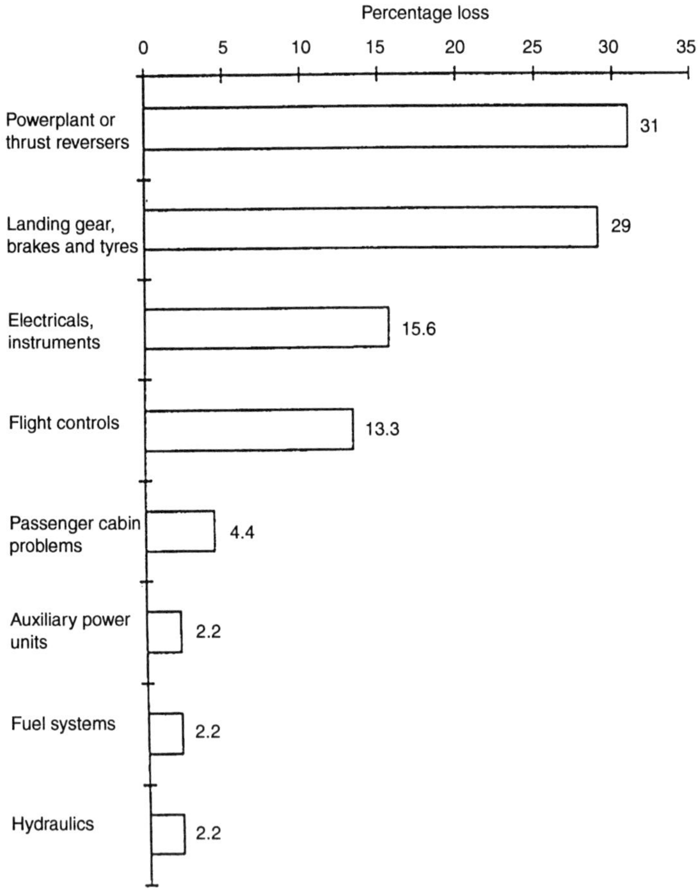

Fig. 1-3 Type of mechanical failure leading to total loss of jet aircraft as a percentage of total failures. Boeing survey. Source: Ref 1

Analysis of Wing Main Spar Cracking*

Following the crash of a Mirage III-0 aircraft, a small crack was detected in a bolt hole in the wing main spar. Because this area was considered to be critical to aircraft safety, the Royal Australian Air Force (RAAF) requested that the crack growth rate during service be determined.

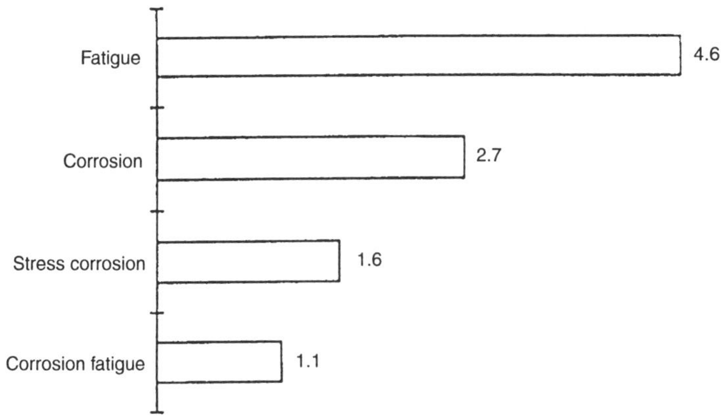

Fig. 1-4 Percentage incidence of major failure mechanisms in aircraft. Source: Ref 2

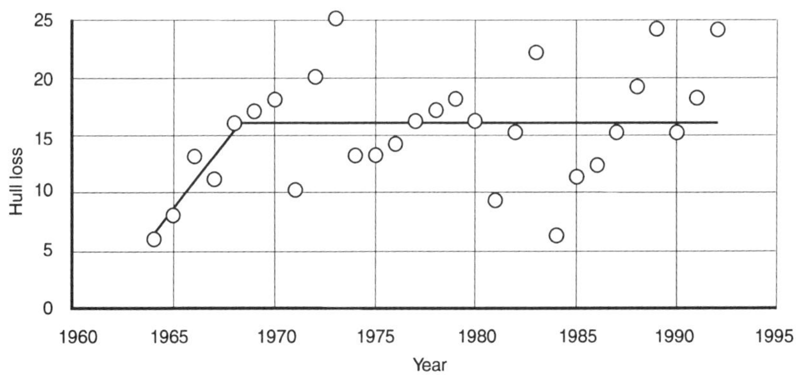

Fig. 1-5 Passenger jet aircraft: hull loss 1964–1992. Source: Ref 2

---

*Adapted from N.T. Goldsmith and G. Clark, in *Handbook of Case Histories in Failure Analysis*, Vol 1, K.A. Esaklul, Ed., ASM International, 1992, p 9–11

At the time, the Mirage III-0 was the main fighter aircraft in service with the RAAF. The results of a full-scale fatigue test indicated the possibility of cracking in spar bolt holes. The crashed components were examined to check for cracking in these locations. The defect was in the main spar of the delta wing. The crash was apparently caused by engine failure. The fatigue crack in the bolt hole was located during a later analysis.

The Mirage III-0 wing main spar (Fig. 1-6) is machined from a large forging of AU4SG, an aluminum-copper alloy similar in composition to 2014. The spar specification indicated that the material was to be heat treated to the T6 condition.

After the crack was discovered, examination of other spars in service confirmed the presence of similar cracking in many aircraft. The particular crack examined was selected because it appeared to be the best preserved following the crash.

Visual Examination. The bolt hole in which cracking was discovered was located in the lower rear flange of the main wing spar. Cracking was observed on the lower aft side of the bolt hole. There was clear evidence of fatigue progression markings.

Eddy current examination of the bolt hole revealed crack indications in the lower portion of the lower flange in the aft section of the spar (Fig. 1-7). Other aircraft in service were examined using the same technique which revealed many other cracks. An array of small cracks was visible along the edge of the bore of the hole (Fig. 1-8).

Fig. 1-6 Lower surface of a main spar from the starboard wing of a Mirage III-0. The row of fastener holes along the flange is used to attach the wing skin.

Fig. 1-7 Detailed view of the lower rear flange of the spar near the root end. Arrow indicates the location of the crack in the failed bolt hole.

Macrofractography. A view of some of the cracks is shown in Fig. 1-9. Note the arrangement of cracks along the bore of the hole. The location of the array of cracks appeared to have been determined by the general stress situation, whereas the precise

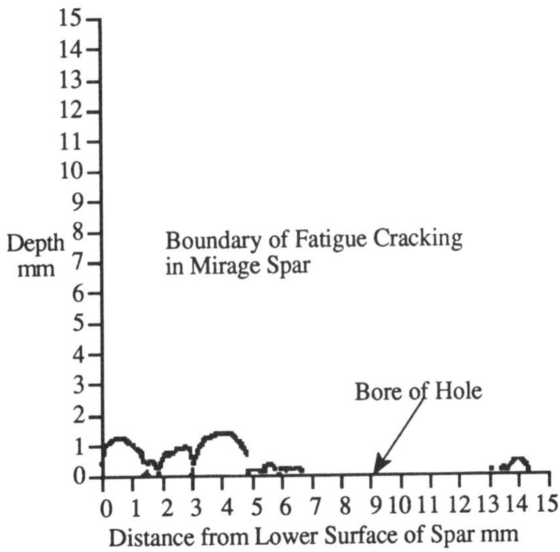

Fig. 1-8 Diagram of the array of cracks found in the bore of the hole after it was opened

Fig. 1-9 SEM micrograph showing the main array of cracks in the bore of the hole. Note the traces of other cracks in the bore, which have not been opened for examination.

location of each crack appeared to have been determined by local discontinuities, such as surface imperfections normally found in machined holes and inclusions typically present in the material.

High-Powered Light Microscopy/Fractography. Examination of the fatigue crack surfaces using high-powered light microscopy revealed many regions of fatigue striations, which occasionally were large. The larger striations could be followed around the crack front and appeared to be present on all the larger crack surfaces.

Scanning Electron Microscopy/Fractography. Examination at high magnification revealed fine and coarse fatigue striations, which are typical of fatigue cracking under a service load spectrum (see Fig. 1-10). It was observed that examination of striations was significantly affected by accelerating voltage. Lower voltages (2 kV) revealed more superficial detail, whereas higher voltages (10 kV) displayed the striations with less apparent contrast and were also more affected by characteristic changes in electron emission caused by gross changes in topography. The optimum voltage (in this case, 5 kV) appears to depend on the scale of the topography to be imaged with good contrast. Fatigue striations usually have only small relief, and lower values of accelerating voltage are therefore more appropriate.

Metallography. The microstructure of the material was typical of an aluminum-copper alloy forging in the T6 condition. Crack origins were located at small discontinuities in the bore of the hole. Superficial corrosion damage obscured fine detail on the crack surface in the area of the origins. Crack paths were transgranular, frequently intersecting the groups of intermetallic compounds normally present in this material.

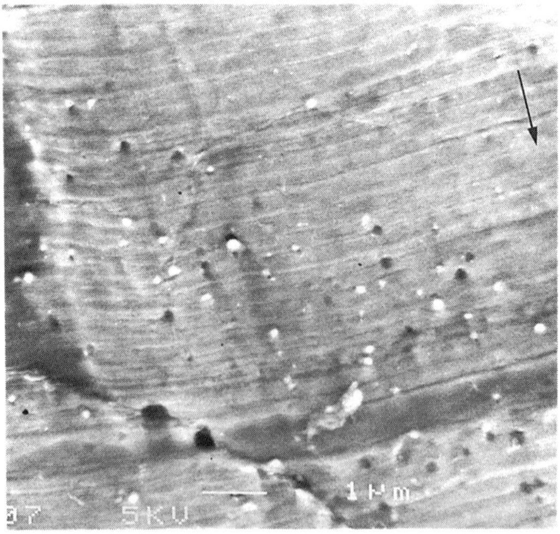

Fig. 1-10 SEM micrograph showing fine and coarse fatigue striations on the crack surface. Arrow indicates crack growth direction.

Chemical analysis of the material showed that it conformed to the specification for AU4SG.

Simulation Specimens. As a result of the discovery of cracking in the full-scale fatigue test, a refurbishment program was established with the goal of prolonging the safe operating life of the spar. As part of this program, many tests were performed on specimens representing the spar. Some tests produced cracks at the same location as the cracks in the spar under examination. These samples were therefore examined to obtain details of crack growth. The flight-by-flight test loading sequence was available. A crack growth curve was obtained by measuring the position of successive appearances on the fracture surface of the mark, produced by a unique peak load in the testing sequence. The resulting crack growth curve was used to calibrate a crack growth model.

Calibration of Crack Growth Model. It was necessary to determine the relationship between the fatigue crack growth produced by the sequence of loads applied in the simulation specimen and the growth associated with the sequence for the crashed aircraft. The principal difficulty was that crashed aircraft data were in the form of "g" exceedances recorded by a series of accelerometers, whereas laboratory tests were conducted using a flight-by-flight sequence that was representative of the fleet.

To model the effect of the accelerometer system in the aircraft in terms of how the recording system modified the real load sequence, a synthetic accelerometer system was simulated using software and used to process the fatigue test sequence. This process truncated real loadings in the same way that actual service loads would have been truncated in the aircraft. The resulting truncated load history was then used as input to a computer program. This program constructed a crack using a table of values of crack advance per cycle versus the stress-intensity factor obtained for the material by standard fatigue testing procedures. This modeled crack growth was performed retrogressively, beginning with the final crack depth and the last load applied in the test. When crack growth was estimated for one complete sequence, estimated depth was compared with actual measured depth on the laboratory fatigue test specimen. If these values were not equal, the calculation could be repeated using a scaling factor for the stress until the crack increments matched. The process could then be repeated for the next crack increment. In this way, a table of values of scaling factors was constructed, with a scaling factor appropriate to each crack depth range. A typical set of these values is plotted in Fig. 1-11.

These factors were then used with the accelerometer data from the crashed aircraft to estimate the development of the crack that had formed during service. The growth of the crack was again estimated retrogressively, and resulting predictions of fatigue markings were compared progressively with real fracture surface markings in order to prevent the rapid accumulation of errors. The results of this process are shown in Fig. 1-12 as a correlation between crack growth and flight numbers.

Discussion. The use of simulation specimens permitted real crack growth rate data to be determined for the configuration of interest. It also allowed accelerometer data from the crashed aircraft to be used in a crack growth model to establish a correlation between real fatigue progression markings and flight load events.

Conclusion and Recommendations. Many spars in the fleet were found to be cracked. The results of the investigation revealed the reasons for crack growth rates in service. This in turn allowed a program of hole refurbishment to be safely implemented. Refurbishment was based on reaming the holes and using interference-fit bushings.

Fatigue Fracture of Aircraft Engine Compressor Disks*

Several compressor disks in military fighter and trainer aircraft gas turbine engines cracked prematurely in the bolt hole regions. The disks were made from precipitation-hardened AM355 martensitic stainless steel.

Critical gas turbine rotating components, such as compressor disks, spacers, cooling plates, and turbine disks, are subjected to cyclic stresses during engine startup and shutdown, as well as major throttle excursions during flight maneuvers. The cyclic stresses can exceed the yield strength of the material at stress-concentration sites, such as bolt holes, bores, blade dovetail slots, or rim serrations (Fig. 1-13), and thus lead to low-cycle fatigue cracking. The cyclic lives of these components are usually established

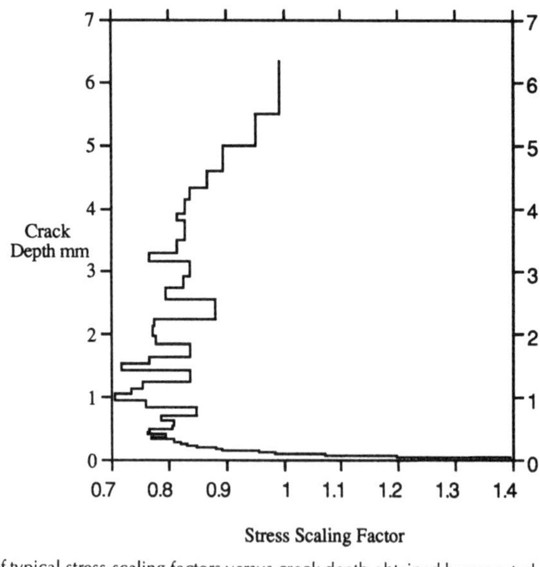

Fig. 1-11 Plot of typical stress-scaling factors versus crack depth obtained by repeated computation of incremental crack depths using the actual loading history modified by a synthetic "fatigue meter"

---

*Adapted from A.K. Koul and R.V. Dainty, in *Handbook of Case Histories of Failure Analysis*, Vol 1, K.A. Esaklul, Ed., ASM International, 1992, p 241–250

on the basis of a probabilistic crack initiation criterion for a known fracture-critical location where, typically, 1 in 1000 components (or, in the case of a smaller fleet, 1 in 741 parts) will form a 0.8 mm (0.03 in.) crack during the lifetime of an engine fleet. This crack size is usually considered detectable with high reliability by the most commonly used inspection technique—liquid penetrant inspection.

In some engines, such as those considered in this case study, the cyclic lifetimes of critical components may be based on a dysfunction criterion, where either 1 in 1000 or 1 in 741 components will initiate and propagate a crack considerably longer than the 0.8 mm (0.03 in.) initiation criterion. (It should be noted that dysfunction crack size is not a critical crack length for catastrophic failure. It is factored down from such a length for a number of reasons, such as to prevent exceedance of vibration allowables in cracked components.) These cyclic lives are calculated analytically using databases generated on material coupons or through spin-pit or accelerated engine testing for an assumed engine mission profile. If the actual mission profile deviates from the assumed mission, a mission severity factor (MSF) is then incorporated to estimate the cyclic lives of the components. The components are regularly inspected at specified engine overhaul intervals to check for any signs of premature cracking.

The dysfunction lives of the baseline engines and the MSF for the engines under investigation were regularly updated by the engine manufacturer on the basis of field experience. After almost 14 years of engine operation, severe bolt hole cracking of disks and spacers in the eight-stage compressor train assembly of all engines, including the baseline, was observed during scheduled overhaul inspections. It became clear in the early part of this investigation that the actual flying mission was considerably more severe than that assumed in the original life calculations. Consequently, the dysfunction life of the baseline component was reduced and the MSF for the engines under investigation was increased. Therefore, the allowable dysfunction life of the disk was drasti-

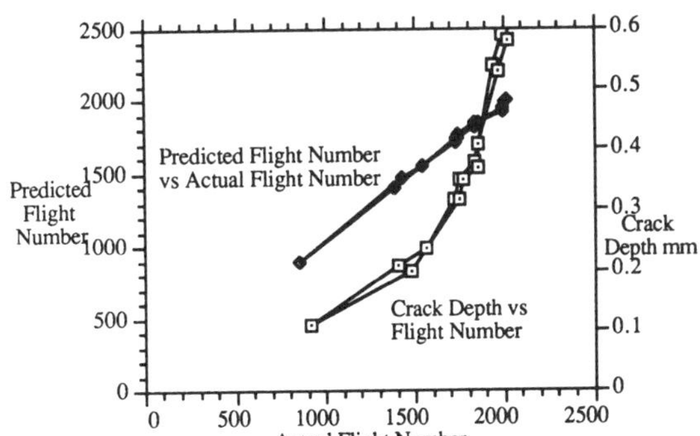

Fig. 1-12 Plot of incremental crack depth obtained by using "fatigue meter" data and stress-scaling factors. Positions of large striations are plotted against the number of the flight that was predicted to have produced them. Also shown is a plot of "predicted" versus "actual" flight numbers at which large loads were experienced. Note the slope of this relationship, which approaches 45°.

cally reduced (Table 1-1). (It should be noted that the dysfunction life of the compressor disk was calculated on the basis of a 4.5 mm, or 0.18 in., crack length by the engine manufacturer.) In all disks, the cracks had initiated in the bolt hole region and propagated radically inward toward the bore of the disk.

Limited data on flying altitude, power lever angle, speed, and number of engine revolutions per minute had previously been recorded for 11 different flights, each of which used a separate set of maneuvers, but where an individual flight did not constitute a complete mission. Depending on aircraft operation role, it was estimated that the average number of low-cycle fatigue cycles per mission hour varied from three to five (Ref 3).

Table 1-1 History of dysfunction lives and mission severity factors for the compressor disk

Date	Baseline dysfunction life, h	Mission severity factor	Dysfunction life incorporating mission severity, h
1970	3000	1.0	3000
1978	4500	0.94	4787
1979	4400	0.94	4680
1981	3200	1.80	1777
1983	3200	1.4	2285

Fig. 1-13 Compressor disk

The disks and spacers contained 40 equally spaced tie bolt holes. Regions containing these bolt holes were known to be the fracture-critical locations in all compressor disks and spacers.

Based on the number of components retired in the cracked condition, it was determined that the fifth-stage compressor disk was the most crack-prone disk in the compressor train (Table 1-2). Therefore, most of the experimental and analytical work was conducted on this part.

Visual Examination. Long cracks in a number of disk bolt hole regions were visible with the naked eye. No evidence of component distortion or component discoloration (indicative of overheating) was noted.

Nondestructive Evaluation. The disks had been inspected at the engine overhaul and maintenance facility at regular intervals by trained inspectors certified at level II, using the MIL-STD-6866 liquid penetrant inspection technique. In some disks, cracks as long as 14 to 19 mm (0.55 to 0.75 in.) were found (Fig. 1-14) This was unusual, because the disks had been inspected several times using liquid penetrants, which should detect 0.8 mm (0.03 in.) cracks with high probability. However, none of these cracks had resulted in a catastrophic failure, indicating that the disks were capable of tolerating cracks considerably longer than the dysfunction crack length of 4.5 mm (0.18 in.) assumed by the engine manufacturer.

A nondestructive evaluation demonstration program on a statistically significant number of components was carried out to determine the sensitivity and reliability of liquid penetrant inspection. Ten fifth-stage compressor disks that were known to contain service-induced bolt hole cracks were examined to determine the probability of detection (POD) of the MIL-STD-6866 technique. Details of the demonstration program are provided in Ref 4 to 7.

The POD curve, along with a 95% confidence bound for the liquid penetrant technique, is presented in Fig. 1-15. The crack length values at 90% POD and at a 95% confidence level are presented in Table 1-3, along with the maximum crack size missed. The 90/95 POD value given in Table 3 clearly suggests that it is quite possible to miss cracks as long as 6 mm (0.24 in.) when using liquid penetrants. The longest crack missed

Table 1-2 Total number of disks cracked in compressor train assembly between 1976 and 1984

Stage	No. cracked
1	0
2	0
3	0
4	2
5	109
6	19
7	81
8	6

was approximately 4 mm (0.16 in.). These results indicated that long cracks had in fact been missed during regularly scheduled inspections.

Surface Examination. Bolt hole regions were cut from a number of cracked disks. To examine the fracture surfaces, the cracks were pried open (Fig. 1-16). Scanning electron microscopy (SEM) of the bolt hole cracks clearly established that a low-cycle fatigue mechanism was responsible for initiating thumbnail (Fig. 1-17) or corner-type (Fig. 1-17) cracks. In most specimens, cracks initiated near the subsurface primary particles—probably carbides or carbonitrides. Multiple crack initiation sites were also found (Fig. 1-17). Results of a detailed analysis of the crack initiation sites in approximately 400 bolt hole specimens are summarized in Table 1-4. These results indicate that most bolt hole specimens contained through-cracks with multiple crack initiation sites.

Fatigue striation spacings were measured on the fracture surfaces of nine cracks ranging in length from 4 to 19 mm (0.16 to 0.75 in.). Typical results for a 14 mm (0.56 in.) long crack are shown in Fig. 1-18. Average striation spacings for crack lengths less than 1 mm (0.04 in.) were 0.5 μm (20 μin.). For crack lengths between 1 and 10 mm (0.04 and 0.4 in.), the striation spacings were relatively constant at a value of approximately 1.1 μm (44 μin.). At crack lengths greater than approximately 10 mm (0.4 in.), the spacings increased sharply to about 2.5 μm (100 μin.). As indicated in Fig. 1-18, over a crack length from 1 to 10 mm (0.04 to 0.4 in.), the striation spacing and thus the crack propagation

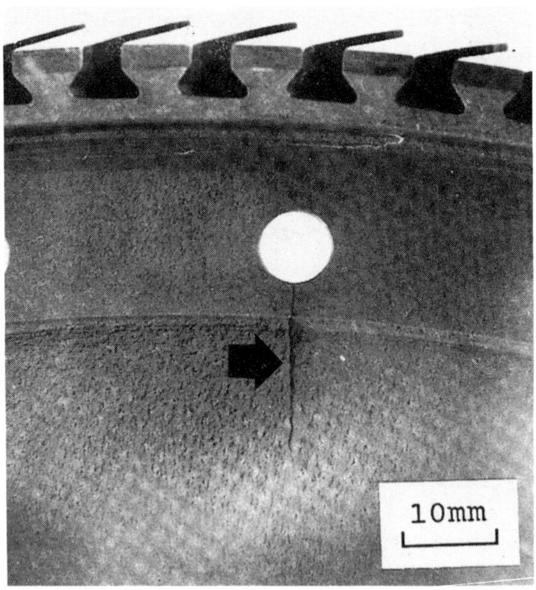

Fig. 1-14 Micrograph showing 19 mm (0.75 in.) long crack in retired compressor disk

rate appeared to be relatively constant. It was clear that the dysfunction crack length was in fact 10 mm (0.4 in.), rather than the 4.5 mm (0.18 in.) length assumed by the engine manufacturer.

An important observation concerning the fracture surfaces was the tendency for a banded or recurring striation pattern to form, as indicated by arrows in Fig. 1-19. This recurring striation pattern consisted of discrete regions of finely spaced striations separated by a single, widely spaced striation. Most often, the number of striations within a band appeared to be about five; in rare cases, about ten striations were found grouped together. It is reasonable to assume that each discrete band represents the increment of crack growth that occurred during a single flight, and that the fine striations represent flight maneuvers and changes in the engine rotational speed.

Metallography. Radial sections were cut from the cracked disks and were further divided into five subsections, from the bore to rim of the disks. The sections were polished and etched to reveal the martensitic structure and the prior-austenite grain boundary structure (Fig. 1-20). There were no obvious differences in the microstructures of the bore and rim sections. This was not unexpected, because the maximum service temperature that this component experiences (~200 °C, or 390 °F) is not severe enough to cause microstructural degradation.

However, these results did confirm that the disks had not been overheated during engine operation. In cracked bolt hole sections, considerable crack tip branching was observed in several cases (Fig. 1-21). The reasons for this behavior were not clear.

Plastic replicas were prepared from crack-tip locations of the etched bolt hole specimens and examined using a transmission electron microscope. It was found that

Table 1-3 Crack detection capabilities of liquid penetrant inspection

Item	Crack size mm	Crack size in.
90/95 POD	6.40	0.26
Longest crack missed	3.98	0.16
Smallest crack detected	0.50	0.02

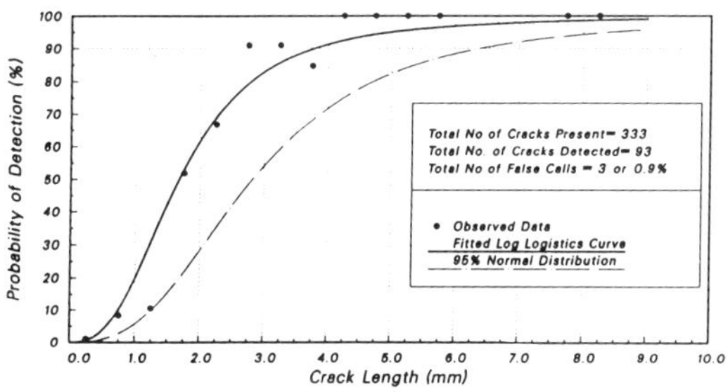

Fig. 1-15 Probability of detection of bolt hole low-cycle fatigue cracks in compressor disks using liquid penetrant inspection

significant changes had occurred in precipitate morphology around the crack-tip regions that experience severe cyclic straining. Although the precipitate particles away from the crack tip were spherical and uniformly distributed (Fig. 1-22) these precipitates appeared elongated along what may have been planes of intense shear within the plastic zone (Fig. 1-22) The appearance of precipitate shearing at approximately 45° to the direction of crack growth supports this speculation (Fig. 1-22). The microstructural changes in the crack-tip region indicated that the crack-tip stress intensities were not small.

Hardness. The room-temperature hardness measurements taken from the service-exposed disks were found to be within the specifications of the engine manufacturer for the AM355 material (see Table 1-5), that is, above 37 HRC.

Table 1-4 Classification of bolt hole crack initiation sites

No.	Initiation sites	% of total cracks
1	Single	11.3
2	Corner	20.4
3	Multiple	69.9

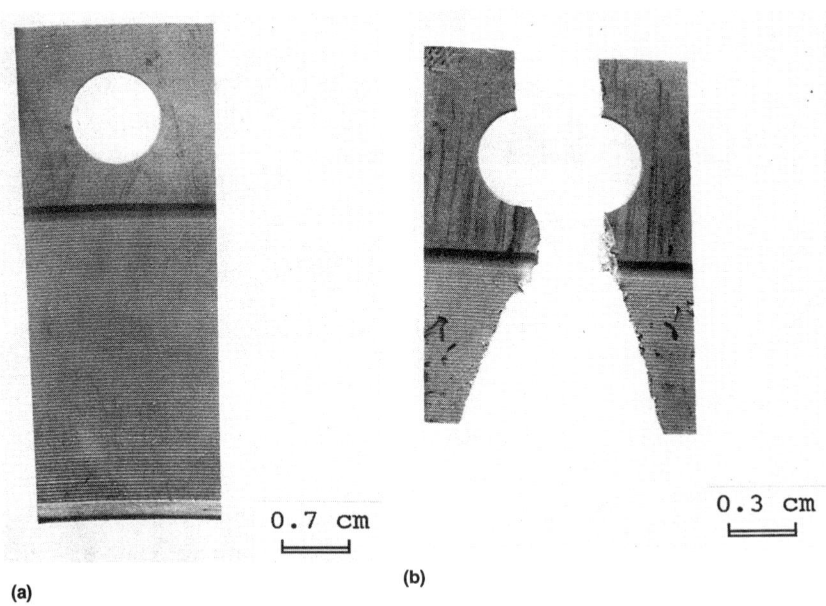

Fig. 1-16 Destructive procedures followed for prying cracks open and measuring crack length. (a) Typical bolt hole coupon. (b) Coupon with V-notch prior to fracturing specimen through the crack

Tensile Properties. Tensile tests were conducted on specimens machined from cracked disks that conformed to ASTM specifications. The tensile properties of the

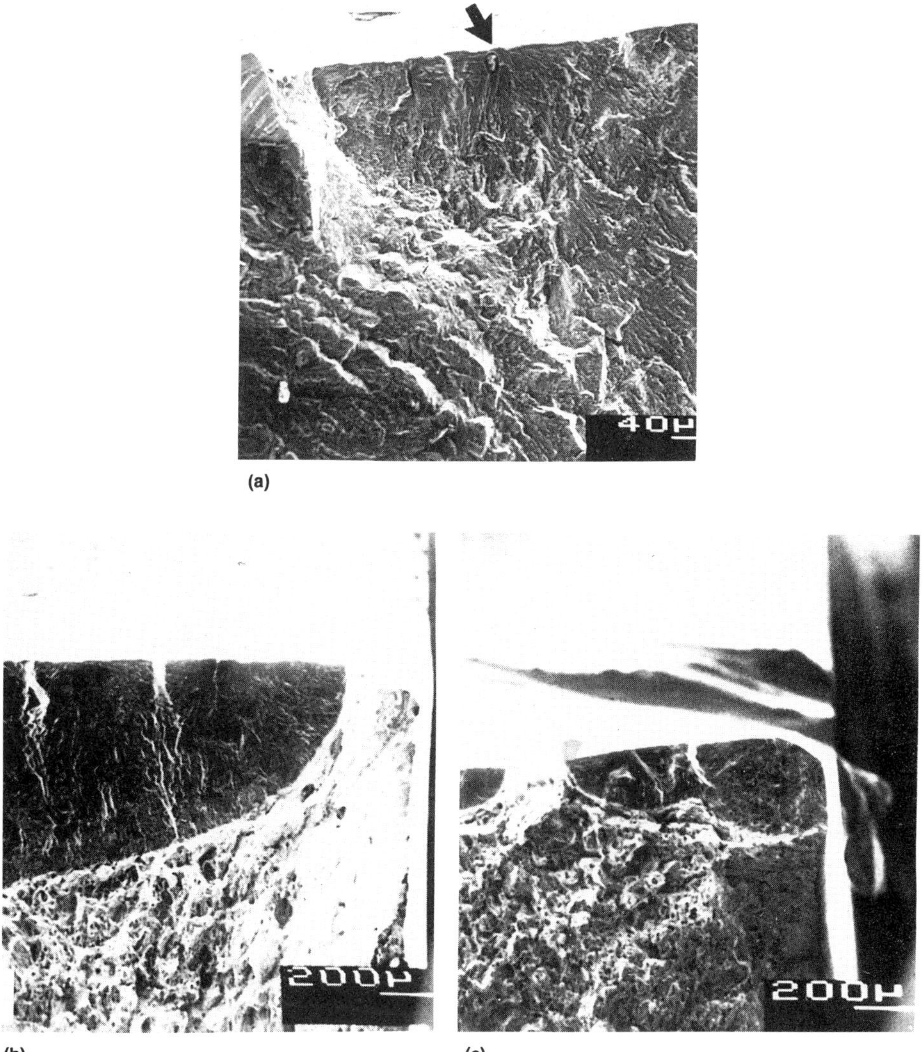

Fig. 1-17 (a) Typical thumbnail crack. (b) Corner-type crack. (c) Example of multiple crack initiation sites

cracked disks were well within the specifications prescribed for the disk material (Table 1-6).

Analytical Stress Analysis. A preliminary two-dimensional finite-element stress analysis for the disk was performed. The details of the mesh geometry and the model

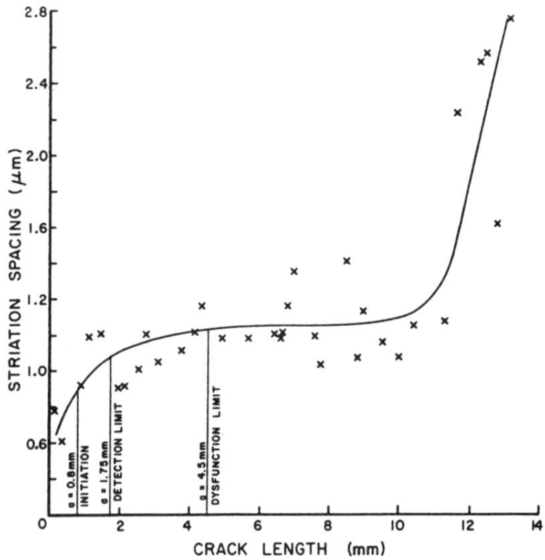

Fig. 1-18 Results of striation spacing measurements of long bolt hole crack

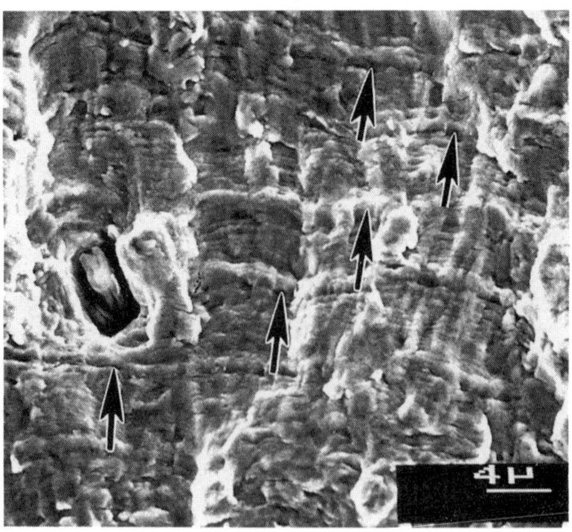

Fig. 1-19 Banded striation pattern observed in service-induced cracks

information are provided in Ref 3. The effects of blade mass, aerodynamic loads, bolt and washer mass, and interactions caused by the spacers on either side of the disk were not considered in these computations. The actual values of *K* were obviously underestimated because of these assumptions. It is of interest to note, however, that *K* remains constant over a significant portion of the crack length (Fig. 1-23). This is consistent with the fractographic results of the cracked disks described in the section "Surface Examination."

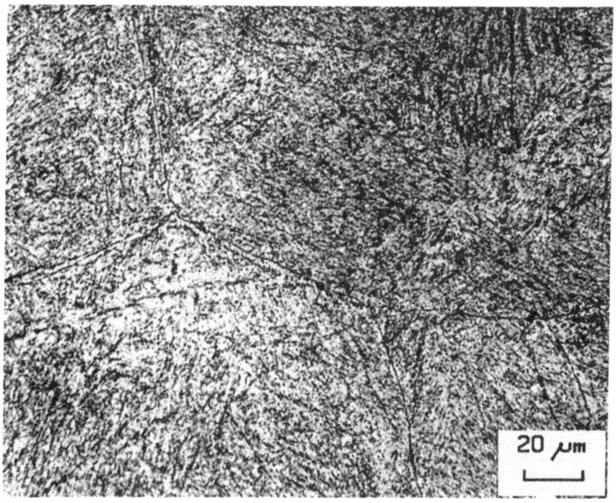

Fig. 1-20 Typical etched microstructure of compressor disk material

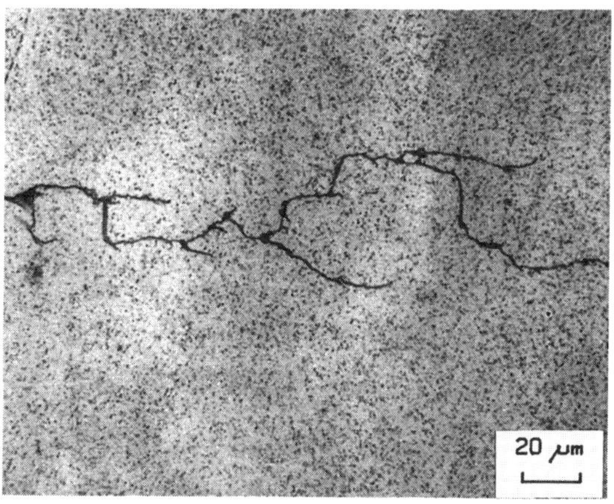

Fig. 1-21 Crack-tip branching of bolt hole crack

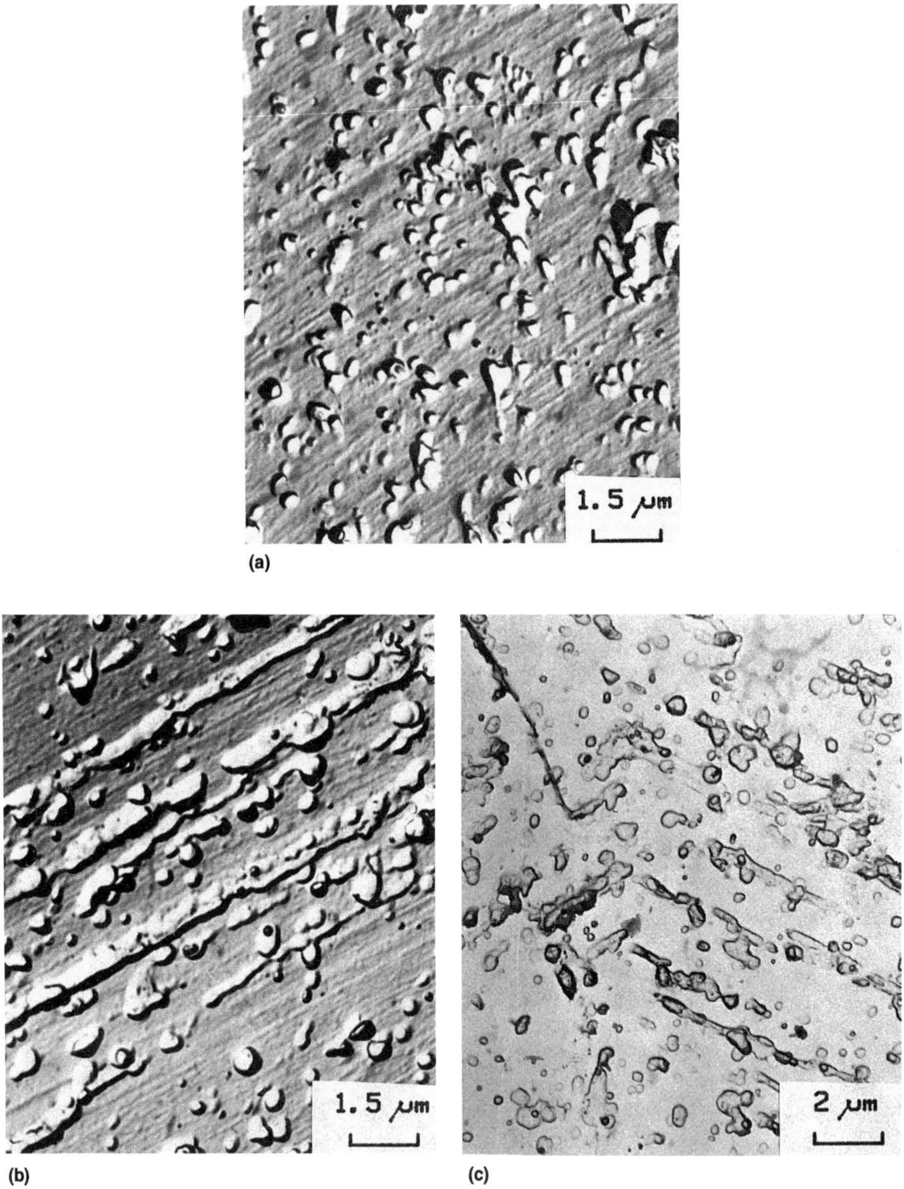

Fig. 1-22 Precipitate morphologies at crack tip in compressor disk. (a) Away from crack tip. (b) and (c) Near crack tip

Experimental Stress Analysis. The cyclic plastic zone radius (*Ry*) ahead of the crack tip was measured for a number of bolt hole cracks to experimentally estimate the ΔK values. It is well known that during cyclic loading, a small zone of reversed plastic yielding called the cyclic plastic zone can be found, along with a larger monotonic plastic zone. The areas over which cyclic hardening or softening occurred as a result of plastic yielding was revealed by a series of microhardness measurements at the crack tip, where indentations were made at 0.05 mm (0.002 in.) space intervals at a load of 50 g, as shown in Fig. 1-24.

Table 1-5 Chemical composition of a typical disk

Element	Composition,%	
	Typical disk	Specification requirement (a)
Carbon	0.11	0.1 – 0.15
Manganese	0.86	0.5 – 1.25
Silicon	0.21	NR
Phosphorus	0.012	NR
Sulfur	0.011	NR
Chromium	15.6	15 – 16
Nickel	4.3	4 – 5
Molybdenum	2.53	2.5 – 3.25
Nitrogen	...	0.07 – 0.13
Iron	bal	bal

(a) NR, no requirement

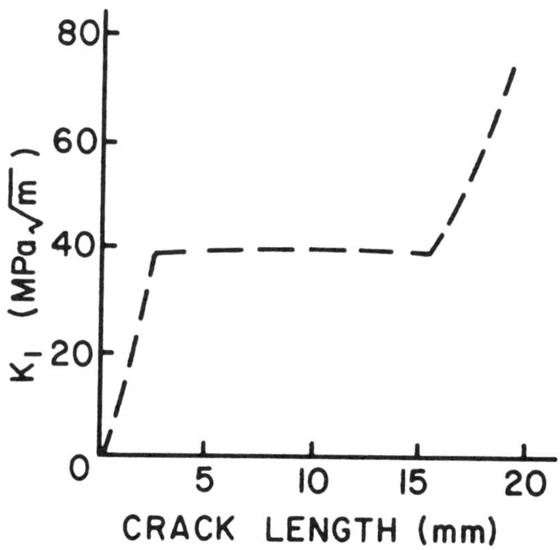

Fig. 1-23 Results of two-dimensional finite-element analysis, showing variation of stress-intensity factor with crack lengths

In the AM355 compressor disk investigated, a large hardened zone was located very close to the crack tip (Fig. 1-25). In a number of bolt hole specimens containing crack lengths ranging from 2 to 7 mm (0.08 to 0.28 in.), the size of plastic zones did not vary significantly, indicating that ΔK was indeed constant and Ry was approximately 0.45 mm (0.018 in.). The ΔK was calculated using the relation:

$$Ry = C_1 \left(\frac{\Delta K}{\sigma_{ys}} \right)^2 \qquad \text{(Eq 1)}$$

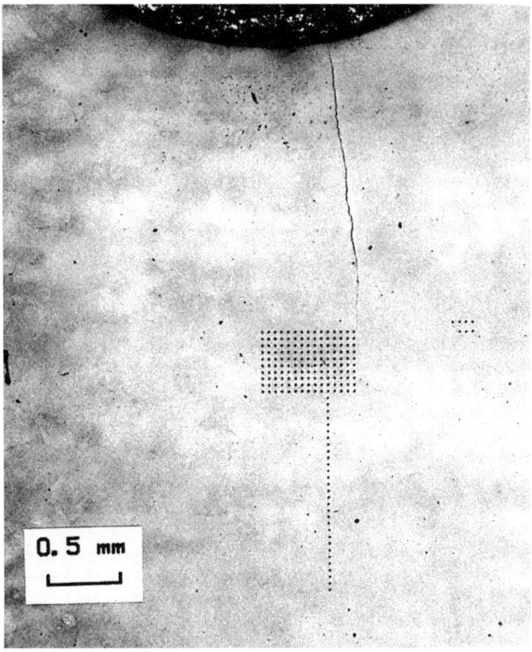

Fig. 1-24 Grid pattern of microhardness measurements used to determine plastic zone size

Table 1-6 Tensile properties of compressor disks

	Yield stress		Ultimate tensile strength		Elongation,
Material	MPa	ksi	MPa	ksi	%
Minimum specifications for forging	1070	155.2	1170	170	...
Disk sample 1	1129	164	1272	184	11.0
Disk sample 2	1112	161	1247	180.8	10.0
Disk sample 3	1071	155.3	1248	181.0	10.0

where σ_{ys} is the yield strength and C_1 is a constant that is dependent on material and type of loading. Assuming plane stress loading conditions because the disks are relatively thin, and using a value of $\frac{1}{2}\pi$ for C_1 and 1100 MPa (160 ksi) for σ_{ys}, the value of ΔK was computed to be 58 MPa$\sqrt{m}$ (53 ksi$\sqrt{in.}$). This operative ΔK value is not very small, relative to the ΔK_c value ranging from 85 to 90 MPa$\sqrt{m}$ (77 to 82 ksi$\sqrt{in.}$) for AM355 material (Ref 3).

Fracture Mechanics Predictions. Deterministic (worst-case) fracture mechanics calculations were performed to determine the safe inspection interval (SII) values for the disks. Assuming that a bolt hole crack of a given length (a_i) is missed during liquid penetrant inspection, allowing it to propagate up to a dysfunction length (a_d) at known ΔK values, the SII values were calculated using the Paris equation:

$$\frac{da}{dN} = C\,(\Delta K)^n \qquad \text{(Eq 2)}$$

where C and n are material constants. Under constant ΔK conditions, Eq 2 can be rearranged to predict the SII through the following equation:

$$\text{SII} = N = \frac{a_c - a_i}{C(\Delta K)^n} \qquad \text{(Eq 3)}$$

In the present case, Eq 3 is valid for crack lengths ranging from about 1 to 10 mm (0.04 to 0.4 in.).

The upper bound of the material constant C and the value of the constant n were determined experimentally. The results are presented in Ref 6. Upon using the maximum crack length value missed by the liquid penetrant technique (~4 mm, or 0.16 in.) for a_i from Table 1-3 instead of the 0.8 mm (0.03 in.) value used in most fracture mechanics calculations, an a_d value of 10 mm (0.4 in.) from Fig. 1-18 instead of the 4.5 mm (0.18 in.) value suggested by the engine manufacturer, a C value of 3.497×10^{-12} m/cycle (11.5×10^{-12} ft/cycle), an n value of 3.16, and a ΔK value of 58 MPa$\sqrt{m}$ (53 ksi$\sqrt{in.}$) in Eq 3, the SII was predicted to be 4592 cycles.

An alternative approach involved using quantitative fractographic results from service-induced cracks and the results of the demonstration program, where:

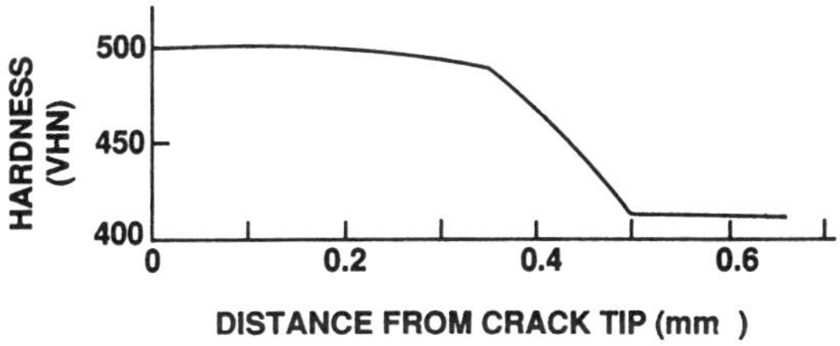

Fig. 1-25 Typical results of microhardness measurements at crack tip of compressor disk bolt hole crack

$$\text{SII} = \frac{a_d - a_i}{\text{Average striation spacing}} \quad \text{(Eq 4)}$$

Using an a_d value of 10 mm (0.4 in.), an a_i value of 4 mm (0.16 in.) (the maximum crack length value missed by liquid penetrant inspection) and the average striation spacing value of 1.1 μm (44 μin.), the SII was predicted to be 5454 cycles. This SII value is approximately 19% higher than the SII predicted on the basis of fracture mechanics calculations. It is suggested that an upper-bound C value used to calculate SII through Eq 3 represents the worst-case scenario, whereas the striation spacing data used in Eq 4 represent an average value; therefore, the SII calculated on the basis of quantitative fractography results is probably closer to the mean SII.

Simulation Tests. Spin-pit simulation tests were conducted on fully bladed disks in vacuum. Details of the spin-pit test apparatus are given in Ref 8. Two loading sequences (constant amplitude and variable amplitude) were used, as shown in Fig. 1-26(a) and (b). The variable-amplitude loading sequence was derived on the basis of revolutions per minute data obtained on different flights comprising different flight maneuvers. Two disks were cycled using the constant-amplitude loading sequence at levels between 50 and 100% of the maximum engine speed. A third disk was cycled using the variable-amplitude sequence. For disk 3, a 1500-cycle marker block (same loading sequence as that shown in Fig. 1-26a) was applied at the beginning of the test in an attempt to form a distinguishable fractographic region between the service-induced fracture areas and the variable-amplitude spin-pit fracture areas. Marker blocks were further introduced intermittently once the cracks in the disks started propagating radically inward during the variable-amplitude spin-pit tests (Ref 8). Liquid penetrant inspection was used to monitor the length of the propagating cracks in all disks subjected to spin-pit testing.

The constant-amplitude loading fatigue crack growth data for a bolt hole crack in disk 2 are presented in Fig. 1-27. The spin-pit tests showed that the crack propagation rates remained relatively constant up to a crack length of approximately 10 mm (0.4 in.). A typical fatigue striation pattern observed during spin-pit crack growth is shown in Fig. 1-28. Using striation counting methods, the fatigue crack growth curves for the spin-pit cracks were also determined.

A typical result is shown in Fig. 1-29, where crack length values measured by liquid penetrant inspection are superimposed for comparison. Again, quantitative fractography results of spin-pit cracks indicated that the fatigue crack growth rates were nearly constant over a wide range of crack lengths. The average striation spacing within the constant fatigue crack growth rate portion of Fig. 1-29 was on the order of 0.7 μm (28 μin.). This value was somewhat lower than 1.1 μm (44 μin.), the average striation spacing value determined for the service-induced cracks (Fig. 1-18). Therefore, spinpit testing loading conditions were not as severe as those experienced by the disks during service.

Typical fatigue crack growth data for disk 3, tested under the variable-amplitude loading sequence, are presented in Fig. 1-30. In plotting these data, it was assumed that each throttle excursion in the mission profile represented one loading cycle. It is important to recognize that the amount of crack growth would be a function of the specific cycle in the variable-amplitude sequence. Consequently, direct comparison of the fa-

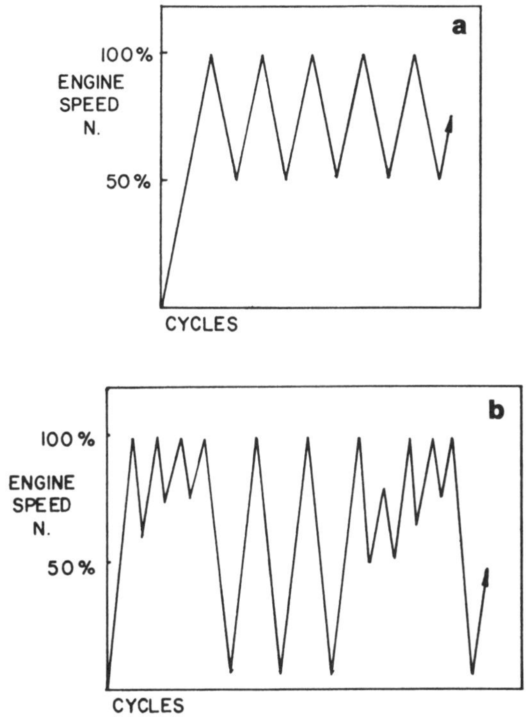

Fig. 1-26 Loading cycles used in spin-pit testing. (a) Constant-amplitude idle-max-idle duty cycle. (b) Variable-amplitude duty cycle

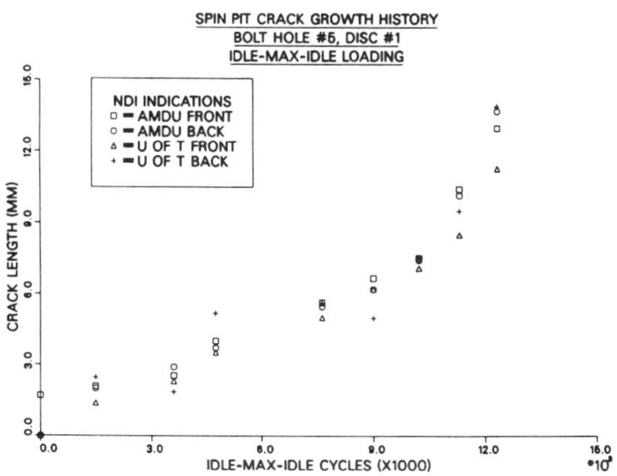

Fig. 1-27 Spin-pit crack growth history of bolt hole crack in disk 1 using idle-max-idle loading

tigue crack growth data for the constant-amplitude and variable-amplitude tests was difficult. Quantitative determination of the fatigue crack growth curves, using a striation counting method, proved difficult because of poorly defined fatigue fracture topography and problems associated with correlating the randomly spaced striations (Fig. 1-31) with the loads applied from the spectrum. In addition, the demarcations between the marker block bands and variable-amplitude fracture regions were not clearly discernible. It was thus impossible to derive any meaningful quantitative fractographic information from variable-amplitude spin-pit tests.

Discussion. Table 1-7 compares the major findings of this investigation with the predictions of the manufacturer. The field experience clearly indicated that the disks were capable of withstanding long cracks without causing catastrophic failures. The quantitative fractography of service cracks, the finite-element stress analysis, and the plastic zone size measurements ahead of the crack tip indicated that the crack-tip stress-intensity factor remained constant with increasing crack length up to approximately 10 mm (0.4 in.). The spin-pit simulations further confirmed these observations, because the crack growth rates remained approximately constant up to a crack length of about 10 mm (0.4 in.).

Under a damage-tolerance-based maintenance methodology, components would be inspected after a predetermined SII, the cracked components would be retired, and the uncracked components would be returned to service for another SII. In this manner,

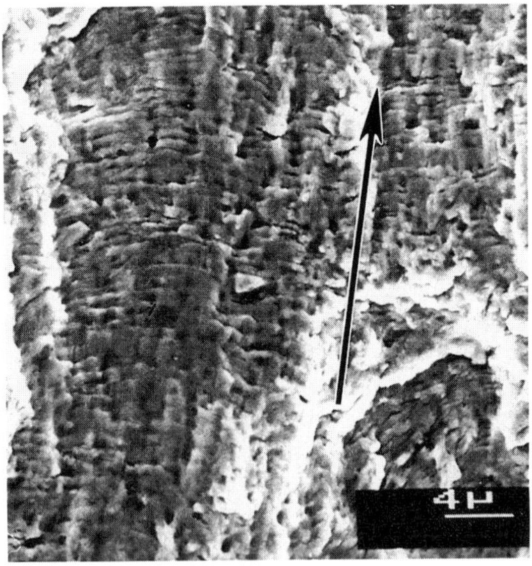

Fig. 1-28 Striations in typical spin-pit fracture area of bolt hole crack, 4.3 mm (0.17 in.) from crack initiation site, using constant-amplitude loading

rather than scrapping all components because of premature cracking of individual disks and spacers, the components would be retired on an as-required basis.

Quantitative fractography calculations performed on service-induced cracks indicate that the average SII is approximately 5454 cycles. The constant-amplitude spin-pit simulation tests, on the other hand, suggest that the average SII is on the order of 6500 cycles if the crack is assumed to propagate from 4 to 10 mm (0.16 to 0.4 in.) in length (compare with Fig. 1-27). The higher SII value, calculated on the basis of spin-pit tests, is also reflected in the smaller average striation spacing width (0.7 μm, or 28 μin.) in spin-pit cracks, whereas average striation spacing was determined to be 1.1 μm (44 μin.) for the service-induced cracks. These observations obviously indicate that the fatigue crack growth rates of spin-pit cracks are lower than those determined for service cracks. This discrepancy probably arises from the fact that spin-pit tests cannot simulate the influence of aerodynamic and inertia loads. As a result, the overall load at the fracture-

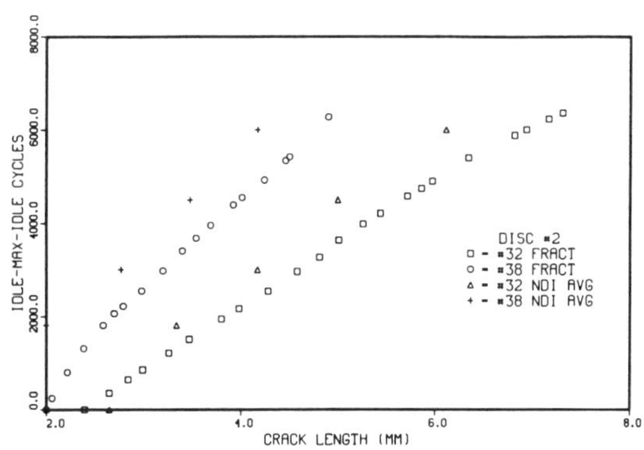

Fig. 1-29 Comparison of fractographic and liquid penetrant crack growth data for two cracks in disk 2 subjected to constant-amplitude loading

Table 1-7 Comparison of previous predictions of different operational and component variables with experiments and analysis

Variable	Engine manufacturer predictions	Predicted through experiments and analysis
Fracture-prone disk	5th stage	5th stage
Fracture-critical location	Bolt hole	Bolt hole
Crack initiation mechanism	Low-cycle fatigue	Low-cycle fatigue
Crack propagation direction	Radially inward	Radially inward
Dysfunction crack length, mm (in.)	4.5 (1.8)	10.0 (0.4)
Detection limit of liquid penetration inspection mm (in.)	0.8 (0.03)	4.0 (0.16)
Operative stess-intensity range, MPa$\sqrt{m}$ (ksi$\sqrt{in.}$)	...	58.0 (53)
No. of low-cycle fatigue cycles per mission hour	2	3-5

critical location in a spin-pit simulation test is lower than that which is operative during service.

Conclusion and Recommendations. The aircraft under investigation were not equipped with airborne data recorders for monitoring the cyclic usage of the engines. It was thus necessary to convert the cyclic SII into flying hours. The SII values obtained through different fracture mechanics calculations are compared in Table 1-8. These calculations indicate that the SII for service-induced cracks is of the order of 900 to 2200 flying hours, depending on the assumed cyclic usage of the engines and the method followed to calculate the SII value.

In order to accurately predict the SII in terms of flying hours, it was recommended that airborne data recorders be used on a number of aircraft. The intent was to monitor different engine usage parameters over a few hundred flights to accurately define an average mission profile for the engines.

It was also recommended that each aircraft be fitted with an airborne data recording system so that the engines could be inspected after a certain amount of cyclic usage, rather than relying on an average mission profile for converting the cyclic usage into flying hours. In this manner, damage-tolerance-based maintenance methodology can be effectively used to retire the cracked disks with little danger of catastrophic failures.

Failed Mixer Pivot Support of Army Attack Helicopter*

A mixer pivot support was found to be broken during a pre-flight inspection. The mixer pivot support is a flight safety-critical component and is part of the rotor support assembly shown in Fig. 1-32. The mixer pivot support fits through the transmission support. If the part had failed during flight, the helicopter would have crashed.

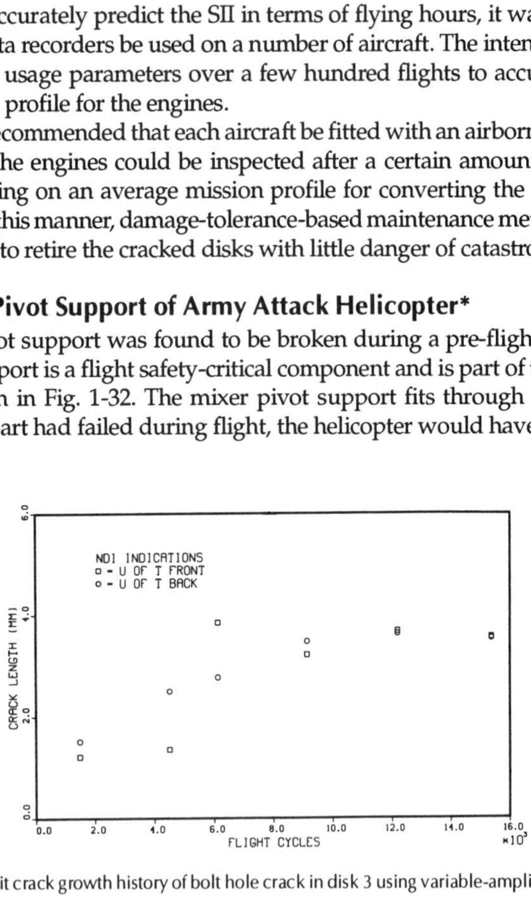

Fig. 1-30 Spin-pit crack growth history of bolt hole crack in disk 3 using variable-amplitude loading

*Adapted from V.K. Champagne, G. Wechsler, and M. Pepi, in *Handbook of Case Histories in Failure Analysis*, Vol 2, K.A. Esaklul, Ed., ASM International, 1993, p 15–18

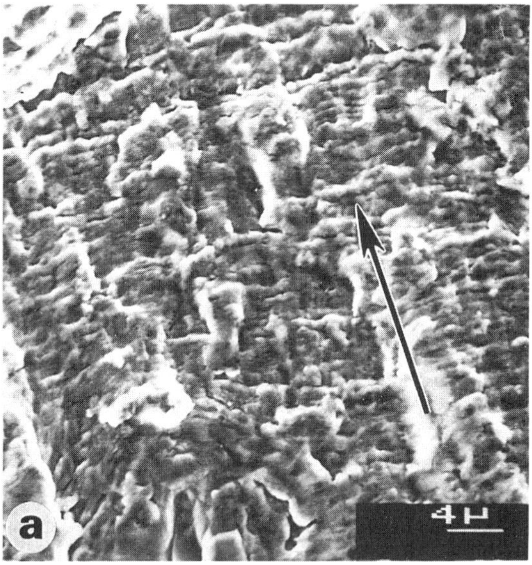

(a)

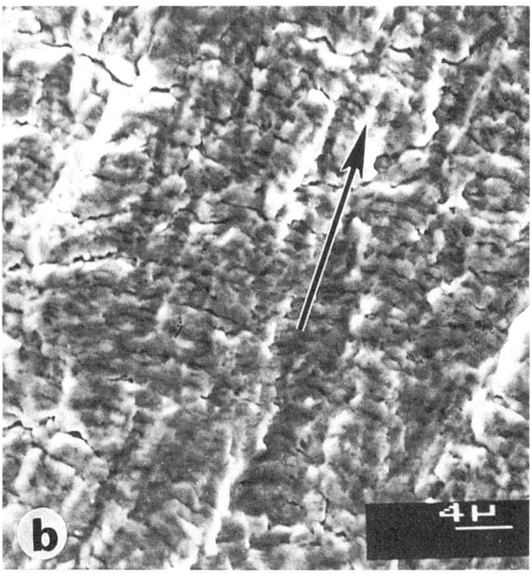

(b)

Fig. 1-31 Typical striation patterns of crack propagated in spin pit in disk 3 using variable-amplitude loading. (a) 3.7 mm (0.146 in.) from crack initiation site. (b) 3.8 mm (1.50 in.) from crack initiation site

The component was in service approximately 1449 days. The original replacement time for the mixer pivot support was 800 days. However, this period was extended to 1440 days due to a lack of spare parts. A second extension of six months beyond 1440 days was granted for the same reason, as long as no surface corrosion was observed upon visual inspection. According to inspection records, no evidence of corrosion was detected on the part. However, the results of this investigation proved otherwise.

The component was specified to be forged from ESR 4340 steel bar stock. A vacuum-plated cadmium coating was applied according to MIL-C-8837. Mixer pivot supports fabricated from the same manufacturer and from the same heat treat/shipping lots as the broken component had not failed during service.

Surface Examination. Figure 1-33 shows the broken component in the as-received condition. The failure occurred at the machined radius. The two broken pieces were labeled Part A (upper half) and Part B (lower half). Part A showed no obvious signs of corrosion or mechanical damage. In contrast, Part B exhibited significant wear on surfaces that contacted the busing. These regions were characterized by dark stains. The cadmium coating appeared to have been almost entirely worn away during service and severe corrosion pitting had occurred within these areas. The extensive corrosion prevented accurate measurement of the surface finish. However, the radius where the fracture occurred was measured and found to be within requirements (3.3 mm, or 0.130 in.). Figure 1-34 is an optical fractograph which reveals the crack origin and direction of propagation. The radial lines and chevron patterns indicated that the crack proceeded from the bottom right of the photograph (as designated by the arrow) up along both sides of the central hole. The fracture appears to have originated at the bottom radius.

Scanning Electron Microscopy/Fractography. Figure 1-35 is a SEM fractograph representative of the morphology found on approximately 90% of the total fracture surface. The mode of fracture was intergranular decohesion, indicative of a brittle fracture. No evidence of fatigue striations were observed, but since these features are difficult to resolve in such high-strength materials, fatigue could not be entirely ruled out as a failure mechanism. Extensive corrosion pitting was found along the edge identified as the crack origin, but the exact point of crack initiation was difficult to resolve due to mechanical damage. Energy dispersive spectroscopy confirmed the by-products of corrosion (Fig. 1-36). Chlorides and particles consisting primarily of silicon and oxygen (sand) were found within pits. Chlorides have been known to accelerate corrosion of carbon steels.

Table 1-8 Comparison of safe inspection interval values predicted by different test methods

Basis of prediction	Safe inspection interval in hours assuming: 3 cycles/ mission hour	5 cycles/ mission hour
Worst-case crack growth rate	1530	918
Quantitative fractography of service cracks	1818	1091
Quantitative fractography of spin-pit cracks	2167	1300

Note: The SII value predicted by the engine manufacturer was 555 h (Ref 3)

Microstructural Analysis. Samples were prepared to examine the microstructure of the component, as well as determine the extent of pitting. The microstructure of fine martensite was consistent with the heat treatment performed on the component. Unusual material defects or large inclusions were not observed. Cracks were noted originating from the bottom of pits, and extending into the base material parallel to the fracture plane, as shown in Fig. 1-37. These cracks appeared to extend in an intergranular fashion.

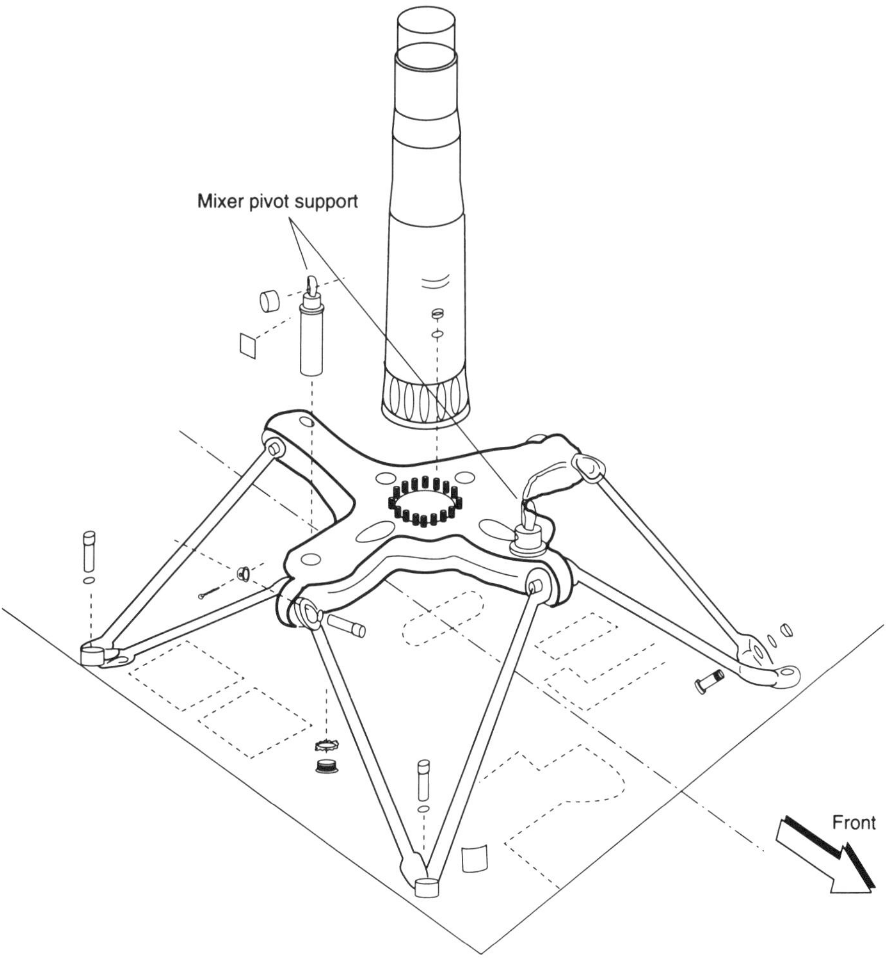

Fig. 1-32 Schematic of the rotor support assembly. Mixer pivot support is shown with an arrow.

Chemical Analysis/Alloy Identification. Samples from the failed mixer pivot support were chemically analyzed, and the results (Table 1-9) met the requirements of ESR 4340 steel.

Hardness measurements performed across a section of the failed component averaged 55.6 HRC, and conformed to the required hardness of 54 to 57 HRC.

Tensile Properties. Tensile specimens were fabricated from the failed component with the intent of obtaining fracture surfaces which could be compared to the fracture surface under investigation. These specimens were pulled to failure and the resultant fracture surfaces were examined using a scanning electron microscope. The tensile specimens displayed a typical ductile cup-cone fracture. The dimpled morphology was indicative of a ductile failure.

Discussion. Severe corrosion pitting occurred along the machined radius of the component and served as a crack initiation site. Desert sand taken from Saudia Arabia and analyzed at ARL revealed high concentrations of chlorides which assisted in the corrosion process. Hydrogen diffused into the high strength material (56 HRC) as a result of the corrosion process and migrated into areas of high stress concentration (crack tip). Evidence for this claim is shown by the tensile tests; when a section of

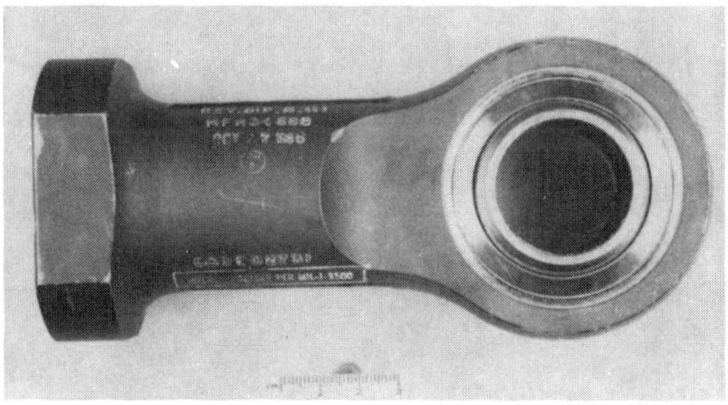

(a)

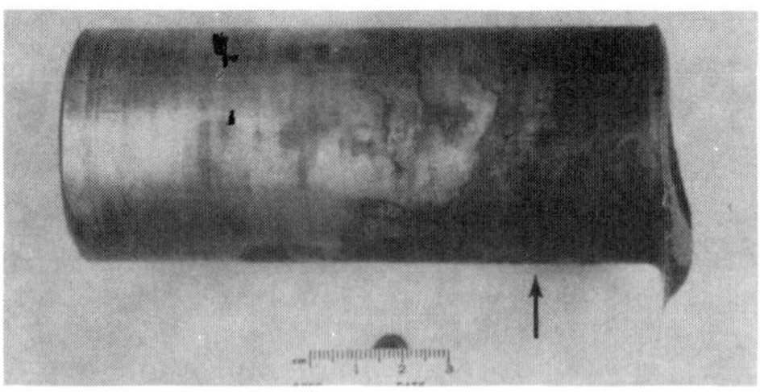

(b)

Fig. 1-33 Macrographs of the two broken halves of the mixer pivot support as received. Region of significant wear is denoted by arrow.

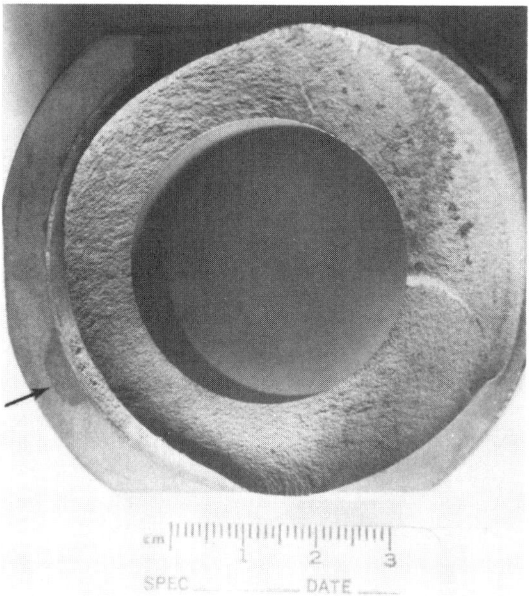

Fig. 1-34 Optical fractograph of Face A (arrow identifies crack origin). Approximately 0.81×

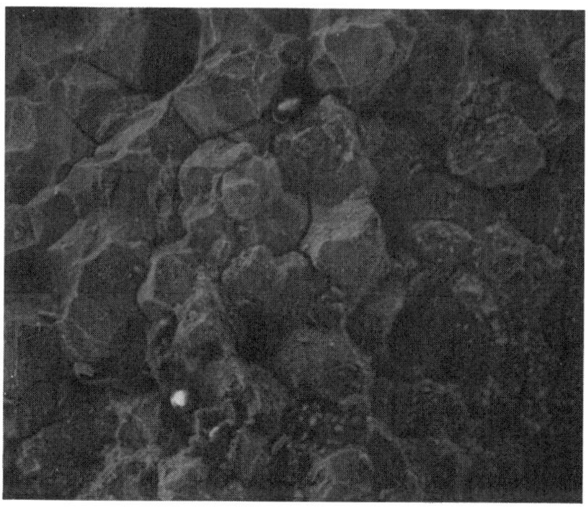

Fig. 1-35 SEM showing intergranular mode of failure. Approximately 350×

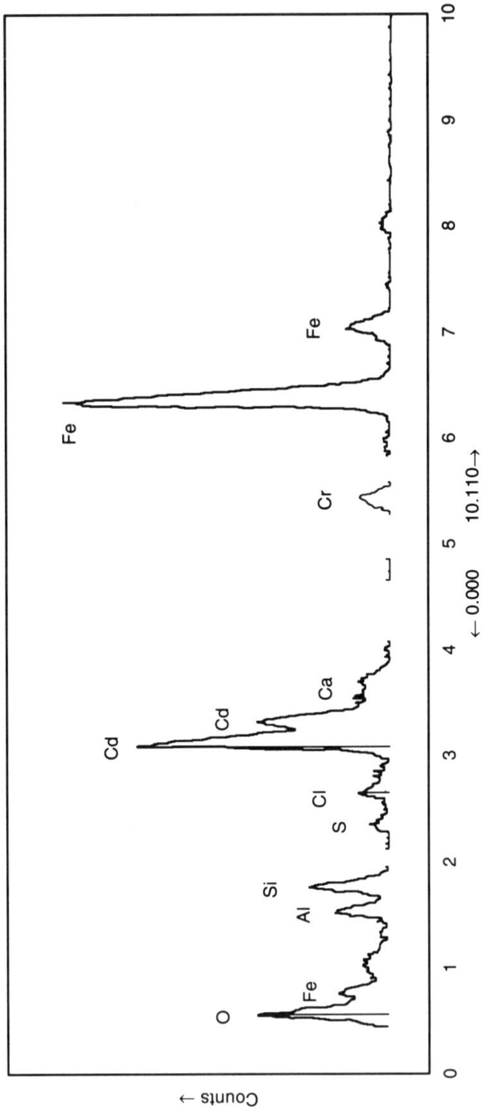

Fig.1-36 EDS spectrum from within a pit adjacent to the fracture

material taken from the failed component was pulled to failure, the resulting fracture surface was dimpled, but the failure mode over 90% of the fracture surface under investigation was intergranular. In addition, the final fracture region on the component (shear lip) also displayed a dimpled topography. This evidence indicates that the material can fracture in a ductile fashion. Hydrogen-assisted cracking occurs in an intergranular fashion in this type of material when heat treated to the hardened condition.

Conclusion and Recommendations. The cracking of the mixer pivot support initiated at the machined radius within a severely pitted region. The fracture did not originate where fretting was most severe. The failure began due to hydrogen charging as a result of corrosion. This condition was aggravated by the stress concentration effects of pitting at the radius and the high notch sensitivity of the material. Failure was hydrogen assisted, and was most likely a combination of stress-corrosion cracking and corrosion fatigue.

It was recommended that the components be removed from service at the first indication of corrosion. Visual inspection with the aid of a magnifying lens could be used to detect corrosion in the field. The component could continue to be utilized when hardened to 54 to 57 HRC as long as there is no visual evidence of corrosion. In this way, the ballistic properties could be maintained. However, a more conservative approach (which would sacrifice some of the ballistic properties of the material) would be to heat treat the component to a softer condition. A softer material would be less notch sensitive, and the inspection intervals could then be longer since the critical crack size would be increased.

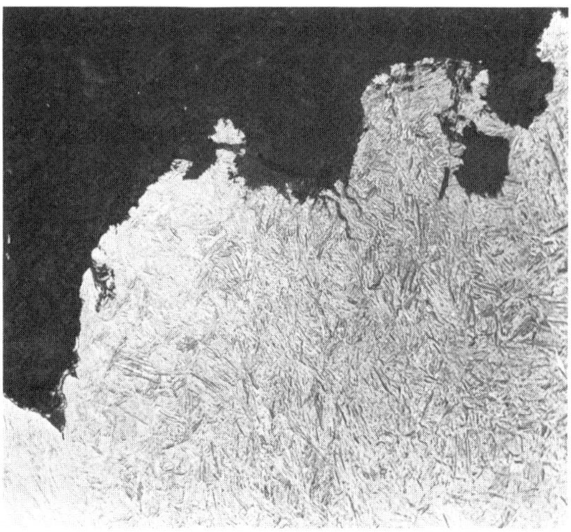

Fig. 1-37 Corrosion pits with cracks extending into to the base material parallel to the fracture plane. Approximately 265×

Failure of a Transport Aircraft Relay Valve Guide*

A steel relay valve guide component operating in the fuel supply system of a transport aircraft broke into two pieces during flight. The valve guide was one of a batch of such components imported fro tail plane pitch control. Valve failure introduces a great risk of fire, which may endanger the safety of the aircraft by the escape of fuel to hot zones around the engine bay. The part was reportedly machined to its entire length by conventional machining—except for two sets of elongated holes, which were precision machined by the electrodischarge process (Fig. 1-38).

The valve guide fractured into two pieces after it had served for only 26 h of flight (Fig. 1-38). The valve was operating at a working load of 20.7 MPa (3000 psi) at the time of failure. The valve guide was manufactured from SAE 52000, a high-carbon chromium ball bearing steel. One of the two broken pieces was selected for fractographic analysis. A section across the origin of failure was metallographically prepared for micro-structural studies.

Visual examination revealed that the valve guide had separated transversely through one of the two sets of elongated holes. The hole bore surface was irregular and discolored.

Macrofractography. Examination of the failed elongated hole under a low-power stereoscopic microscope revealed a rough and corroded bore surface, with clear signs of melting and craters (Fig. 1-39). The fracture had a smooth but rather coarse crystalline surface in a plane at a right angle to the valve axis.

Both elongated holes contained almost identical deep depressions at the bore edge caused by severe burning combined with erosion effects (Fig. 1-39). The fracture surface exhibited a conspicuous dark-shaded, triangular zone adjoining the sunken hole bore edge. Closer observation revealed several faint arclike markings of rather wide spacing running parallel to the profile of the dark-shaded zone. This indicated that the crack had propagated in stages through several lines of arrest in the crack path, having originated at the crater region in the outer corner of the bore surface. All these features suggested typical brittle cracking under low-cycle fatigue.

Fig. 1-38 Rear valve guide, showing transverse fracture through an elongated hole. No plastic deformation was evident around the fracture zone.

*Adapted from A.K Das, in *Handbook of Case Histories in Failure Analysis*, Vol 2, K.A. Esaklul, Ed., ASM International, 1993, p 28–32

Microstructural Analysis. Microexamination of a section across the zone of origin revealed a conspicuous white layer (remelted) in the outer surface of the bore to the maximum depth of 0.035 mm (0.0014 in.). The microstructure also contained a 0.35 mm (0.014 in.) deep overtempered zone just beneath the white layer (Fig. 1-40). Otherwise, the core structure showed satisfactory tempered martensite with a uniform distribution of numerous tiny globular carbides in the matrix, typical of a high-carbon chromium ball bearing steel.

Microexamination also revealed that the fracture was enveloped with a trace amount of white layer to a considerable depth (Fig. 1-40). Moreover, the white layer showed several transverse cracks emanating from the outer bore surface and occasionally penetrating the base of the overtempered zone (Fig. 1-41).

Hardness. A microhardness survey using a 500 g load was performed. A maximum hardness value of 1100 HV in the outermost white layer and a minimum value of 550

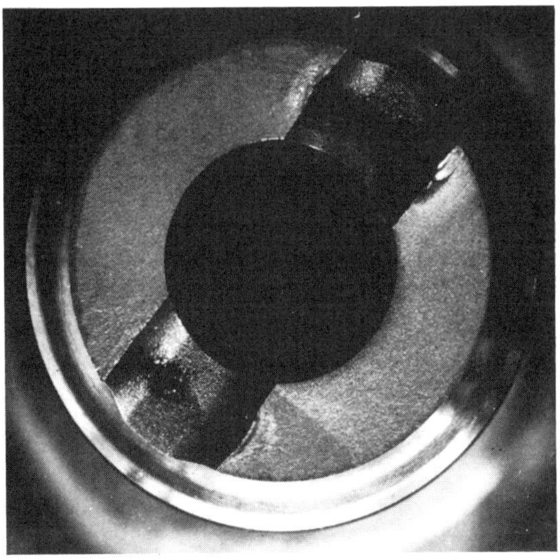

Fig. 1-39 Fractograph of the transverse fracture, showing brittle crack propagation through widely spaced lines of arrest and from the melted crater region on the hole corner

Table 1-9 Chemical analysis

	Composition, wt%										
	C	Mn	Si	P	S	Cr	Ni	Mo	Cu	Al	Fe
HMS-6-1121	0.39 to 0.41	0.60 to 0.80	0.20 to 0.35	0.010 max	0.008 max	0.70 to 0.90	1.65 to 2.00	0.20 to 0.30	0.35 max	0.030 max	Remain
Mixer pivot support	0.39	0.69	0.21	0.008	0.005	0.82	1.99	0.28	0.11	0.012	Remain

HV in the overtempered zone were obtained. The core hardness yielded an average value of 800 HV.

Discussion. Fractographic studies established that the failure occurred by low-cycle fatigue, as evidenced by the presence of several lines of arrest marks in the crack propagation path. Close analysis of the fracture path revealed that the brittle crack originated at the severely eroded zone of craters in the outer corner of the bore surface.

The presence of surface defects, such as discoloration, erosion, and craters, indicated a high-temperature melting phenomenon that occurred during electrodischarge machining, forming a very hard, deep white layer. It was evident from the microstructure

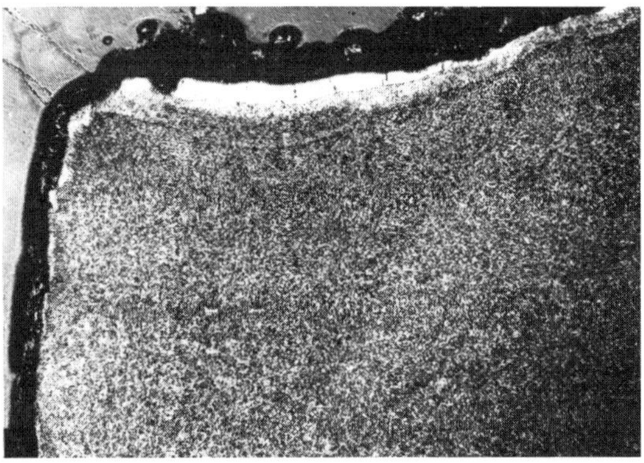

Fig. 1-40 Low-magnification optical micrograph of a section across the crater zone, showing the presence of a conspicuous hard, white layer that seeped into the cracked zone while in the molten condition during electrodischarge machining. An underlying tempered zone exists between the core and the layer.

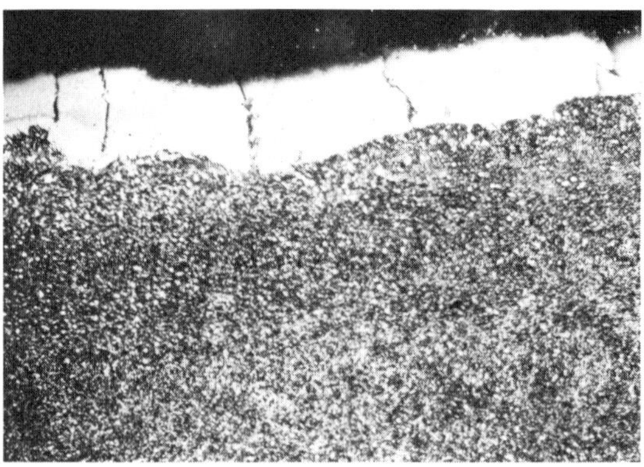

Fig. 1-41 Higher magnification view of the zone in Fig. 1-40, showing parallel transverse cracks in the white layer that occasionally penetrated the tempered martensite matrix

that, because the white layer had seeped into the fracture surface, crack initiation took place during the high-temperature machining stages.

On the other hand, the development of transverse cracks in the outermost bore surface was a direct consequence of drastic quenching of the hard, brittle white layer. Under this condition, any minor stresses during service would lead to premature brittle failure. However, the depth of the white layer (0.035mm, or 0.0014 in.) exceeded the limit normally permissible for finished hardenable steel products (0.025 mm, or 0.001 in.). As such, the excess depth was believed to have contributed to the extreme brittleness of the white layer. All these findings suggested that the process parameters adopted for electrodischarge machining could have been faulty, especially with respect to the maintenance of proper current density. It was also possible that short circuiting led to high-temperature melting.

Conclusion and Recommendations. The brittle cracking of the valve guide was caused by low-cycle fatigue, which was attributed to the presence of a preexisting crack in the outer corner of the bore. This crack resulted from an unduly overheated structure (white layer) formed during machining. Stereoscopic macroexamination as well as borescopic inspection of the electrodischarge machined bore surfaces should be performed before such components be accepted from the exporter.

Fatigue Fracture of a Helicopter Tail Rotor Blade due to Field-Induced Corrosion*

A helicopter tail rotor blade failed and separated during flight. The tail rotor blade spar is the main structural component of the tail rotor blade system and is safety-of-flight critical. During flight, the tail rotor blade failed and the outer section of the blade separated from the aircraft. The crew executed a successful emergency landing. Subsequent inspection revealed that the spar of one tail rotor blade had failed and that the entire section from the spar shank outboard had separated in flight. The departed outboard section was recovered and there was no strike damage evident.

The blade was constructed of 2014 aluminum alloy heat treated to the T-652 condition. The spar shank was hollow and the cavity was filled with lead wool ballast material. The cavity was sealed with a thermoset material.

Visual Examination. The tail rotor blade spar failure location is shown in Fig. 1-42. The failure was located approximately 32.38 cm (12.75 in.) from the inboard end of the blade. The fracture was located within the zone of the spar shank where the lead wool ballast filled the spar, as shown in Fig. 1-43. In addition, the spar cavity was found to contain an unknown fluid. Dimensional analysis revealed no discrepancies.

Surface Examination. Visual examination of the spar fracture surface revealed a flat fracture with beach mark features typical of fatigue. The fatigue fracture initiated at the

---

*Adapted from R.H. McSwain, in *Handbook of Case Histories in Failure Analysis,* Vol 1, K.A. Esaklul, Ed., ASM International, 1992, p 30–32

inner surface of the spar cavity and propagated outward until final overload failure occurred.

Macrofractography. Macrospic examination of the fracture surface revealed a flat area at the origin with a trough-shaped corrosion pit which was 0.43 mm (0.017 in.) long and 0.10 mm (0.004 in.) deep. The crack propagated over half the cross-section of the spar and to the surface of the spar before final overload occurred. Figure 1-44 shows the origin area and lead wool fibers in contact with the spar cavity surface. The corrosion pit at the origin is also shown. The orgin and origin corrosion pit are shown at higher magnification in Fig. 1-45. The corrosion product that filled the pits is lighter, due to charging in the scanning electron microscope beam.

Scanning Electron Microscopy/Fractography. Fractographic analysis by scanning electron microscopy revealed fatigue striations with varying spacings, caused by the complex load spectrum on the blade during crack propagation. The initial 1.10 mm

Fig. 1-42 Failed rotor blade showing fracture shank of the blade. Approximately 0.31×

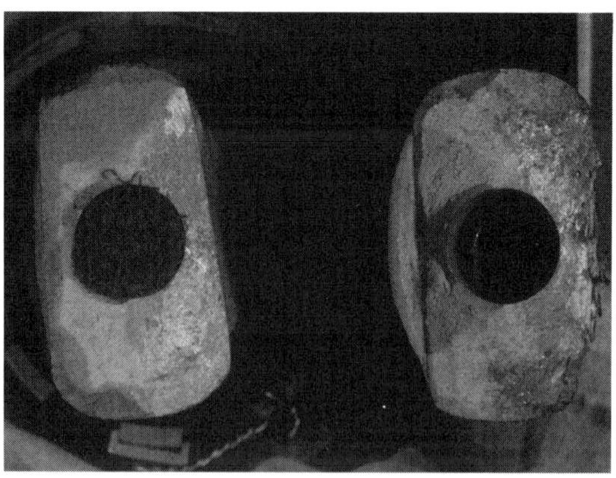

Fig. 1-43 Fracture surfaces in failed tail rotor blade spar shank. Approximately 0.66×

(0.040 in.) of cracking produced a flat fracture surface typical of high-cycle fatigue. The remainder of the fracture surface was rough with evidence of a higher crack propagation rate.

Fig. 1-44 Scanning electron micrograph of tail rotor blade failure origin. Micrograph shows lead wool ballast in aluminum spar bore cavity and fracture which initiated at bore wall. Approximately 13×

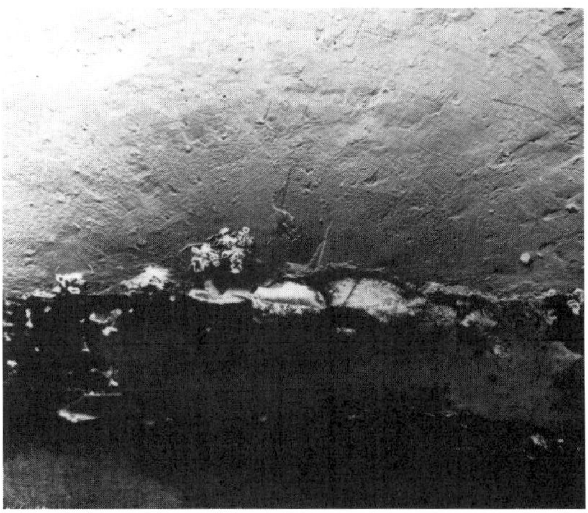

Fig. 1-45 Scanning electron micrograph of tail rotor blade failure origin. Micrograph shows multiple pits at origin with associated corrosion product. Beach marks are shown emanating from pits, typical of a fatigue failure mode. Approximately 63×

Corrosion Patterns. Examination of the inner surface of the spar cavity revealed an unusual pattern of corrosion as shown in Fig 1-46. The same area with corrosion removed is shown in Fig. 1-47.

Microstructural analysis of the spar revealed no material anomalies that would have contributed to the failure. Sections were taken through the failed spar to determine the extent of the corrosive attack. The sectioning revealed spherical pits at the surface of the cavity wall with isolated intergranular attack.

Chemical analysis by X-ray energy spectroscopy revealed a composition consistent with the specified 2014 aluminum alloy.

Hardness tests averaged 84 HRB. The typical hardness of 2014-T652 aluminum is 81 to 90 HRB. The electrical conductivity was also found to be appropriate for the specified material.

Discussion. The fatigue cracking initiated at a corrosion pit on the inner surface of the spar cavity. A fluid substance was found inside the spar cavity and the thermoset plastic seal covering the cavity was broken. The material properties and microstructural features were all consistent with 2014-T652 aluminum. The spar cavity should have been free of moisture or any fluid substance, if the seal had performed as designed.

Conclusion and Recommendations. The tail rotor blade failed by fatigue which initiated from a trough-shaped corrosion pit located at the inner wall of the spar shank cavity. The pitting was a result of galvanic attack caused by entry of moisture into the spar cavity. The moisture served as an electrolyte for galvanic corrosion between the lead wool ballast and the aluminum spar material. Corrosion pits were formed along the

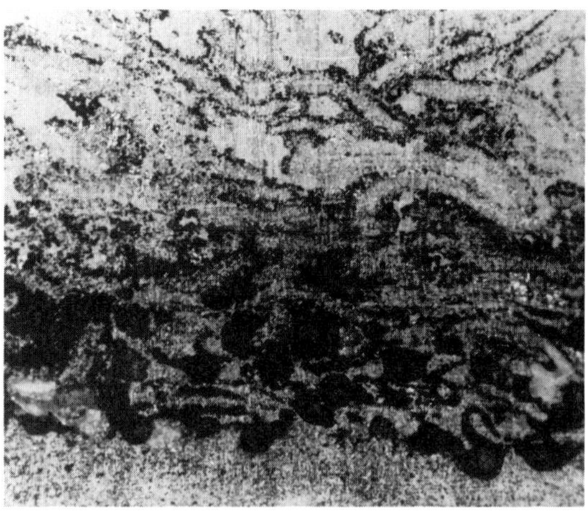

Fig. 1-46 Light micrograph of tail rotor blade shank bore surface. Patterns remaining on surface were result of electrolytic attack produced by lead wool adjacent to aluminum spar in presence of an electrolyte. Black material was found to be corrosion/reaction product filling spherical pits. Approximately 13×

contact points between the lead wool fibers and the spar cavity surface. The moisture entry was due to the failure of a thermoset polymeric seal inside the spar cavity.

The failure could have been prevented by ensuring the integrity of the thermoset polymeric seal, and by barrier coating the wall of the spar cavity with an insulating material. Alternate methods of encapsulating the lead wool would also have prevented the galvanic corrosion, corrosion pitting, and resulting fatigue failure.

Processing-Induced Fatigue Fracture of a Helicopter Tail Rotor Blade*

A helicopter tail rotor blade failed and separated during flight. The helicopter tail rotor blade provides anti-torque flight control as it reacts against the main rotor system. The tail rotor blade spar is the primary structural component of the tail rotor blade assembly and is safety-of-flight critical.

In flight, the aircrew heard a loud bang followed by severe vibrations. The crew executed a successful emergency landing. Subsequent inspection revealed that the spar of one tail rotor blade had failed and that the entire air foil section had departed the aircraft. There was no strike damage evident. The outboard departed section was never recovered.

The tail rotor blade was constructed of 7075 aluminum alloy heat treated to the T7351 condition. The blade had fiberglass pockets.

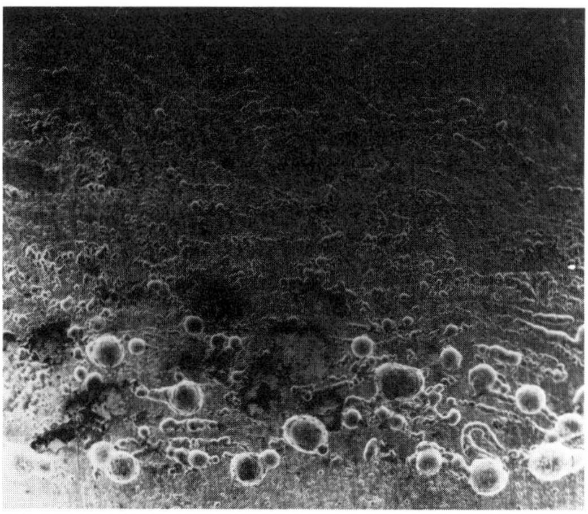

Fig. 1-47 Scanning electron micrograph of tail rotor blade bore surface after cleaning. Area shown is identical to that shown in Fig. 1-46. Removal of corrosion/reaction product revealed spherical pits and etched patterns beneath the black material. Approximately 13×

---

*Adapted from R.H. McSwain, in *Handbook of Case Histories in Failure Analysis*, Vol 2, K.A. Esaklul, Ed., ASM International, 1993, p33–35

Visual Examination. The residual inboard portion of the failed blade spar is shown in Fig 1-48. The fracture was flat and was located approximately 59.7 cm (23.5 in.) from the inboard end of the spar.

Surface examination. The spar fracture surface showed a flat fracture with numerous highly reflective facets present, typical of fatigue, as shown in Fig 1-49. Visual examination also revealed that the flat fracture zone extended over approximately 85% of the spar cross-section before final overload. The arrow in Fig. 1-49 points to a black spot adjacent to the inner surface of the spar where the cracking originated.

Macrofractography. The black spot was the fracture origin, as shown in Fig. 1-50. The black spot encircled the bottom of a pit extending from the inner surface of the spar to a depth of 0.38 mm (0.015 in.). The fracture surface exhibited characteristics of fatigue failure for approximately 85% of its cross-section. The remaining portion of the fracture surface, a darker coarser area, exhibited characteristics typical of overload failure.

Scanning Electron Microscopy Fractography. A low-magnification scanning electron micrograph of the fracture origin is shown in Fig. 1-50. The micrograph shows the corrosion pit located at the fatigue origin. The pit was 0.38 mm (0.015 in.) deep and 0.254 mm (0.001 in.) across at the surface of the spar cavity. No significant amounts of corrosion product were found in the pit. Examination of the other spar cavity surfaces

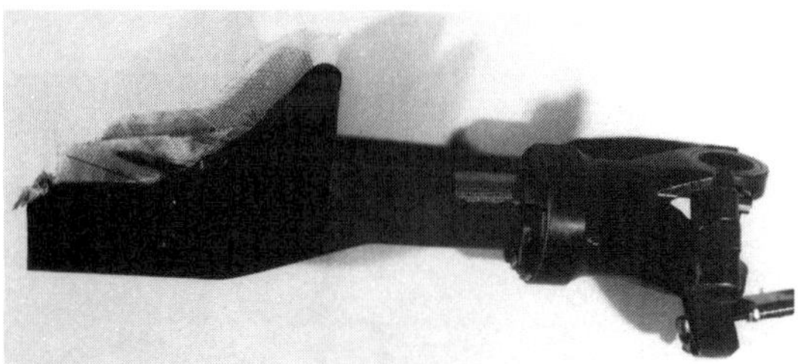

Fig. 1-48 Remaining portion of helicopter tail rotor blade after spar failure and outboard section separation. Approximately 0.14×

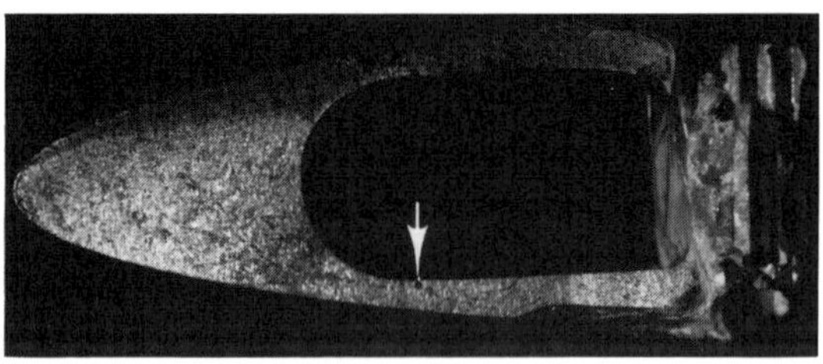

Fig. 1-49 Tail rotor blade spar failure surface showing flat fracture reflective facets, and a dark spot at the arrow. Approximately 0.95×

revealed numerous smaller corrosion pits. All of the pits were filled with corrosion-preventative primer.

Figure 1-51 shows a higher-magnification scanning electron micrograph of the spar fracture surface and shows uniformly spaced fatigue striations.

Corrosion Patterns. Though corrosion pitting was present, there was an absence of significant amounts of corrosion product, and all of the pits were filled with corrosion-preventative primer. Moisture entry into the spar cavity that could have caused corrosion damage was not evident.

The lack of corrosion product in the pits, the presence of primer in the pits, and the fact that the spar had never been reworked, all indicated that the pitting occurred during manufacture, prior to application of the primer.

Microstructural analysis of the spar revealed numerous large intermetallic alloy particles, some of which were exposed to the surface. Those which were exposed had been chemically attacked, resulting in various sized surface pits.

Chemical analysis of the spar material by X-ray energy spectroscopy revealed an alloy composition similar to 7075 aluminum as specified.

Analysis of the fatigue origin pit using X-ray energy spectroscopy showed some residual chlorine and sulfur from the pitting or from later processing.

Hardness tests averaged 72 HRB, below the minimum hardness of 77.5 HRB typically specified for 7075-T7351 aluminum.

The conductivity was found to be 41% IACS, on the higher end of the 38.0 to 42.5% IACS range specified for 7075-T7351 aluminum.

Tensile Properties. The ultimate tensile strength was 427 MPa (62 ksi), below the minimum specified of 475 MPa (69 ksi). The 0.2% offset yield strength was 345 MPa (50 ksi) below the minimum specified yield of 393 MPa (57 ksi). The elongation was 11.3%,

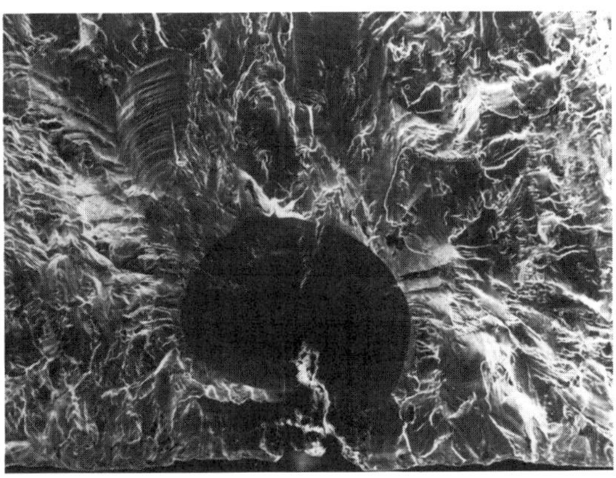

Fig. 1-50 Fracture origin of tail rotor blade spar. Crack propagated from pit that was 0.38 mm (0.015 in.) deep extending from the inner surface of the spar. Approximately 43.7×

satisfactory for the minimum elongation specification of 8%. Hardness, conductivity, and mechanical data all indicated that the spar had been excessively heated after the T7351 condition was established.

Discussion. Further research revealed that the spar had received multiple nickel plating during the manufacturing process. Each plating was removed by acid etching prior to reapplication of the plating. The etching process was the most probable cause for the observed attack of the intermetallic alloy particles and pit formation observed in the microstructural analysis. Records indicated that the failed spar was plated and etched three times before an acceptable plating was achieved. The pits were undetected in the remaining manufacturing operations were covered with corrosion-preventative primer. In addition, the spars were given an unauthorized baking operation at 232 °C (450 °F) for 4 h following each nickel plating operation. The etch pitting became the site for fatigue crack initiation. The reduction in mechanical properties due to the post-plating baking reduced the fatigue strength of the spar. The fatigue crack propagated through 85% of the cross-section of the spar before final failure occurred.

Conclusions and Recommendations. The tail rotor blade failed by fatigue which initiated from a 0.38 mm (0.015 in.) deep acid etch pit located at the inner wall of the spar cavity. The pitting was a result of a multiple plating and plating removal etch process during manufacturing. An unauthorized series of baking operations reduced the mechanical properties of the spar. The reduced mechanical properties contributed to the fatigue process.

Better processing controls during the etching and plating procedures would have lessened the possibility of defective plating and subsequent stripping. The post-plating

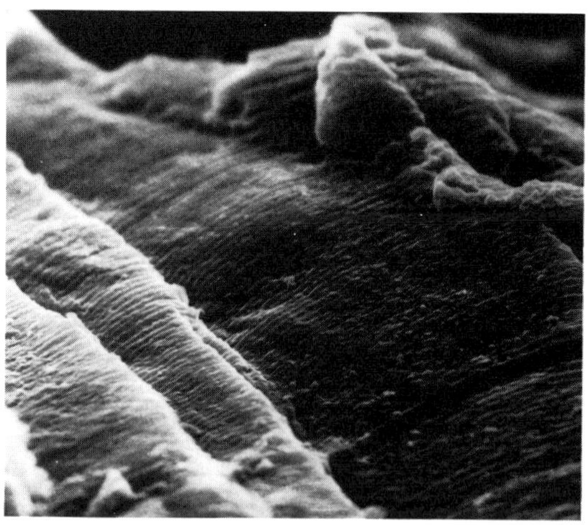

Fig. 1-51 Scanning electron micrograph of fatigue striations on tail rotor blade spar fracture surface. Approximately 3534×

baking which reduced the mechanical properties should not have been performed.

Fatigue Fracture of a Transport Aircraft Crankshaft during Flight*

A steel piston engine crankshaft in a transport aircraft failed catastrophically during flight.

The massive, complex-shaped crankshaft (Fig. 1-52) was machined from a triple alloy steel forging. During flight, the crankshaft experienced a combination of highly complex stresses, including bending and torsion, particularly at the critical pin radius zone. The crankshaft was magnetic particle inspected during periodic major overhauls for the presence of defects or crack indications in the pin radius zone.

One of two crankshafts of a twin piston engine catastrophically fractured into two pieces in an area near the pin radius zone (Fig. 1-53) during flight. The aircraft landed safely with the help of the other engine. The aircraft was maintaining an average flight speed of 2050 rev/min at the time of failure. It had previously undergone three major overhauls involving 3000 h of service.

The crankshaft was manufactured from SAE 4340 steel. Similar crankshafts were reported to have completed their predicted service life satisfactorily. The undamaged portion of the crankshaft was selected for detailed fracture analysis under a low-power stereoscopic microscope. A small portion containing the fracture origin zone was sectioned for in-depth analysis by microexamination and scanning electron microscopy (SEM).

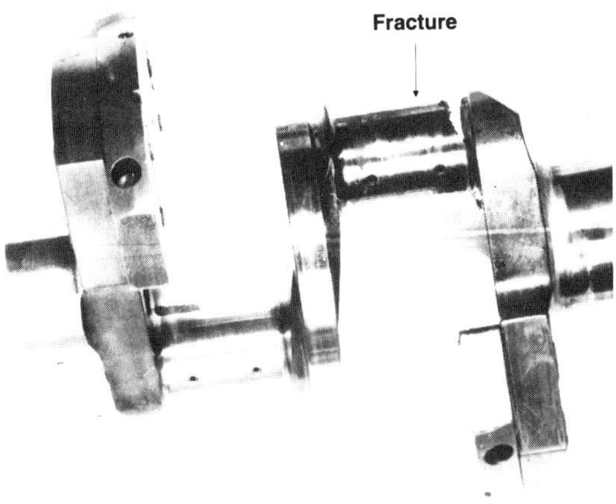

Fig. 1-52 Both portions of the failed crankshaft, showing fracture through the pin radius zone. 0.02×

---

*Adapted from A.K. Das, in *Handbook of Case Histories in Failure Analysis,* Vol 2, K.A. Esaklul, Ed., ASM International, 1993, p 36-38

Visual examination revealed that the failure occurred transversely to the crankshaft axis, with no evidence of plastic deformation around the fracture zone. The fracture features of the broken half on the left side of the crankshaft (Fig. 1-53) were found to be more or less intact, whereas those of the mating surface were completely obliterated by rubbing that occurred after failure.

Surface examination. Examination of the fracture using a low-power stereoscopic microscope revealed a smooth surface with a distinct "eye zone" formed by several conspicuous beach marks around a tiny subsurface defect that served as the origin (black spot, Fig. 1-54). The fracture features were indicative of bending fatigue failure. The smooth zone of the fracture covered slightly more than half the circumference. The remainder exhibited a coarse, irregular surface, typical of overstress failure.

Scanning Electron Microscopy/Fractography. The fracture origin area was examined at higher magnification by SEM to identify the nature of the subsurface defect. The defect was a lone refractory-type inclusion enclosed by a nonmetallic film (Fig. 1-55). The inclusion appeared to be Al_2O_3, as this type of hard refractory inclusion does not undergo deformation by heavy mechanical working, such as forging. However, actual characterization of the inclusion could not be confirmed owing to a lack of appropriate instrumentation at the time of investigation.

Metallography. Microexamination of a section across the fracture in the region adjacent to the subsurface defect revealed that the material was very unclean, with numerous globular nonmetallic oxide inclusions (Fig. 1-56). Otherwise, the microstructure consisted of fine-grained tempered martensite.

Fig. 1-53 Fractograph of the undamaged fracture portion, showing the distinct eye zone of a fatigue crack. 1.11×

Chemical Analysis. The chemical composition of the crankshaft material conformed to SAE 4340 specifications (Table 1-10).

Hardness. Microhardness measurements using a 30 kg load produced an average value of 383 HV. The minimum specified hardness requirement was 360 HV.

Discussion. Fractographic studies established the mode of failure as fatigue under a complex combination of bending and torsional stresses. Its origin was traced to a subsurface hard refractory (Al_2O_3) type of inclusion. The subsurface defect acted as a severe notch for the initiation of fatigue cracking under dynamically loaded conditions. The overall cleanliness of the material was marred by a high content of oxide inclusions.

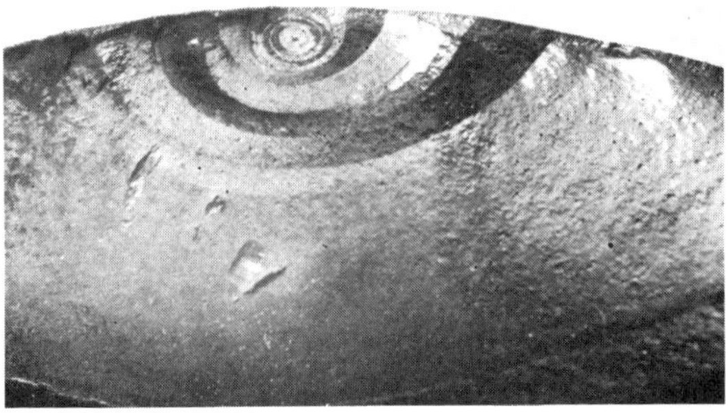

Fig. 1-54 Close up view of the eye zone, showing a tiny black subsurface inclusion that acted as the fracture initiation site. 3.92×

Fig. 1-55 SEM micrograph showing uniformly spaced beach marks around the hard irregular refractory inclusion enclosed in a nonmetallic oxide film

During three successive overhauls, the crankshaft was magnetic particle inspected using the ac method. No crack indication was reported, even though the subsurface crack may have existed in the dormant state. Failure to detect the presence of a defect or crack was attributed to the small size of the inclusion (about 0.01 mm, or 0.0004 in.) and its depth 1 mm (0.04 in.) below the surface. The ac method is insensitive at such depths, and the dc method should be employed instead.

Conclusion and Recommendations. The catastrophic fracture was attributed to the presence of a subsurface defect in the form of a lone hard refractory inclusion in the critically loaded pin radius zone which became the initiation site for a hidden fatigue crack. The uncleanliness of the material further reduced its fatigue resistance.

It was recommended that the supplier exercise utmost care to maintain product quality, including the use of the dc method of magnetic particle inspection.

Table 1-10 Results of chemical analysis

	Composition, %	
Element	Failed crankshaft	SAE 4340 requirements
Carbon	0.41	0.38-0.43
Manganese	0.75	0.65-0.80
Phosphorus	0.015	0.025 (max)
Sulfur	0.025	0.040 (max)
Nickel	1.75	1.65-2.00
Chromium	0.82	0.70-0.90
Molybdenum	0.25	0.20-0.30

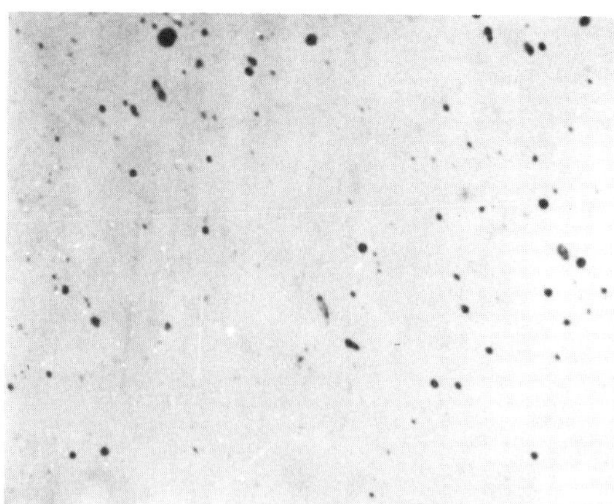

Fig. 1-56 Micrograph showing nonmetallic globular oxide inclusions

Hydrocarbon Processing

Capital value of large process plant losses as a percentage of the nominal value of hydrocarbon produced between 1963 and 1992 (Ref 9) are shown in Figure 1-57. Numbers of losses per billion tons of oil and oil equivalent of gas produced increase at a lower rate (Ref 9), as shown in Figure 1-58, which suggests that the mean capital loss has increased during the time frame reviewed. This is shown to be the case in Figure 1-59.

The causes of disruptive losses in hydrocarbon processing and chemical industries worldwide (Ref 10) and in Europe (Ref 11) are shown in Figure 1-60. Mechanical failure of equipment is the dominant failure type followed by operational error.

The type of equipment in which failures giving rise to large losses is identified in Table 1-11 (Ref 10), with piping highest at $41.9 million USD. The data of Table 1-12 (Ref 11) indicate that mechanical piping failure at 25% of the total 82% of mechanical failures is most common. This predominant failure of piping is further reflected in Figure 1-61, which indicates asset loss risk as a function of equipment type (Ref 12). Corrosion failures in this survey (Ref 11) constitute 18% of the total.

In terms of non-disruptive failure in chemical (Ref 13) and offshore (Ref 14) process plant, the dominant mechanisms of mechanical failure and corrosive failure are shown in Fig. 1-62. Fatigue and mechanical overload and to a lesser extent, poor fabrication appear to be the dominant mechanisms in mechanical failures, which themselves constitute about 50% of the total. Whereas general corrosion, stress corrosion cracking and to a slightly lesser extent, pitting, intergranular and preferential weld corrosion

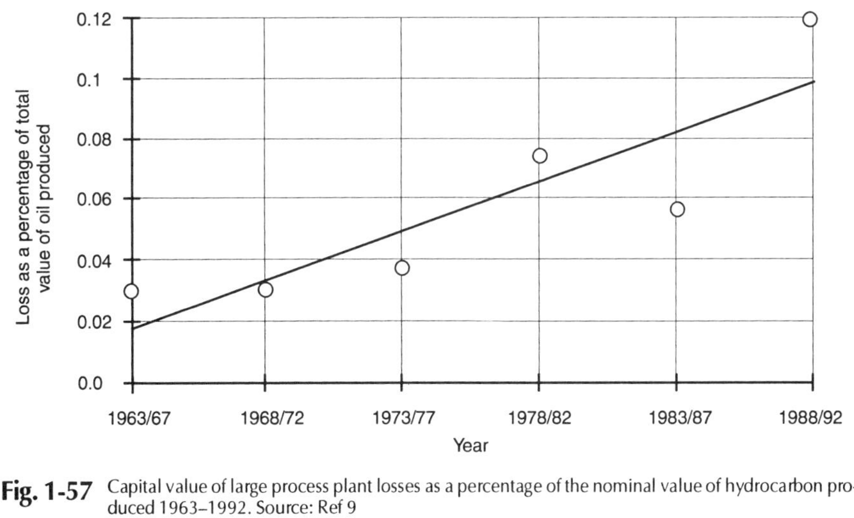

Fig. 1-57 Capital value of large process plant losses as a percentage of the nominal value of hydrocarbon produced 1963–1992. Source: Ref 9

dominate the corrosive failures, which in themselves, constitute the other 50% of the total.

The types of process unit which exhibited the greatest number losses in the time-frame 1962 to 1991 are shown in Table 1-13 (Ref 10). Refineries head the list, since there are more refineries than any other type of process unit at 40% of the total number, but with an average dollar loss of $44.6 million USD. Gas Processing Plant, suffered the greatest dollar loss at $55.5 million USD, however. In the same time-frame, vapor-cloud

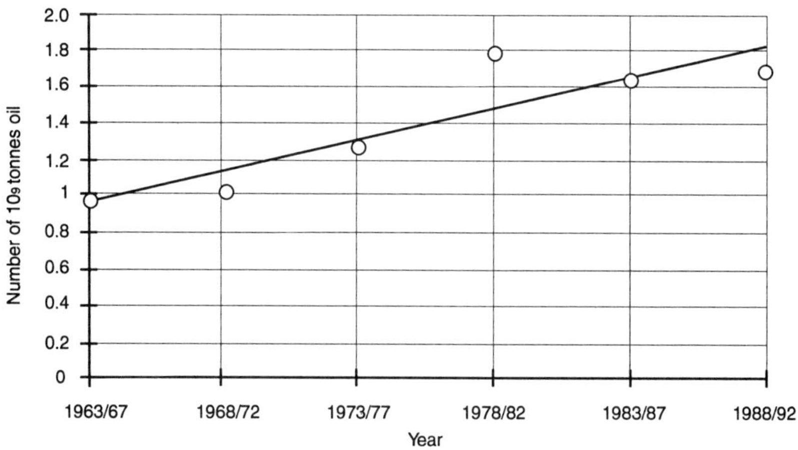

Fig. 1-58 Number of large process plant losses per billion tonnes of oil and oil equivalent of gas produced 1963-1992. Source: Ref 9

Table 1-11 Equipment in which failures leading to large losses in process plant occurred: worldwide survey, all losses

Equipment	% of total number of losses	Average financial loss, U.S. $ million
Piping systems	31	41.9
Tanks	17	40.5
Reactors	13	28.9
Process drums	7	25.5
Marine vessels	6	32.0
Pumps/compressors	5	19.2
Heat exchangers	3	24.0
Towers	3	53.8
Heaters/boilers	1	28.6
Miscellaneous	9	34.7
Unknown	5	25.0

Source: Ref 10

explosions caused the highest annual loss ($59.6 million USD) as shown in Table 1-14 (Ref 10).

It is clear from these data that the overall failure rates in hydrocarbon processing equipment is increasing.

To further elucidate the failure mechanisms and the equipment in which they occur, typical examples of these failures are given in the following section.

Table 1-12 Type of failure and equipment in which failure occurred: disruptive failures in European countries, mechanical and corrosion failures only

	% of total number of losses
Mechanical failures	
Piping	25
Instruments and control systems	9
Valves	6
Machinery	6
Welds	5
Total, **mechanical**	**82**
Corrosion	
Internal	11
External	7
Total, **corrosion**	**18**

Source: Ref 11

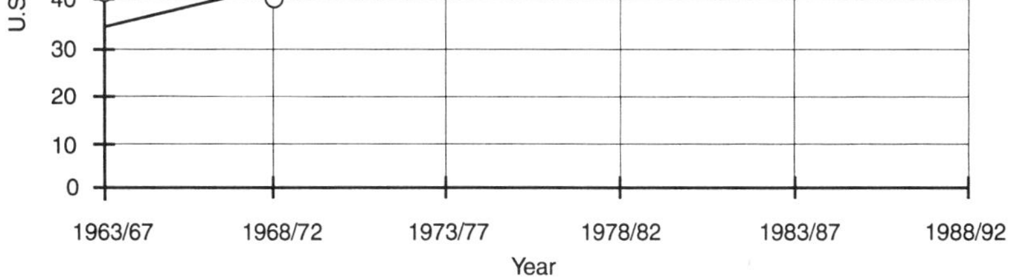

Fig. 1-59 Mean capital value of major process plant losses 1963–1992. Source: Ref 9

Hydrocarbon Processing Service Failures and Analysis

Following are example case histories from *Handbook of Case Histories in Failure Analysis,* Vol 1 and 2.

Hydrotest Failure of a Carbon Steel Pressure Vessel*

A carbon steel pressure vessel failed while being hydrotested in the fabricating shop. The pressure vessel was about 760 mm (30 in.) in diameter and 4.6 m (15 ft) long with a 33.3 mm (1 5/16 in.) wall thickness. It was intended for use at an oil production facility. The vessel had been completely fabricated and was being hydrotested prior to painting and shipping. Upon reaching a pressure of 14 MPa (2000 psi), the vessel fractured, ejecting two large sections of the wall. The temperature of the water being used for the hydrotest was 15 °C (60 °F), and the ambient temperature was about 10 °C (50 °F).

The pressure vessel was fabricated from ASTM A515 grade 70 steel plate, in accordance with the ASME Boiler and Pressure Vessel Code, Section VIII, Division 1. As such, the steel was in the as-rolled condition, and the vessel was not required to be stress relieved after welding. A515 pressure vessel plate is carbon steel made to a coarse grain

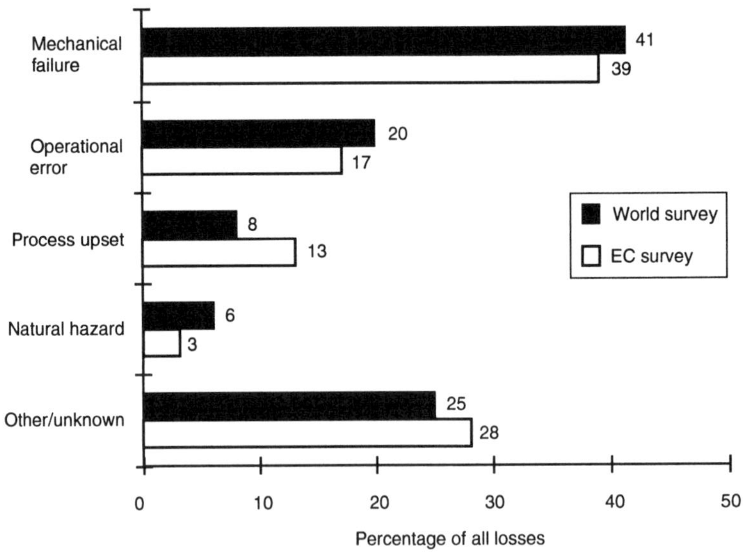

Fig. 1-60 Cause of disruptive losses in hydrocarbon processing and chemical industries worldwide and in Europe. Source: Ref 10, 11

*Adapted from W.R. Warke, in *Handbook of Case Histories in Failure Analysis,* Vol 1, K.A. Esaklul, Ed., ASM International, 1992, p 143–146

practice (silicon killed) and is intended for elevated-temperature use. However, the code permitted the alloy to be used to –30 °C (–20 °F), the minimum design metal temperature of the vessel. Other pressure vessels fabricated at the same time passed the hydrotest and were placed in service.

The two sections of the pressure vessel wall that were ejected during the failure and a section of the wall adjacent to one of these pieces were sent for laboratory failure analysis.

Visual Examination. The side of the pressure vessel is shown in Fig. 1-63, and the two pieces that were ejected are shown in Fig. 1-64. In addition to the cracks around the two pieces, a crack ran up and over one end of the vessel (Fig. 1-65). No bulging or other overall deformation of the vessel was visible.

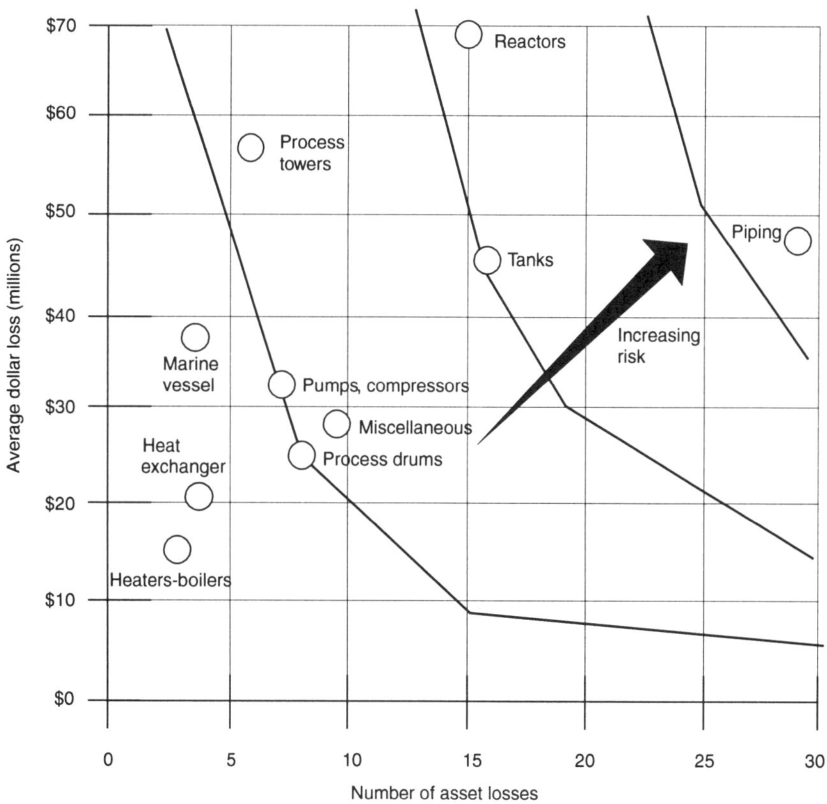

Fig. 1-61 Number of asset losses (frequency). Source: Ref 12

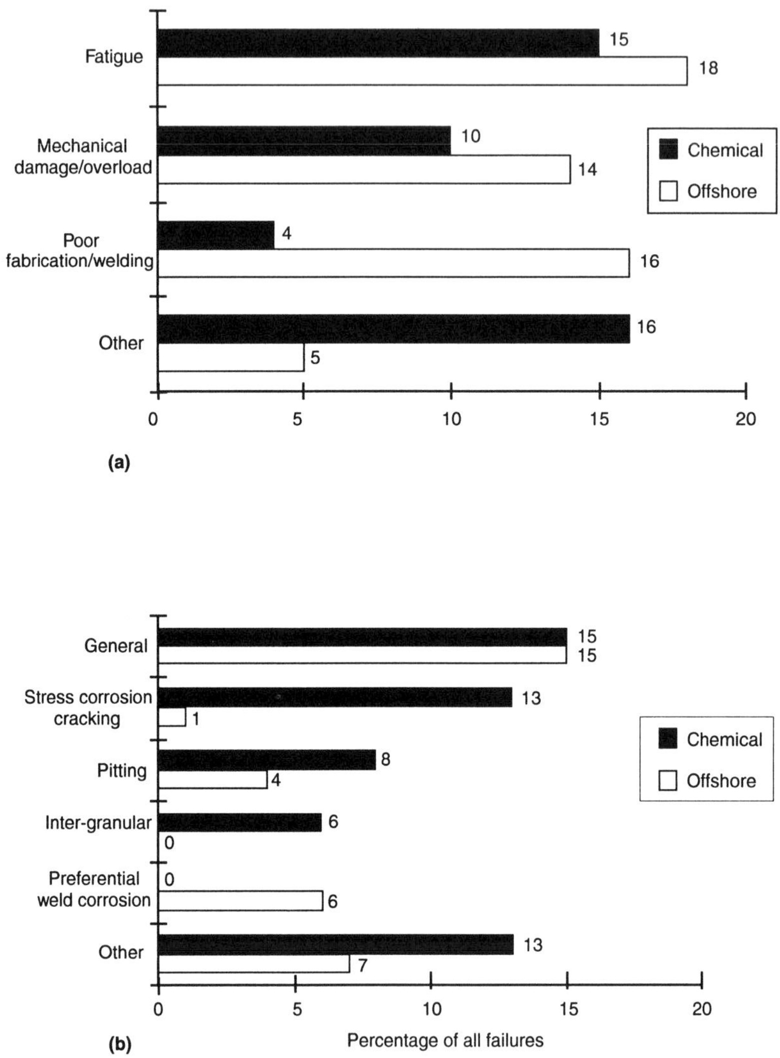

Fig. 1-62 Type of nondisruptive failure in chemical and offshore process plant: (a) mechanical failures (total, chemical 45%, offshore 53%); (b) corrosion failures (total, chemical 55%, offshore 33%). Source: Ref 13, 14

Surface Examination. The fracture surfaces had the general characteristics of brittle fracture, in that they were flat (no shear lips), perpendicular to the plate surface, and covered with chevron marks. The chevron marks enabled the propagation direction of each crack to be determined, as indicated by the arrows in Fig. 1-65. They point to an origin at the toe of one of the nozzle welds on the inside diameter of the vessel. Figure 1-66 shows specimen A (Fig. 1-65), which contained this nozzle; the location of the origin is indicated.

Macrofractography. Figure 1-67(a) illustrates the fracture origin, and a small dark surface flaw is visible at the convergence of the chevrons. This preexisting thumbnail crack was about 1.5 mm (0.06 in.) deep and 15.2 mm (0.60 in.) long and was located at the toe of the outermost pass of the nozzle weld. Figure 1-67(b) shows the inside diameter plate surface of the mating section of the vessel wall. The rough globular area

Table 1-13 Types of process unit in which major losses occurred: worldwide survey, 1962–1991

Type of plant	% of total	Average capital loss, U.S. $ million
Refineries	40	44.6
Petrochemical plant	35	47.0
Terminals	11	28.1
Gas processing plant	6	55.5
Other	8	31.4

Source: Ref 10

Table 1-14 Type of incident causing major losses in process plant: worldwide survey, 1962–1991

	Number	%	Average loss, U.S. $ million
Type of loss			
Fire	62	36	36.1
Vapor cloud explosions	59	35	59.6
Internal and other explosions	43	25	33.6
Other	6	4	24.7
Total	**170**	**100**	**43.2**

	Fire	Vapor cloud explosion	Internal explosion	Other
Type of loss by type of plant; percentage of total				
Refineries	48	31	15	6
Petrochemical plant	17	37	46	0
Terminals	44	28	22	6
Gas processing plant	40	60	0	0
Other	50	36	7	7

Source: Ref 10

at the origin was an arc strike where a welder had started the arc when making the nozzle weld.

Microstructural Analysis. A metallographic section was prepared through the arc strike at the origin; the results are shown in Fig. 1-68. The base plate microstructure was a ferrite-pearlite mixture with an ASTM grain size of about 4 to 5, as expected for A515 grade 70 steel. The arc strike exhibited martensitic resolidified and heat-affected zones (HAZs). The large inclusions near the surface were identified by energy-dispersive analysis as aluminum silicate and manganese-iron silicate containing calcium and titanium. Because the cover passes of the nozzle weld were made with AWS 7024 weld rod, the inclusions were found to be welding slag from the cover passes. Consequently, the arc strike must have been made when one of the final passes was made.

Material and Weld. A sample of the steel plate containing the point of origin was chemically analyzed, and the results (Table 1-15) met the requirements of the ASTM specification.

Hardness. The hardness of the steel plate was measured as 187 HB. The resolidified and HAZs had hardnesses ranging from 31 to 33 HRC, when converted from microhardness data.

Tensile properties of the plate were determined using 12.83 mm (0.505 in.) diam tensile bars. Although the results (Table 1-16) met the requirements of the ASTM specification, the tensile strength was very near the upper limit for this material.

Impact Toughness. Standard Charpy V-notch impact tests were conducted on the plate; the results are shown in Fig. 1-69. The 20 J (15 ft ·lb) transition temperatures were 65 and 50 °C (150 and 120 °F) for the longitudinal and hoop (rolling) directions, respectively. The toughness at the hydrotest temperature in the longitudinal orientation of the vessel (i.e., in the direction of the fracture at the origin) was 6.8 J (5 ft ·lb).

Fracture toughness was not determined experimentally, but was estimated from the Charpy toughness, assuming that the critical stress-intensity factor is roughly 15.5 times

Fig. 1-63 Pressure vessel immediately after the hydrotest failure

the square root of the Charpy value on the lower shelf. The value for K_{Ic}, at the failure temperature, obtained in this manner was 37.4 MPa$\sqrt{m}$ (34 ksi$\sqrt{in.}$).

Analytical Stress Analysis. The nominal axial stress (roughly perpendicular to the fracture surface at the origin) in the vessel wall corresponding to the internal pressure at the time of failure was calculated to be 79 MPa (11.5 ksi). However, several factors existed that would have resulted in the presence of a much higher stress at the origin. First, the vessel had not been postweld heat treated or stress relieved. Consequently, the radial residual stresses resulting from thermal contraction following nozzle welding would have to be added. Second, the weld toe angle was very steep and as such would have caused a stress concentration factor of perhaps 4 to 5. Finally, the multiaxial stress state resulting from the overall and local geometries could have allowed the stress at the origin to approach the uniaxial yield strength, without being relieved by plastic flow. In view of these factors, a local stress level of 415 MPa (60 ksi) was assumed to exist at the origin at the time of failure.

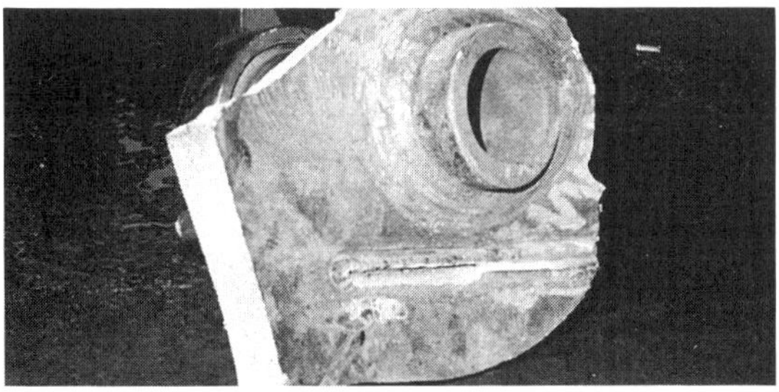

(a)

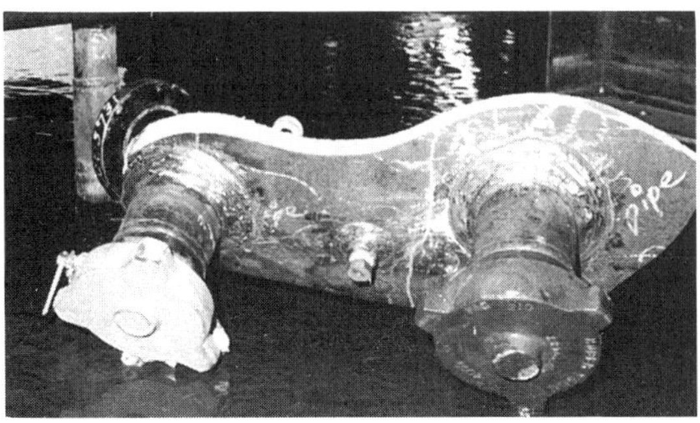

(b)

Fig. 1-64 Fragments of the vessel wall that were ejected at the time of failure

Fracture Mechanics Predictions. A fracture mechanics calculation was made using the above estimate of the stress and the measured preexisting crack size at the arc strike. A Q value of 0.75 was used for the 10:1 ratio of crack length to depth. A value of the stress intensity at failure of 33 MPa$\sqrt{m}$ (30 ksi$\sqrt{in.}$) was calculated.

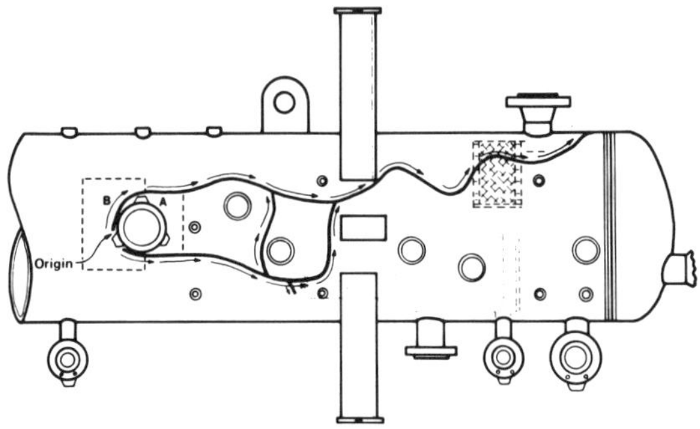

Fig. 1-65 Schematic of the failed pressure vessel showing the location of the origin and the path of the propagating fracture. A and B indicate sections of the vessel selected for laboratory examination.

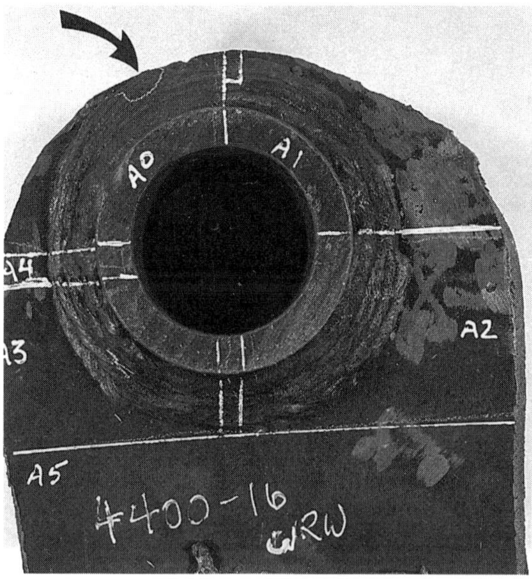

Fig. 1-66 Inside surface of specimen A (Fig. 1-65), showing the fracture origin at the toe of the nozzle weld (arrow)

Discussion. The chemical analysis and tensile tests on the steel from which the pressure vessel was fabricated met the requirements of A515 grade 70. The fabrication and testing of the vessel were in accordance with the ASME Boiler and Pressure Vessel Code in effect at the time. However, the toughness of the steel was low, which is not unusual for as-rolled A515 plate. A rough estimate of the fracture toughness of the steel

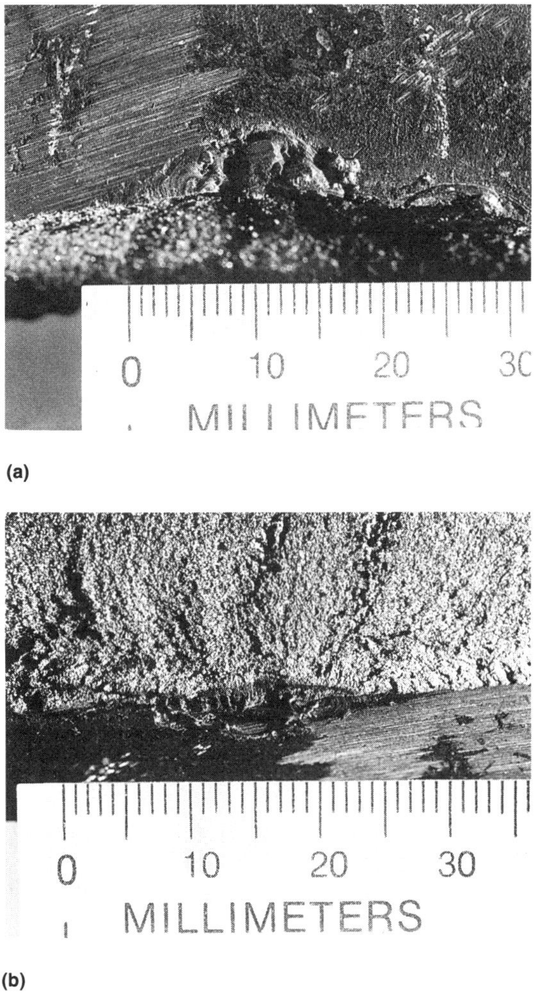

Fig. 1-67 Macrographs of the fracture origin showing (a) the thumbnail crack on the fracture surface and (b) the arc strike on the inner surface of the vessel

and of the stress intensity at a small flaw at the toe of a nozzle weld showed a remarkable degree of agreement, within 10%. This agreement indicated that, under the conditions existing at the time of the hydrotest, the flaw could have initiated a brittle fracture of the vessel, as observed.

Conclusion and Recommendations. The hydrotest failure of this pressure vessel was the result of a preexisting crack in an arc strike at a nozzle weld on the inside of the vessel. Although the crack was only 1.5 mm (0.06 in.) deep and 15.2 mm (0.60 in.) long, it was located in a highly stressed area, and the material was of insufficient toughness to resist rapid crack propagation.

Failure could have been prevented by either eliminating the flaw, lowering the stresses, or increasing the toughness of the steel. Arc strikes are common in welded fabrication and are a source of problems on occasion. Because of the low heat input and

Table 1-15 Results of chemical analysis

	Composition, %	
Element	Plate	A515 grade 70 requirements (a)
Carbon	0.29	0.33 (max)
Manganese	0.82	0.90 (max)
Phosphorus	0.019	0.035 (max)
Sulfur	0.021	0.040 (max)
Silicon	0.21	0.15-0.35
Nickel	0.13	NR
Chromium	0.17	NR
Molybdenum	0.04	NR
Copper	0.10	NR

(a) NR, no requirement

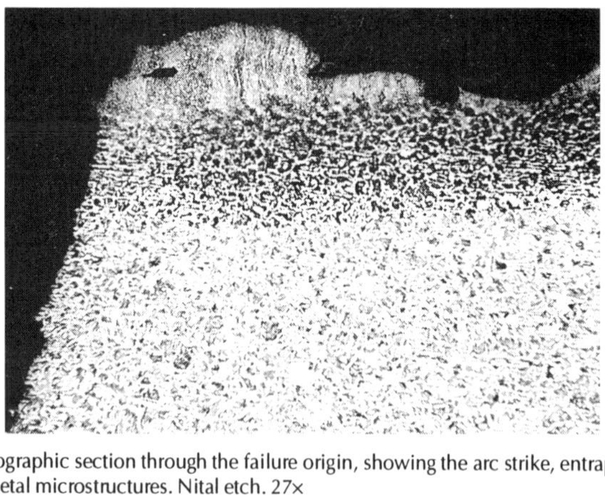

Fig. 1-68 Metallographic section through the failure origin, showing the arc strike, entrapped slag, HAZ, and base metal microstructures. Nital etch. 27×

the quenching action of the cold steel plate, high hardness and microcracking are not uncommon. Welders often weld back over their arc strikes, or subsequently grind them out. In the confined quarters of the inside of this pressure vessel, however, the arc strike could have been easily overlooked. Postweld heat treatment would have lowered the residual stresses, and a less steep weld toe would have reduced the stress concentration at the toe. The toughness of the steel could have been increased by normalizing, but a better course of action would be to use A516 grade 70 steel. Specification A516 requires that the steel be produced by a fine grain practice, which is normally met by aluminum deoxidation. Normalized A516 steel usually has excellent toughness.

Stress-Corrosion Cracking of CMn Steel in a CO_2 Absorber in a Chemical Plant*

The source of cracking in the circumferential weld of a JIS-SM50B carbon-manganese steel pipe used in a CO_2 absorber was investigated; the absorber had been in service for 18 years. The seam had been weld-repaired twice, and the repair welds had been locally stress relieved.

Table 1-16 Results of tensile testing

Property	Plate	A515 grade 70
Yield strength (0.2% offset), MPa (ksi)	400 (58.2)	262 (38.0) (min)
Ultimate tensile strength, MPa (ksi)	615 (89.2)	483 to 620 (70.0–90.0)
Elongation, % MPa	25.5	17.0 (min)
Reduction in area, %	56.5	...

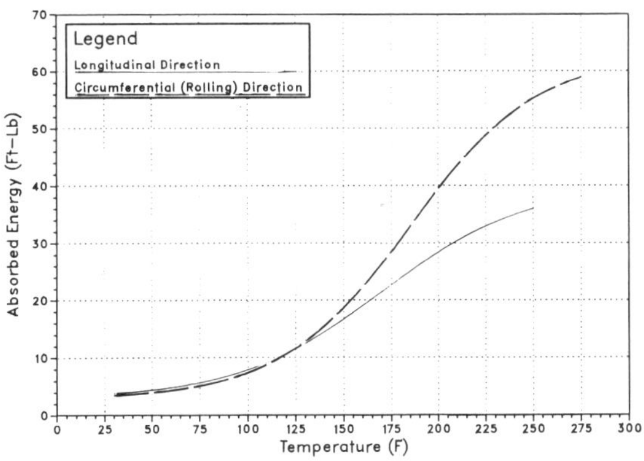

Fig. 1-69 A515 grade 70 pressure vessel Charpy V-notch impact transition curves

*Adapted from S.K. Chauhan, in *Handbook of Case Histories in Failure Analysis,* Vol 1, K.A. Esaklul, Ed., ASM International, 1992, p 191–193

A CO_2 absorber is used to remove CO_2 from the process gas so the gas can be reused in the chemical plant. Carbon dioxide is removed by passing it through a solution (Table 1-17) flowing downward from the top of the vessel (Fig. 1-70). The process gas is allowed to rise in the solution. The solution absorbs CO_2 from the process gas, and the gas exiting the top of the vessel is free of CO_2. The temperature of the solution during this process is approximately 125 °C (260 °F).

Stress corrosion is caused by the presence of external or internal stresses in addition to the corrosive medium. The following condition produced these factors in this case:

- After welding, the circumferential seam was heat treated by local stress relieving, which does not ensure complete removal of residual stresses.
- The seam had been repaired twice without being stress relieved, which added considerable residual stresses to the weld and nearby region.
- The solution passing through the vessel contained CO_2-CO-H_2O, KHCO, and Cl^- ions, which induce stress-corrosion cracking (SCC) in carbon steels at temperatures greater than 100 °C (212 °F).

The vessel was 37.4 m (123 ft) long, 45 mm (1.8 in.) thick, and 3.9 m (13 ft) in outside diameter and was made of carbon-manganese steel conforming to Japanese specifica-

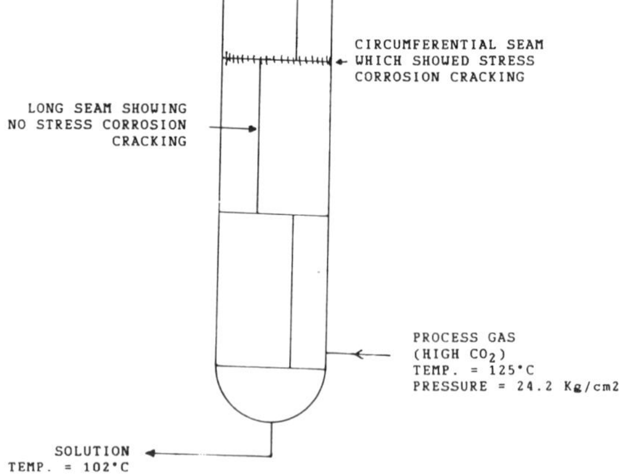

Fig. 1-70 Absorber vessel used to remove CO_2 from the process gas

tion JIS-SM50B. It was welded with E-7018 AWS-classified welding electrodes. The longitudinal seams of the same vessel showed no tendency toward SCC. However, these weld seams had been furnace stress relieved.

The vessel was shortened to remove the cracked circumferential weld and welded again. Two plates, 850 × 120 × 45 and 250 × 100 × 45 mm (35 × 5 × 1.8 and 10 × 4 × 1.8 in.) from the cut portion, with weld in the center, were sent for evaluation.

Visual Examination. Weld seams on one side of each plate were clearly visible. No cracks were visible after cleaning the plates with acetone.

Radiography. The plates were ground flat to facilitate radiographic examination. Both plates were checked for internal cracks. Several branched cracks in the weld and heat-affected zone (HAZ) of the longer plate were reported. These cracks were parallel to the weld and branched severely into the base metal.

Liquid penetrant testing was conducted on both surfaces of the plates. A large number of branching cracks were noticed in the weld and HAZ. However, these cracks were seen only on the inner surface of the vessel.

Ultrasonic examination revealed a few cracks in the weld as well as in the HAZ. These cracks were 2 to 3 mm (0.08 to 0.12 in.) deep measured from the inner surface.

Microstructural Analysis. One section of the small plate was polished, etched with 3% nital, and examined at 5× magnification. All cracks were in the HAZ of the weld, extending and branching into the base metal just adjacent to this zone. The depth of the cracks was 10 to 12 mm (0.4 to 0.5 in.). Two repair beads, one 10 to 12 mm (0.4 to 0.5 in.) deep and another 2 to 3 mm (0.08 to 0.12 in.) deep, were also seen adjacent to the original weld (Fig. 1-71). Uniform grain structure was observed in the base metal, with coarser grains visible in the HAZ.

Crack Paths. The cracks propagating into the base metal followed a predominantly transgranular path (Fig. 1-72).

Chemical Analysis. The compositions of the parent metal and the weld metal are listed in Table 1-18. Both compositions are within the specifications of JIS SM50B and E-7018, respectively.

Hardness measurements were taken on the base metal, original weld, two repair welds, and their HAZs using the Vicker's diamond pyramid method at a load of 10 kg (~22 lb). The results were within specification and are given in Table 1-19.

Table 1-17 Solution used to remove CO_2 from the process gas

Components	Value
K_2CO_3, %	18
Diethylene amine (DEA), %	2.7
KHCO, %	13.8
As_2O_3, %	0.91
Aqueous K_2 CO3, %	27.5
Chloride, ppm	97.8
Iron, ppm	19.1
Suspended matter, ppm	11.9

Impact Toughness. Charpy V-notch impact testing was conducted at 0 °C (32 °F) and at room temperature to determine the toughness of the weld, HAZ, and base metal. Specimens were taken from the top and bottom of the weld, HAZ and base metal. The results met the material specifications and are given in Table 1-20.

Simulation Tests. To evaluate the susceptibility of the weld to SCC, two U-bent specimens (80 × 20 × 3 mm, or 3.2 × 0.8 × 0.12 in.), machined from the same material with weld in the center and bent to a radius of 5 mm (0.2 in.), were exposed to the same operating conditions inside the vessel (ASTM G30). These specimens were checked with liquid penetrant prior to insertion in the vessel. After 1 year, the specimens were removed and examined. Liquid penetrant examination of these specimens also showed multidirectional cracks in the weld and the HAZ.

Discussion. Nondestructive testing revealed severe branching of the cracks on the inner side of the plates. This type of cracking is generally representative of SCC. Microexamination of the cracks revealed that propagation was transgranular, further supporting the possibility of SCC. The results of the simulation test confirmed this mode of failure.

The vessel was in service at a temperature of 125 °C (255 °F) in the presence of carbonate ions and CO_2-CO-H_2O. In this environment, carbon steel is very susceptible to SCC.

All the cracks were progressing through the HAZ of the weld. It appears that the cracks originated in this zone and subsequently propagated into the base metal and weld. The weld joint under consideration was locally stress relieved on site, while other joints were stress relieved inside a furnace. The vessel was also repaired to a depth of 10 to 12 mm (0.4 to 0.5 in.) once and to an additional 2 to 3 mm (0.08 to 0.12 in.) the following

Table 1-18 Chemical analysis results

	Composition, %								
Material	C	Mn	Si	S	P	As	Sn	Pb	Sb
Weld	0.049	0.62	0.56	0.028	0.017	0.0014	0.0025	0.001	nil
Parent metal	0.18	1.47	0.35	0.022	0.018	0.0021	0.0030	0.001	nil
JIS-SM50B	0.18 (max)	1.50 (max)	0.55 (max)	0.04 (max)	0.04 (max)	...	...	...	...

Table 1-19 Hardness results

Location	Hardness, HV
Original weld (inner side)	176-181
Original weld (outer side)	180-187
First repair weld (10-12 mm deep)	206-221
Second repair weld (2-3 mm deep)	276-279
HAZ of first repair weld	195-206
HAZ of second repair weld	203-209
HAZ of original weld	161-170
Base metal	150-160

year. After these repairs, no stress relieving was done. This further enhanced the stress conditions in the HAZ and aggravated the SCC.

Conclusion and Recommendations. The basic cause of failure was stress corrosion caused by locked in residual stresses resulting from improper and inadequate stress relieving.

To avoid SCC the welding seams should be furnace heat treated at a temperature of 600 to 640 °C (1110 to 1180 °F) for a minimum of 1 h per inch of section thickness.

Cracking of Inconel 800H in a Steam Methane Reformer Furnace*

Maintenance shutdown inspections regularly showed cracks in the welded outlet pigtail connections of a steam-methane reformer furnace.

The catalytic steam reforming process uses light to medium hydrocarbons in a certain ratio to steam in order to produce different synthesis gases (i.e. methanol,

Table 1-20 Charpy V-notch impact test results

	Energy absorbed, J (ft · lbf)					
	Base metal		HAZ		Weld	
Temperature	Top	Bottom	Top	Bottom	Top	Bottom
0 °C (32 °F)	60 (44)	54 (40)	46 (34)	52 (38)	156 (115)	136 (100)
	54 (40)	45 (33)	...	...	...	...
Room temperature	...	...	...	...	187 (138)	122 (90)

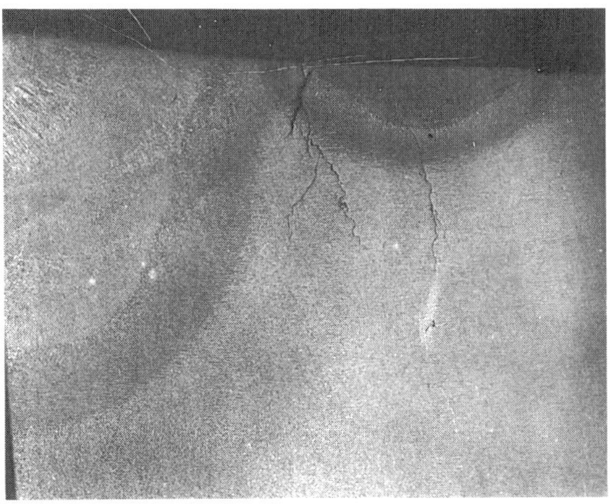

Fig. 1-71 Macrophotograph showing branching crack. 5×

*Adapted from S.R. Gertler, in *Handbook of Case Histories in Failure Analysis,* Vol 2, K.A. Esaklul, Ed., ASM International, 1993, p 168–171

hydrogen, towngas, CO-CO_2, oxoalcohols, or reduction gas). The heart of the steam reforming process is the primary reformer furnace.

The design consists of two parallel cells, each fired on both side walls and containing two staggered rows of vertical tubes. In the radiant section, tube skin temperature of about 850 °C (1562 °F) is maintained by multi-sidewall radiant burners mounted in vertical tiers of horizontal rows. An induced draft fan in the exhaust stack breeching moves the combustion air.

An endothermic catalytic reforming process runs in the catalyst-filling tubes. Gas (2.07 to 2.41 MPa at 450 to 550 °C, or 330–350 psig at 842 to 1022 °F) flows from an inlet manifold system in the top of the furnace through inlet pigtails at the tops of the reformer tubes. At the bottom, tubes are welded to an assembly consisting of a reducer joined to a pipe called to bottom pigtail. The bottom pigtails are welded to multiple collection headers, each of which "tees" to a main transfer line. The top tube ends are flanged for ease of catalyst addition and removal, and the inlet pigtails are accessible and easily pinched off in case of tube leaks. The exit pigtails are in a bottom trough shielded from heat radiation, and are inaccessible while the reformer is in service. The tubes are supported at the top by concrete counterweights.

The components had seen about 50,000 h (5.7 years) of service at about 850 °C (1562 °F). Tubes were originally manufactured in accordance with fabricator specifications, which closely adhered to Sections IX and VIII of the ASME Boiler and Pressure Vessel Code. Pigtail cracking is an industry-wide problem (Ref. 15, 16).

A specimen of tube-to-pigtail connection was cut from each of three tube assemblies removed due to unrelated stress rupture of the tubes. Two of the specimens appeared to have original, unaltered weldments. One had evidently been previously repaired. Two specimens had shown indications of cracking when inspected using dye penetrant.

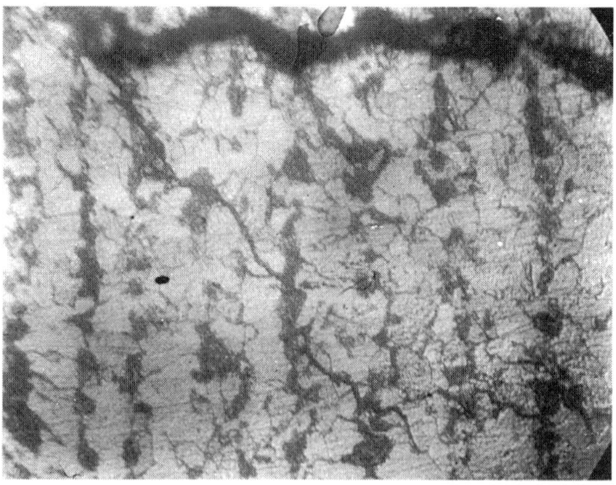

Fig. 1-72 Microphotograph of the crack, showing transgranular path through the base metal. 500×

Visual Examination. A schematic of the tube/pigtail/header assembly is shown in Fig. 1-73. In addition to the subject cracking at the reducer-to-pigtail weld, cracking was also seen at the pigtail-to-weldolet and weldolet-to-header welds. Pigtails were often discovered to be deformed, as shown in Fig. 1-74.

Dye Penetrant Testing. All systems components were abrasively blasted with ground walnut shells prior to inspection and selection of specimens. The intent was to provide a clean test surface without any deformation which might affect detectability. Solvent-based visible dye penetrant was utilized.

Microstructural Analysis. In general, precipitates were observed in significantly varying distributions and morphologies. Both pipe and weld heat-affected zone (HAZ) microstructures showed extensive grain boundary precipitation, in addition to stringers and fine matrix precipitates (Fig. 1-75). Grain boundary carbide morphology was noted as both globular and "film-like." Matrix precipitates were often seen as long carbide stringers, a result of the pipe forming process.

The exteriors of all Inconel 800H pipe samples showed intergranular oxidation, presumably due to the depletion of protective chromium at grain boundaries (Fig. 1-76). Intergranular oxidation was noted to penetrate 0.38 to 1.27 mm (15 to 50 mils) from the surface. These could be translated to rates of 0.064 to 0.203 mm (2.5 to 8 mils) per year. These rates are consistent with those of 800HT exposed to a furnace environment (Ref 17).

Crack Origins/Paths. Samples containing small, but detectable reducer-to-pigtail cracks showed intergranular cracks originating at weld toes and filled with oxidation product. Determination of the cracking mechanism was precluded by the oxidation damage (Fig. 1-78).

Weldments. All weldments exhibited high degrees of secondary precipitates with original fabrication welds exhibiting higher apparent levels than repair welds. The original weld(s) exhibited relatively deeper penetration by intergranular corrosion. This difference was attributed to the orientation and morphology of the columnar weld grains, which in this case offered a path of lesser resistance. Root welds showed more equiaxed microstructures as compared to the primarily columnar microstructure of cover passes.

Material and Weld. The compositions of the three submitted pigtail samples met the specifications for Inconel 800H. Original tube-to-pigtail weld specified Inconel 82 for GTAW (TIG) root pass and Inco-Weld A electrode for SMAW cover passes. Considering iron dilution effects, it appeared the analyzed weldments met the specifications of these filler metals.

SEM/EDS analysis showed base metal grain boundary precipitates to be primarily chromium carbides, but some titanium carbides were also observed. Most apparent stringer compounds were titanium carbides (and titanium cyanonitrides), but chromium carbides were also observed. SEMS/EDS analysis of an original weldment showed a secondary phase in the cover pass rich in nickel, niobium, and silicon. This phase varied in terms of the relative niobium, and silicon. This phase varied in terms of the relative niobium content. Niobium is also an effective carbide stabilizer.

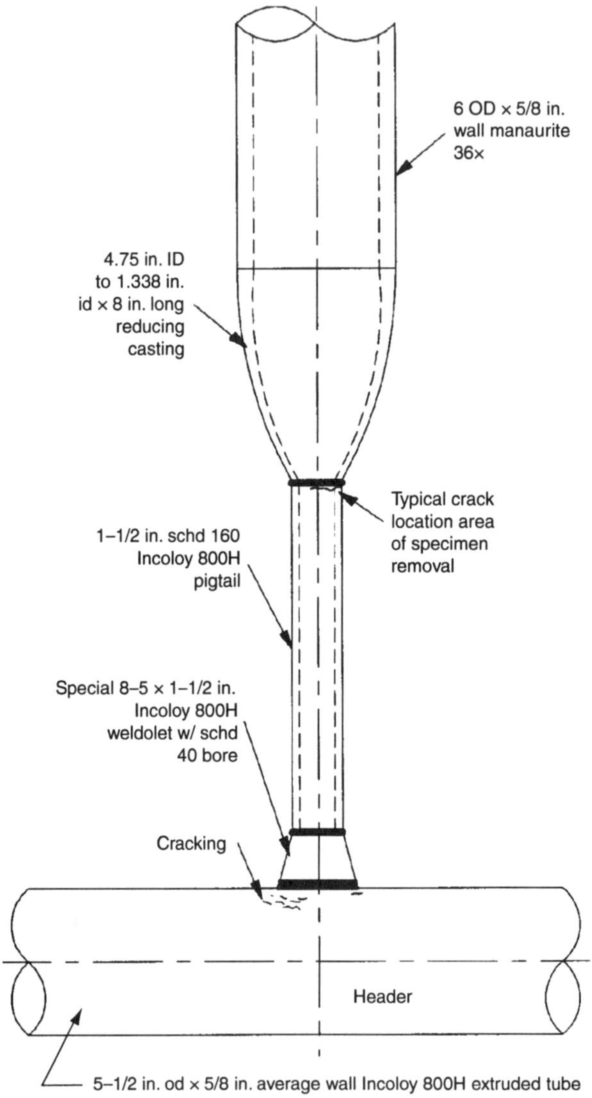

Fig. 1-73 Schematic of pigtail connection

Corrosion Deposits. Grain boundary oxidation products were identified through EDS as consisting mostly of chromium with smaller levels of aluminum, silicon, titanium, manganese, iron, and nickel.

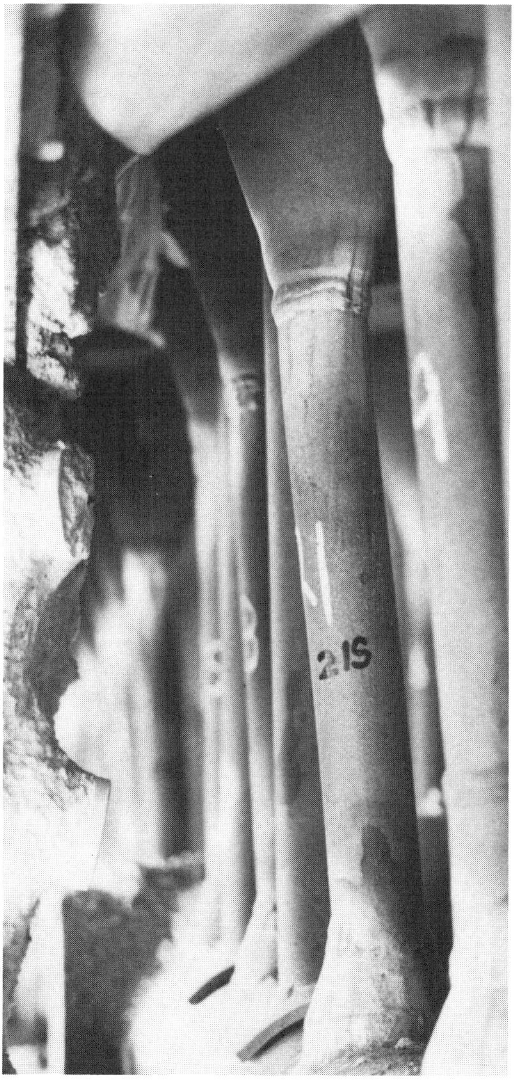

Fig. 1-74 Close-up of actual pigtail showing deformation due to applied moment

Discussion. Metallurgical changes can occur as a result of high-temperature exposure. Such changes can strongly influence performance characteristics. Conditions such as intergranular oxidation and creep voiding are detrimental and irreversible. Oxidized grain boundaries are believed contributory to crack formation and growth and it has been suggested that carbides may participate in fatigue cracking processes (Ref 18). Carbides are necessary for high-temperature properties; however, in the improper form or morphology, their presence can be detrimental. Constitutional liquation, the separation of constituents out of a mixture due to the application of heat, occurs near the HAZ or partially melted regions, and can result in cracking. Brittle grain-boundary precipi-

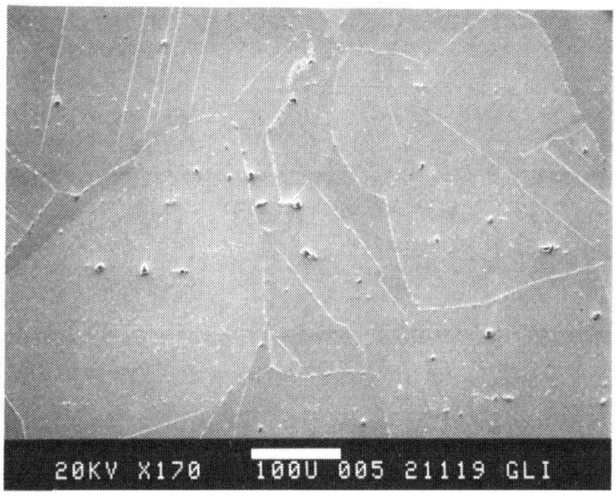

Fig. 1-75 SEM photograph showing grain boundary and matrix precipitates

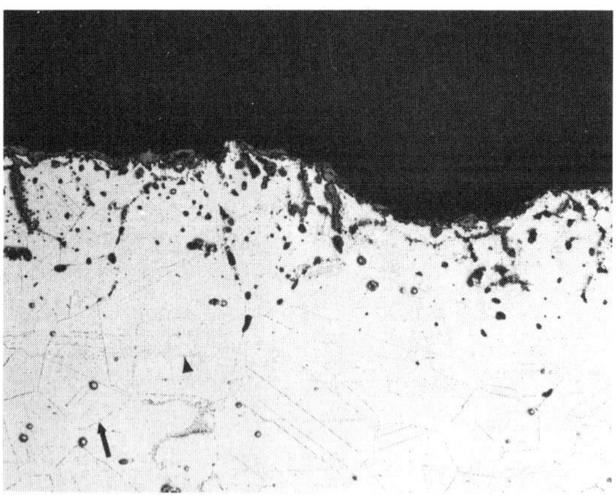

Fig. 1-76 Photomicrograph showing intergranular oxidation, grain boundary precipitates, and matrix stringers. 63×

tates (chromium and titanium carbides) and oxidation products are known contributors to post-weld cracking of Fe-Ni-base superalloys. Such precipitates can be put back into solution through solution-anneal heat treatment.

Conclusion and Recommendations. The damage mechanism is assumed to be synergistic. Thermally-driven tube distortion is the biggest contributor, as it imposes cyclic, moment stresses which are highest at the subject connection. Welding and high-temperature operation result in: increased creep and intergranular oxidation rates; precipi-

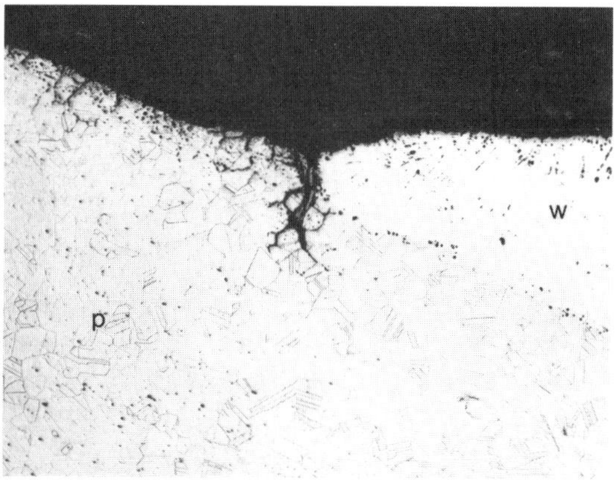

Fig. 1-77 Photomicrograph showing crack at toe of weld. Note voids along fusion zone and intergranular corrosion within base metal and weld metal. 15.75×

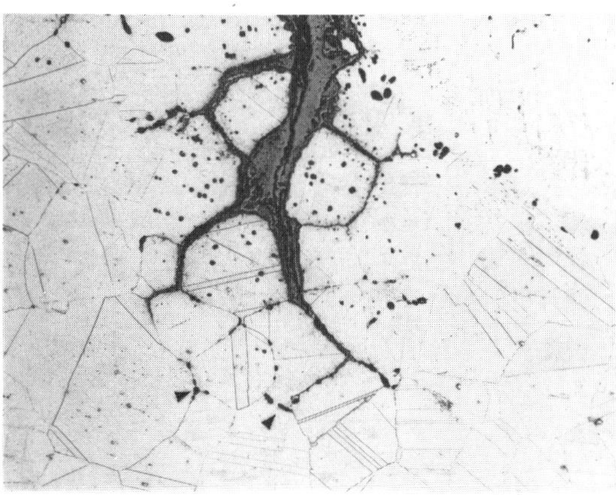

Fig. 1-78 Photomicrograph showing close-up of crack at toe of weld. Note precipitates within weld metal 63×

tation of brittle chromium and titanium carbides at grain boundaries; introduction of high residual stresses; changes in mechanical properties; and stress risers at weld defects and weld transitions. All of these factors contribute to accelerated deterioration and reduced repair weldability.

It was recommended that the moment stresses be reduced and/or materials and components be changed. Refinements in welding procedures and implementation of preweld/postweld heat treatments were also recommended.

Failure of Carbon Steel Superheater Tubes*

The tubes from the superheater section of a two-drum boiler failed while in service. The tubes were identified as being the "old tube" (tube 1) and "new tube" (tube 2). Tube 1 was an original equipment tube and was in service for 19 years. Tube 2 was in service for approximately five years.

The superheater tubes had different nominal outer diameters and nominal wall thicknesses. Tubes 1 and 2 measured 44 mm (1.75 in.) and 50.8 mm (2.00 in.) in diameter, respectively. The wall thicknesses of tube 1 and the failed end of tube 2 measured 4.5 mm (0.176 in.) and 3.8 mm (0.150 in.) thick, respectively. The wall thickness at the end of tube 2 opposite the failure measured 6.0 mm (0.235 in.). These tubes were steam bearing tubes for a 6.2 MPa (900 psig) boiler.

Tube 1 and the failed end of tube 2 were fabricated from low-carbon steel consistent with ASTM A192 specifications. The non-failed end of tube 2 was fabricated from ASTM A213 Grade T22 seamless ferritic alloy steel superheater tubing.

Several other superheater tubes in a boiler of identical design and manufacturer, at the same plant, had failed two years prior to this failure by the same apparent mechanism. Portions of the superheater in the present boiler were retubed five years ago. The two submitted tubes were in parallel in the superheater section at the time of failure. Two selections approximately 406 mm (16 in.) long were submitted for laboratory examination.

Visual Examination. The external surface of the superheater tubes is shown in Fig. 1-79. The failures in tube 1 consisted of two ruptures in the middle of localized bulges. The two failure sites are indicated as "a" and "b" in Fig 1-79. The ruptures were oriented approximately 90° apart with respect to the tube circumference, and were separated by approximately 70 mm (2.75 in.). The tube 2 sample consisted of two tubes welded together with the failure entirely within the carbon steel portion. The tube 2 failure was a thin-lipped rupture displaying a high degree of deformation. The low-alloy portion of tube 2 had no failures associated with it.

Surface Examination. The tube 1 outer diameter surface exhibited two ruptures parallel to its longitudinal axis. Wall thinning to 0.76 mm (0.030 in.) and secondary creep cracks were observed at both tube 1 failure sites (Fig. 1-80 and 1-81). Secondary creep

---

*Adapted from J.P. Ribble, in *Handbook of Case Histories in Failure Analysis*, Vol 2, K.A. Esaklul, Ed., ASM International, 1993, p 177-180

cracks and wall thinning to 0.51 mm (0.020 in.) were observed at the tube 2 failure lip (Fig. 1-82).

The internal surface of tube 1 was covered with a thin layer of yellow-brown deposits over a tightly adherent black scale. The internal surface of tube 2 was covered with a thin layer of tightly adherent black magnetite scale.

Metallography. Representative sections from select areas of both tubes were prepared for metallographic examination. Creep voids were observed along the fracture lip b of tube 1 (Fig. 1-83). The microstructure observed at the mid-wall location of the tube 1 b failure and opposite failure a was in a very advanced stage of spheroidization (Fig. 1-84 and 1-85). In addition, isolated graphite nodules were also observed in the microstructure.

Elongated grains were observed at the failure lip of tube 2 (Fig. 1-86). The microstructure at the mid-wall location of the tube 2 failure consisted of ferrite and reformed pearlite (Fig. 1-87). The presence of transformation products (reformed pearlite) at the failure lip indicated that this portion of tube 2 experienced temperatures above its lower transformation temperature, 727 °C (1340 °F), at the time of failure. Opposite the tube 2 failure, the mid-wall microstructure consisted of spheroidal carbides and occasional islands of pearlite in a ferrite matrix (Fig. 1-88). The mid-wall microstructure observed in the low-alloy portion of tube 2 that did not fail consisted of a fine dispersion of spheroidal carbides (Fig. 1-89).

Material Analysis. Portions of tube 1 and 2 from either side of the weld joint were tested for chemical analysis using optical emission spectroscopy (OES) to determine the alloy composition. The results (Table 1-21) confirmed that tube 1 and the portion of tube 2 containing the failure were fabricated from low-carbon steel consistent with ASTM A192 tubing specification. The similarity of the chemical composition of the failed tube sections suggest they were the same heat of steel. Results of the alloy analysis on the non-failed portion of tube 2 (Table 1-22) indicated that it was fabricated from ASTM A213 Grade T22 tubing.

Internal Deposit Analysis. Scanning electron microscope-energy dispersive spectroscopy (SEM-EDS) was used to determine the elemental composition of the internal deposits on both tubes. The results (Table 1-23) revealed significant quantities of calcium, silicon and sodium species in tube 1. The same species were also present in tube 2 but in smaller quantities. This indicated that some carryover of boiler water was taking place in the superheater section.

Discussion. Heavy oxidation and longitudinal cracking parallel to the failures in both tubes indicated a creep rupture mechanism. Chemical analysis revealed that tube 1 and the failed section of tube 2 were fabricated from plain carbon steel. The non-failed side of tube 2 was fabricated from ASTM A213 Grade T22 low-alloy steel tubing. In steam environments, the generally accepted oxidation limit for plain carbon steel is 454 °C (850 °F), while for Grade T22 it is approximately 580 °C (1075 °F). A spheroidized microstructure with graphite nodules indicated that tube 1 was operating at a temperature significantlty above its oxidation limit and below its lower transformation temperature of 727 °C (1340 °F) for an extended period of time.

Transformation products in the failed region of tube 2 indicated that the failure region experienced a peak temperature in excess of 727 °C (1340 °F). However, secondary creep cracks around the failure and creep voids along the rupture lip indicated that it experienced long-term overheating effects prior to rupture. The presence of a transformed microstructure in the carbon steel portion of tube 2 made an assemment of the degree of long-term overheating impossible. The superior oxidation resistance and microstrucural stability of Grade T22 tubing, via its chromium and molybdenum

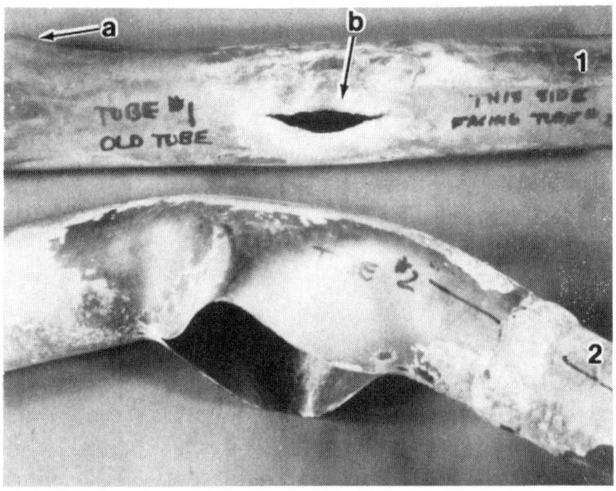

Fig. 1-79 The submitted superheater tube sections. 0.23×

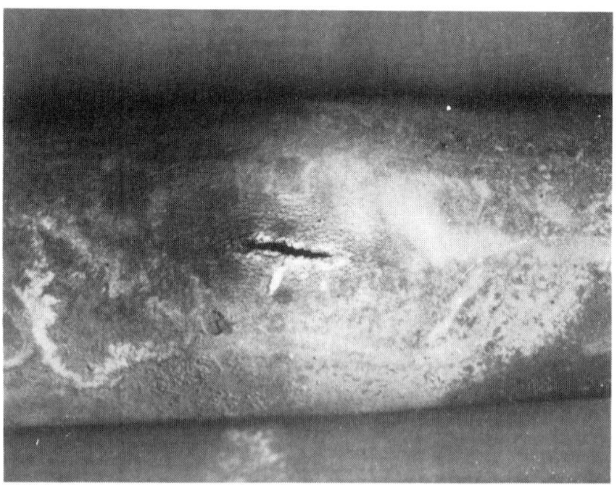

Fig. 1-80 Tube 2 failure site a (see Fig. 1-79). Note the heavy oxidation and longitudinal creep cracks. 0.60×

additions, resulted in the virtual absence of signifcant scale and fine dispersion of spheroidal carbides in the microstructure.

Conclusion and Recommendations. The tube 1 failure was caused by creep mechanism via long-term overheating. Microstructural analysis indicated that the tube 1 peak metal temperatures were greater that 454 °C (850 °F) but less than 727 °C (1340 °F). Secondary creep cracking indicates long-term overheating of tubes 1 and 2 was likely caused by an excessive heat flux and/or insufficient steam distribution. The high-temperature event coinciding with the violent rupture of tube 2 most likely resulted from a

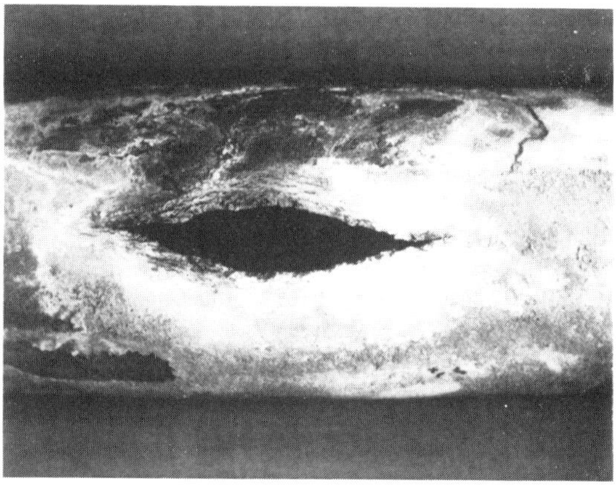

Fig. 1-81 Tube 1 failure site b (see Fig. 1-79). Note the heavy oxidation and longitudinal creep cracks. 0.60×

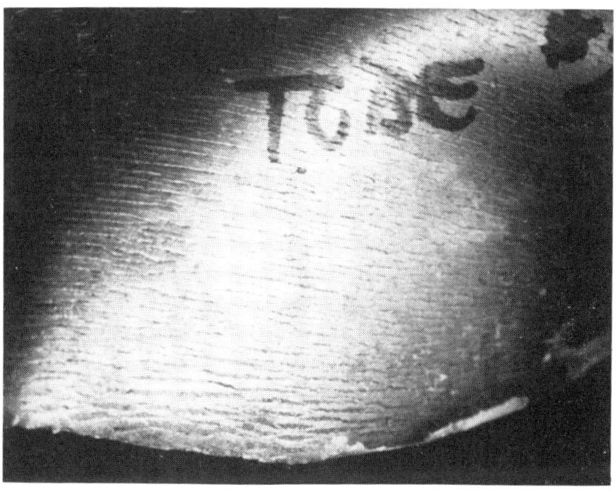

Fig. 1-82 The failure edge of tube 2. Note the secondary creep cracks. 1.06×

sudden loss in steam distribution through this tube circuit. Although SEM-EDS analysis identified some carryover deposits on the internal surfaces, only a small amount of transported deposit was present. Internal deposits were not a primary cause of the tube failures.

A review of steam flow in the superheater should be conducted to ensure adequate steam distribution in the circuit. The entire superheater should be upgraded to ASTM A213 Grade T22 tubing.

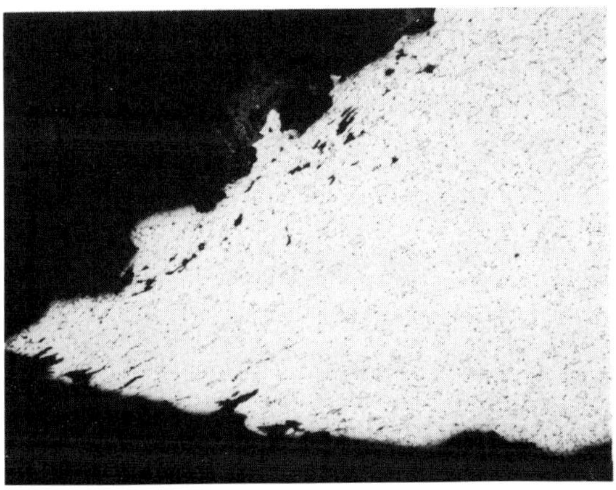

Fig. 1-83 Photomicrograph of the b failure lip in tube 1. Note the creep voids along the fracture surface. Nital etch, 38×

Fig. 1-84 Photomicrograph of the mid-wall microstructure observed in the tube 1 b failure area shown in Fig. 1-83. Note the spheroidization and large graphite nodule. Nital etch, 285×

Alloy analyses of the submitted tube samples indicates the overheating failures occurred in low-carbon steel sections. The portion of tube 2 that did not contain the failure tested as being consistent with ASTM A213 Grade T22 boiler tubing specification. The similarity of the chemical composition of the failed sections suggests these sections were installed at the same time. When the superheater was retubed, the entire superheater assembly should have been upgraded to Grade T22 material.

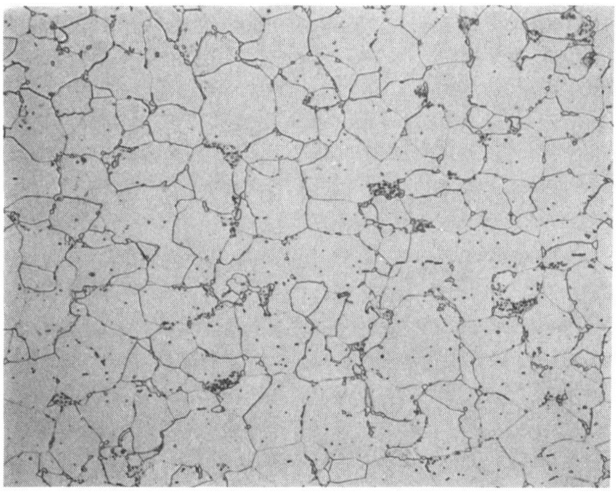

Fig. 1-85 Photomicrograph showing the mid-wall microstructure opposite the failure of Fig. 1-83 and 1-84. Note the spheroidization. Nital etch, 285×

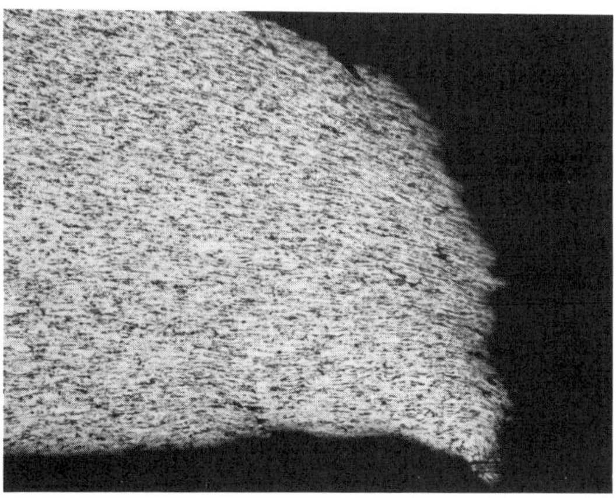

Fig. 1-86 Photomicrograph of the microstructure taken at the failure lip of tube 2. Note the creep voids and elongated grains. Nital etch, 57×

Failure of a Pipe Slip-on Flange*

A 356 mm (14 in.) diam slip-on flange cracked. The specification called for an integral forging or a casting. Instead, the slip-on flange had been fabricated by welding and machining a ring insert within a flange with a larger internal diameter.

The cracked slip-on flange was received in one piece. It had two diametrically opposite cracks located on the pipe and gasket sides.

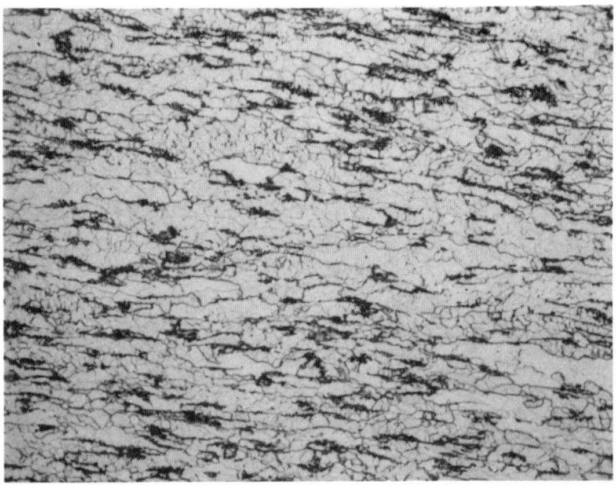

Fig. 1-87 Photomicrograph of the mid-wall microstructure at tube 2 failure lip. Note the reformed pearlite in the ferrite matrix. Nital etch, 285×

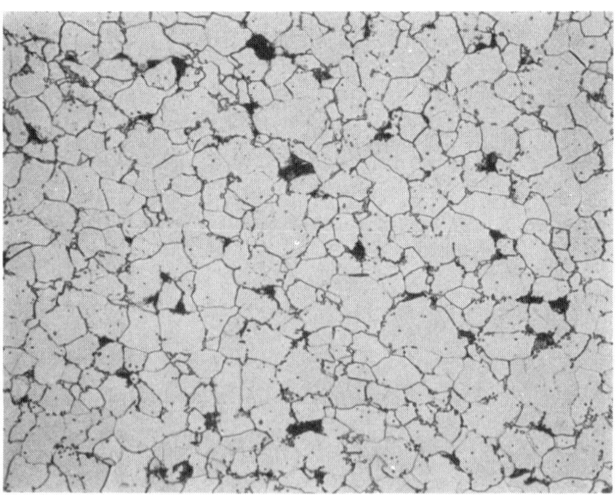

Fig. 1-88 Photomicrograph of the mid-wall microstructure opposite the tube 2 failure. Note the spheroidal carbides and pearlite islands. Nital etch, 285×

Nondestructive Evaluation. The flange was ultrasonically tested from the outside diameter. The indications were rather puzzling, because the reflections were consistently from a cylindrical surface about 95 mm (3.75 in.) from the OD. Dry magnetic particle examination confirmed that the cracks extended more or less continuously in a circular path. Nondestructive evaluation was concluded with dye penetrant tests, as shown in Fig. 1-90 to 1-93. It was quite evident that the flange was not a single-piece

Table 1-21 Alloy analyses of failed tube sections

	Composition, %		
Element	Tube 1	Tube 2	ASTM A192 specification
C	0.11	0.12	0.06 to 0.18
Mn	0.49	0.46	0.27 to 0.63
P	0.006	<0.005	0.035 max
S	0.026	0.25	0.035 max
Si	0.02	0.01	0.25 max
Cr	0.03	0.03	ns
Ni	0.07	0.07	ns
Mo	0.02	0.02	ns
Cu	0.14	0.11	ns
Fe	rem	rem	rem

ns, not specified

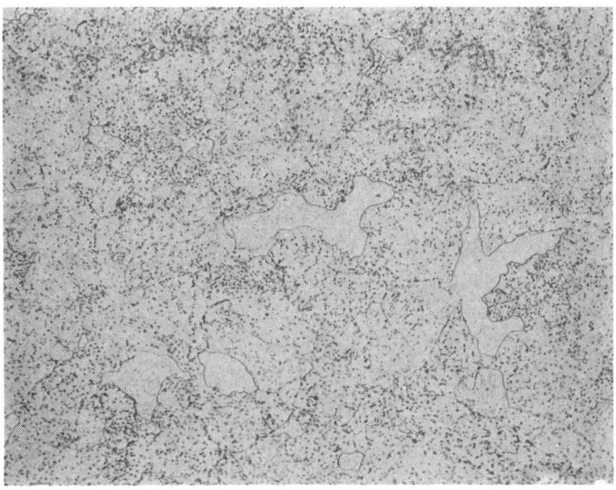

Fig. 1-89 Photomicrograph of the tube 2 mid-wall microstructure on the failure side, in the non-failed portion opposite the weld. Nital etch, 285×

*Adapted from M.D. Chaudhari, in *Handbook of Case Histories in Failure Analysis,* Vol 2, K.A. Esaklul, Ed., ASM International, 1993, p 185–186

component. A large inside diameter flange was reduced to a 356 mm (14 in.) ID unit by welding a 19 mm (.075 in.) thick ring insert. The welded faces had been machined.

Metallography. Figure 1-94 shows the radial macrosection of the flange. Note that the insert ring is welded to the main flange. When another similar piece was further sectioned to retrieve specimens for mounting and polishing, the insert separated from the main flange. The general microstructure (ferritic-pearlitic) is shown in Fig. 1-95.

Chemical Analysis/Identification. Table 1-24 presents results of chemical analysis by optical emission spectrometry.

Table 1-22 Alloy analysis of non-failed section of tube 2

Element	Composition, %: Non-failed section	Composition, %: ASTM A213 Grade T22 specification
C	0.11	0.15 max
Mn	0.49	0.30 to 0.60
P	0.016	0.30 max
S	0.017	0.30 max
Si	0.27	0.50 max
Cr	2.30	1.90 to 2.60
Ni	0.22	ns
Mo	1.02	0.87 to 1.13
Cu	0.09	ns
Fe	rem	rem

ns, not specified

Fig. 1-90 Gasket side of the flange

Mechanical properties. The hardness of the main flange was measured at three sites, with readings of 69, 70, and 72 HRB, for an average hardness of 70 HRB. Three readings of the insert measured 76, 78, and 80 HRB, for an average of 78 HRB. No further work was deemed necessary.

Conclusion. It was concluded that the flange failed at a weak welded joint. The flange did not conform to the ANSI specification, which called for an integral component.

The defective flange was one of a large shipment of similar components imported by the customer. A warning was issued to the end users of the flanges, which were then inspected nondestructively for conformance to specifications.

Mercury Liquid Embrittlement Failure of 5083-O Aluminum Alloy Piping*

A through-wall crack occurred in a butt weld of a 5083-O aluminum alloy piping system in an ethylene plant. The failure mode was identified as mercury liquid metal embrittlement, and 226 of the more than 400 aluminum alloy butt welds in the system were ultrasonically inspected for cracking.

Table 1-23 Analyses of the internal deposits

Element	Composition, % Tube 1	Composition, % Tube 2
Fe	47.1	83.5
Ca	22.2	5.5
Si	16.6	4.5
Na	9.8	4.6
Cl	0.7	...
S	1.0	<0.5
Al	1.7	...
Mn	0.9	...
Cr	...	<0.5
Mg	...	1.6

Fig. 1-91 Close up view of the gasket side, showing the crack indication

Aluminum alloy piping, such as alloy 5083-O, and brazed alloy 3003 aluminum plate-fin heat exchangers are commonly used in hydrocarbon and natural gas processing plants because of their excellent cryogenic and heat-transfer properties, and because of the economics of construction. Process temperatures in ethylene manufacturing plants are as cold as –170 °C (–270 °F), thus making aluminum alloys the preferred materials of construction. Normally, the gaseous and liquid hydrocarbons in the aluminum equipment are not corrosive. The process streams are dry, and the organic chemicals do not react with the aluminum alloys.

The ethylene plant in which the piping failure occurred was designed to process distilled feedstocks that are expected to have low levels of mercury. The plant operated

Table 1-24 Results of chemical analysis

Material	Composition, %											
	C	Mn	P	S	Si	Cu	Sn	Mo	Cr	Mo	Al	V
Main flange steel	0.23	0.62	0.024	0.038	0.24	.027	0.016	0.094	0.079	0.026	0.006	0.003
Welded insert steel	0.26	1.02	0.017	0.039	0.28	0.063	0.003	0.050	0.046	0.032	0.008	0.000
ANSI/ASTM A 105 (a)	0.35	0.60 to 1.05	0.040	0.050	0.35	...	...	...	...	...	...	...

(a) This specification is listed for reference only.

Fig. 1-92 Pipe side of the flange

*Adapted from J.J. English, in *Handbook of Case Histories in Failure Analysis*, Vol 2, K.A. Esaklul, Ed., ASM International, 1993, p 207–213

for 5 years using these distilled feedstocks. In the fifth year of operation, the plant began processing an unrefined feedstock. Shortly thereafter, mercury was found in an in-line type 304 stainless steel pipe strainer just upstream from the location where the subsequent piping failure occurred, namely, the inlet piping to an aluminum plate-fin heat exchanger. Analysis of the various barge shipments of new feedstock showed the presence of up to 40 ppb Hg by weight.

The piping and heat exchangers contained inside a 3.7 m (12 ft) diam by 23 m (76 ft) high cold box were constructed of aluminum alloys. The connecting piping outside the cold box was constructed of type 304 stainless steel. The inlet piping to a brazed

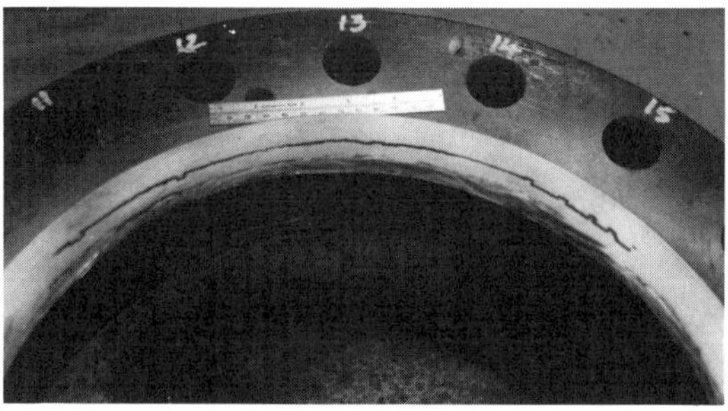

Fig. 1-93 Close up view of the pipe side, showing the crack indication

Fig. 1-94 Macrograph of a radial section etched in 50% HCl for 15 min

aluminum plate-fin exchangers within the cold box was constructed of 5083-O aluminum alloy and operated at –72 °C (–98 °F), which is below the melting point of mercury (–39 °C, or –38 °F). Mercury is not known to embrittle aluminum while it is in the solid state. Therefore, the history of the plant was examined to determine when the area of failure exceeded the melting point of mercury. It was found that the failed piping was heated above the melting point of mercury for a few hours in 1985 and 1986, as well as during a plant shutdown in 1987. An effort had men made by the manufacturing plant to keep the equipment refrigerated below –40 °C (–40 °F) during the 1987 plant shutdown. However, this effort was not successful, and a leak was detected in the piping within 12 h after the piping exceeded the –40 °C (–40 °F) temperature.

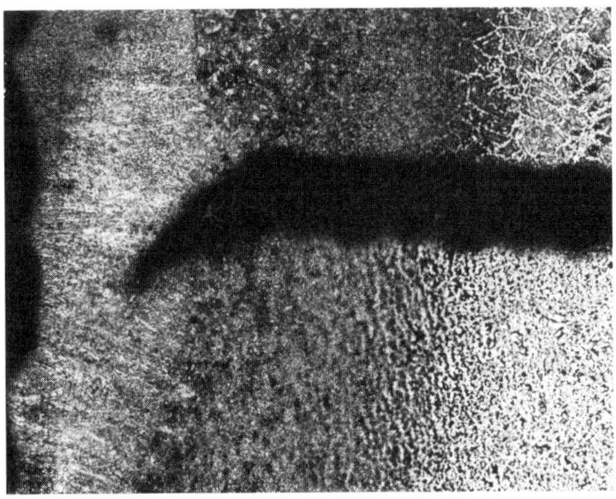

Fig. 1-95 General microstructure in the main flange (lower half), insert (upper half), and weld metal (left edge). Nital etch. 15.75×

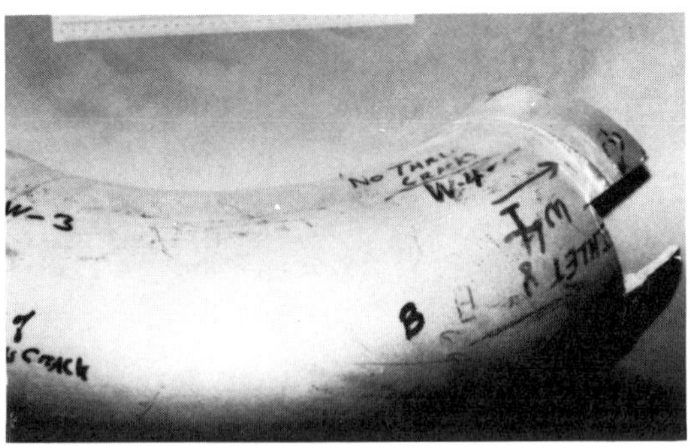

Fig. 1-96 Pipe section and the through-wall crack in weld W-3

Aluminum alloy 5083-O was the specified seamless pipe composition, and the weld filler metal was 5183. The ASTM designation of the alloy piping is not known. Either the gas tungsten arc or gas metal arc welding process was used to make the butt weld where the failure occurred.

The through-wall fracture was located in a 200 mm (8 in.) diam butt weld between a straight section of piping and a 90 ° elbow. Power hacksaw cuts were made in the straight section of piping 75 to 100 mm (3 to 4 in.) from the butt weld to remove the pipe

Fig. 1-97 Interior view of the elbow and the white residue in the flow-impingement area

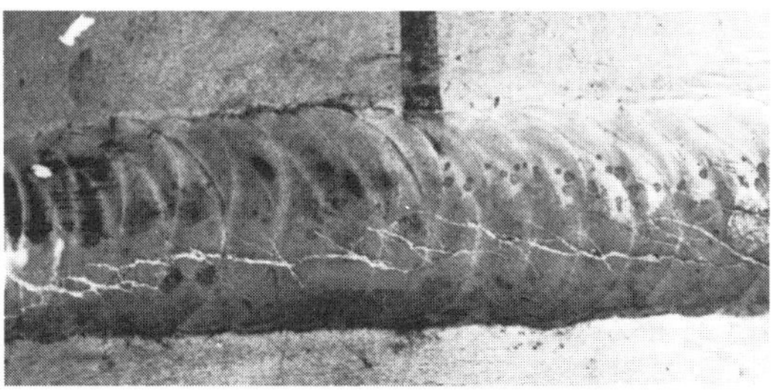

Fig. 1-98 Weld cap at the through-wall crack location, showing branched cracking. The weld cap had a gray stain where mercury had contacted the surface.

section for examination. No water or cutting lubricant was used. The weld containing the through-wall crack was designated W-3.

During the ultrasonic inspection of 226 aluminum alloy piping welds associated with the cold box equipment, a second weld was found in a 200 mm (8 in.) diam pipe section that had branched cracking indications. No crack like indications were visible on the exterior surface of this weld. This particular area of the ethylene manufacturing process normally operated at a temperature above the melting point of mercury. A 0.6 m (2 ft.) long pipe length containing this weld was removed for examination. The weld designated W-19.

Visual Examination. Figure 1-96 shows the 200 mm (8 in.) diam elbow section containing failed weld W-3. Figure 1-97 shows an interior view of the elbow. Note the white residue on the flow impingement surface of the elbow.

Figure 1-98 shows the branched cracking that was readily visible in the weld cap of the 200 mm (8 in.) diam butt weld. A gray stain formed on the exterior surface of the piping where the mercury had escaped from the cracks and wetted the aluminum. This through-wall crack extended along approximately 35 to 40% of the circumference of the weld.

Figure 1-99 shows weld W-19, in which branched cracking was identified by ultrasonic inspection. As noted earlier, no cracks were visible on the exterior surface of the weld. A high-speed cutter was used to excavate the weld to determine whether mercury was present. The freshly cut surface turned a gray color and fine white oxide whiskers began to grow from the cracks exposed by the excavation. A backing ring had been used in the fabrication of this weld.

One of the visual characteristics of mercury degradation is the reactivity of aluminum when the mercury-contaminated surface is first exposed to moist, ambient air. The aluminum turns a gray color, and a voluminous white gamma alumina is observed growing from the freshly exposed aluminum surface.

Radiography. Radiographic examination of the welds was not an effective nondestructive method for identifying mercury-induced cracking. Experiments conducted by an inspection service showed that the tight cracks produced by mercury liquid metal embrittlement did not appear when radiography was performed with an iridium

Fig. 1-99 Weld W-19 after it was excavated to detect the presence of mercury

radioisotopic source (Ref 19). Iridium is the radioactive source used for field radiography.

A complication in the radiographic interpretation occurred because of the gas tungsten arc welding process used to make many of the welds. Deficiencies in the welding procedure resulted in tungsten inclusions in the welds as well as tungsten residue on welding backup rings from arc strikes. These tungsten-rich areas had the same high-density characteristics as mercury on the X-ray film and were first misinterpreted as mercury contamination. When mercury was suspected as having been detected by radiography, these pipe weld areas were heated with a gas torch and radiographed again. If mercury was present in the suspect area, the mercury would have been redistributed by the heat. Many false leads, but no mercury regions were detected when radiography was used to assess the other welds in the equipment. Radiographic examination was helpful when used in combination with ultrasonic examination to try to distinguish original fabrication defects from mercury-induced degradation.

Ultrasonic examination was effective in locating planar weld defects. Of the more than 400 butt welds present in the aluminum piping systems, 226 were ultrasonically inspected. Significant planar defects were found in 89 of these welds. This high number of defects was attributed to the fact that the welds were not radiographically examined during the original construction of the piping systems.

Of the 89 welds containing planar defects, one also showed branch cracking in the weld and in the base metal next to the weld. No visible cracks were present on the exterior of the piping at this weld location. This is the weld (W-19) that was excavated in the field and found to contain mercury-induced cracks. A proprietary ultrasonic procedure similar to the Electric Power Research Institute intergranular crack detection procedures was used to detect the branch cracking.

Surface Examination. The through-wall crack sectioned out of the pipe/elbow weld (W-3) and opened by applying a bending force to the pipe side of the weld. The crack

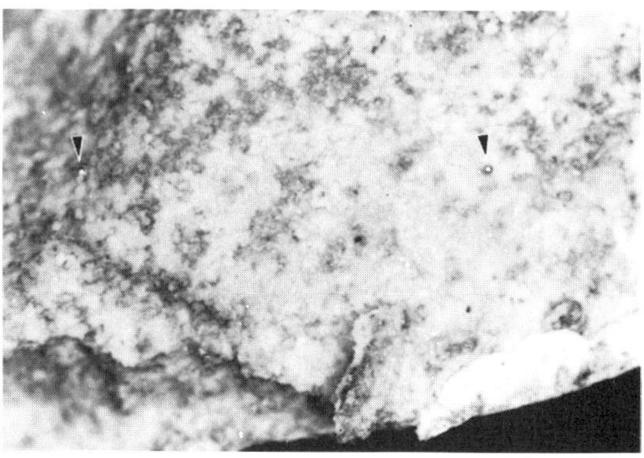

Fig. 1-100 Micrograph of fracture surface at weld W-3 before removal of the white oxide deposit. Note the mercury droplets. 10.26×

opened easily, exposing a fracture surface covered with a white, powdery deposit. Small, reflective, metallic-appearing droplets were observed throughout the white deposit. These droplets are indicated by arrows in Fig. 1-100 and were identified by energy-dispersive spectroscopy (EDS) as mercury.

Scanning Electron Microscopy/Fractography. A small section of the elbow side of the weld fracture was removed for examination by scanning electron micropscopy (SEM). The white deposit, aluminum oxide, obscured most of the details of the fracture surface. This deposit was removed by rinsing the surface with concentrated nitric acid, which is known not to affect the fracture appearance (Ref 20).

Figure 1-101 shows the general appearance of the through-wall crack. The crack originated on the inside (hydrocarbon side) of the pipe (to the right of the arrow in Fig 1-101). This area was actually the weld root-pass surface in the inside of the pipe. Figure

Fig. 1-101 SEM micrograph of the through-wall crack. 14.25×

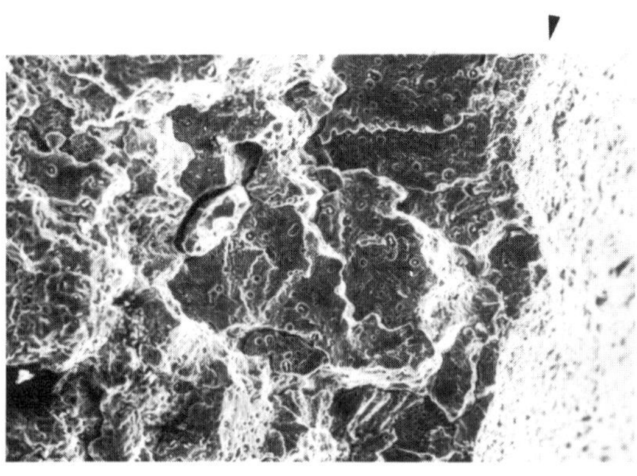

Fig. 1-102 SEM micrograph showing intergranular cracking near the inside surface of the pipe weld.67×

1-102 shows the intergranular path of the cracking in the early propagation region. A number of pits were visible on the grain facets. The nature and extent of the pitting indicated a corrosion reaction of the mercury with the aluminum.

Figures 1-103 and 1-104 show the fracture surface of the through-wall crack at increasing distances from the inside surface. In each case, the fracture appearance was typical of mercury-induced cracking of aluminum (Ref 20). The cracking was primarily intergranular, penetrating the entire wall of the pipe. No shear lip or ductile ligament could be identified near the outer surface of the pipe. The pitting corrosion was heaviest

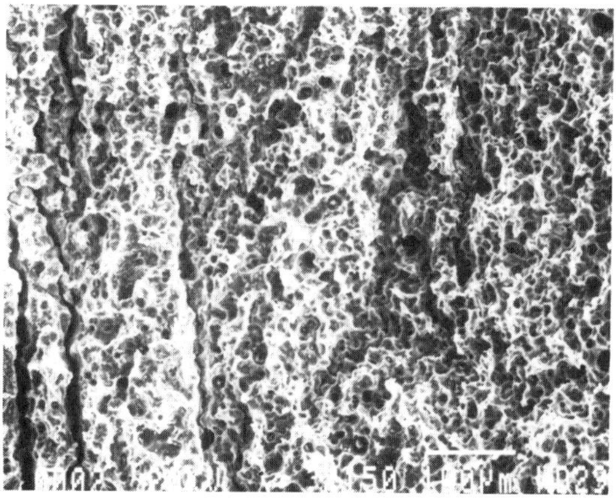

Fig. 1-103 SEM micrograph showing the through-wall fracture near the interior surface of the pipe. Noting pitting. 93×

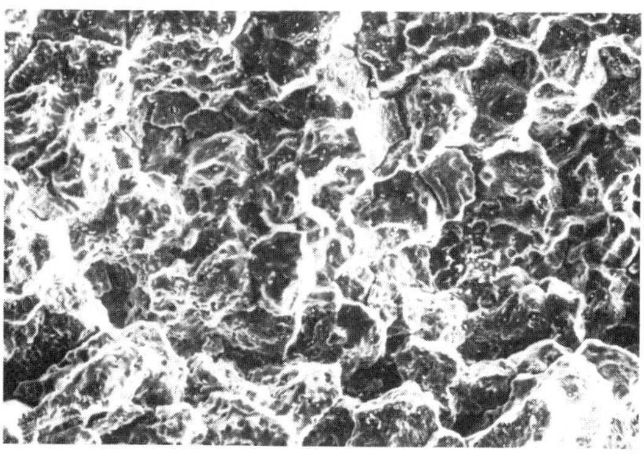

Fig. 1-104 SEM micrograph showing the through-wall crack near the exterior surface of the pipe. Note the less severe pitting compared with Fig. 1-103. 93×

near the inner surface of the fracture, then decreased as the fracture progressed toward the exterior surface (the outside diameter) of the pipe. This suggested that the crack did not progress as a singular event from the inside to the outside diameter.

Scanning electron micropscopy was not performed on weld W-19, which was excavated by high-speed cutting in the field. The appearance of this weld is described further in the next section.

Crack Origins/Paths. Two metallographic sections were prepared form Weld W-3, which contained the through-wall crack. One section was a full cross section of the through-wall crack (Fig. 1-105). The crack was primarily intergranular and extended from the toe of the weld root pass (inner surface) to the weld cap. Secondary cracks branched off the primary crack into the weld metal. A large number of delamination or exfoliation cracks were present in this section, propagating into the base metal (pipe) perpendicular to the weld axis. One of these exfoliation cracks is identified in Fig. 1-105

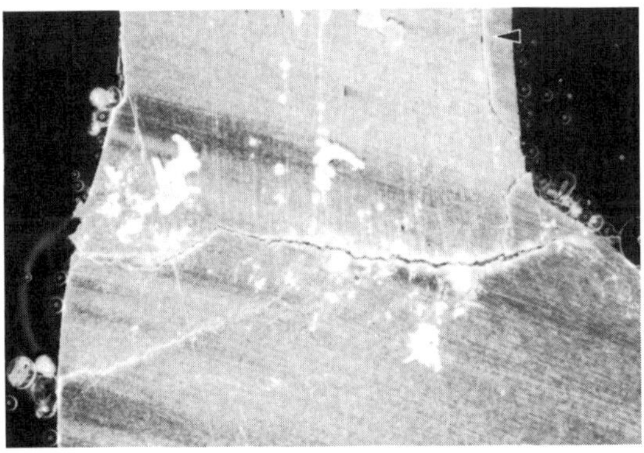

Fig. 1-105 Metallographic cross section at the through-wall crack location

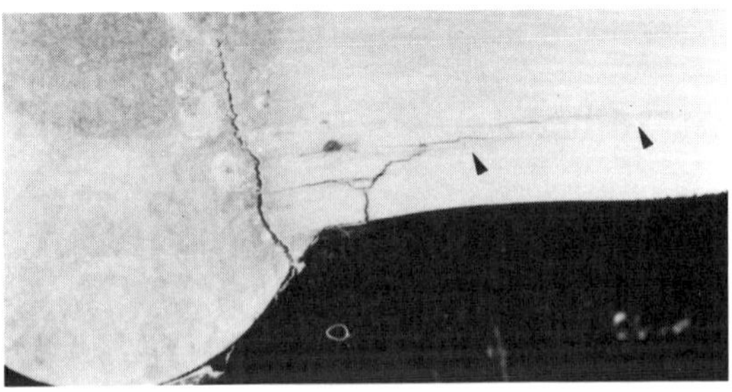

Fig. 1-106 Cross section of weld W-3 near the termination of the through-wall crack. Arrows indicate exfoliation cracks. 3.42×

by an arrow. Shortly after polishing and etching of the primary, secondary, and exfoliation cracks, all the cracks were present in this section, propagating into the base metal (pipe) perpendicular to the weld axis. One of these exfoliation cracks is identified in Fig. 1-105 by an arrow. Shortly after polishing and etching of the primary, secondary, and exfoliation cracks, exuded the white residue characteristic of an Al-Hg-O_2-H_2O interaction.

The second metallographic section was removed near the end of the through-wall crack. In this location, the crack had not propagated completely through the weld metal (Fig. 1-106). Several exfoliation cracks were present near the inside surface of the pipe. Figure 1-107 is a higher-magnification view of this region of exfoliation. Figure 1-108 shows the crack tip of one of the cracks, it the intergranular path, and the considerable amount of branching.

Metallographic examination of the base metal next to weld W-19, which had been excavated in the field, revealed exfoliation cracking. This cracking was perpendicular to

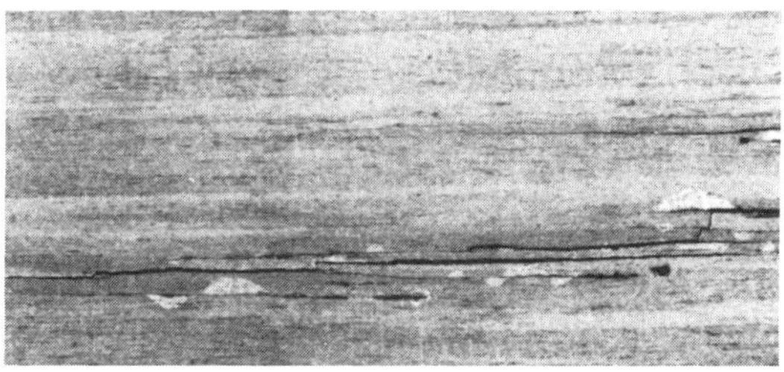

Fig. 1-107 Micrograph showing the exfoliation region on the pipe. 14.25×

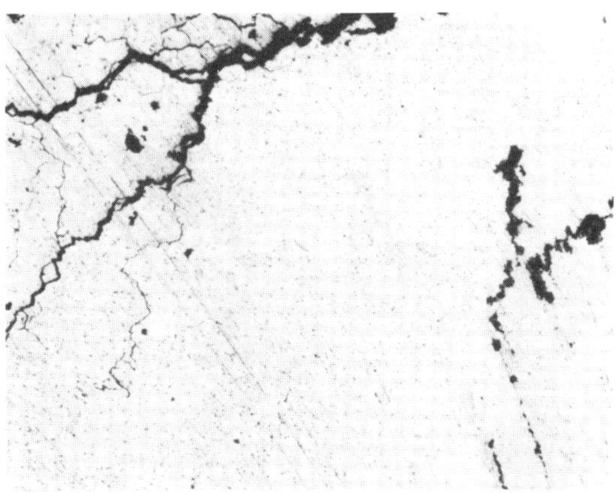

Fig. 1-108 Micrograph of a crack tip, showing branching and an intergranular mode of fracture. 30.5×

the weld axis and parallel to pipe surface. The condition extended about 38 mm (1.5 in.) from the weld fusion line. Many of these cracks contained a light-colored phase (Fig. 1-109), later identified as a hard, chromium-manganese-rich intermetallic compound. Figure 1-110 shows this segregation.

Chemical Analysis/Identification. Results of chemical analyses of the pipe, elbow, and weld metals, as determined by x-ray fluorescence, are given in Table 1-25. Both the pipe and elbow were within the specification for aluminum alloy 5083. The weld metal composition corresponded to the specification for 5183 filler metal.

Corrosion Deposits. White deposits from the inside surface of the pipe/elbow section were dissolved in acid and analyzed by atomic absorption. The composition of the white deposit was as follows: 26.4% Al, 1.84%Hg, and 226 ppm Zn. This white residue was identified as aluminum oxide, which resulted from the reaction of an aluminum-mercury solid solution with atmospheric oxygen and/or water vapor. Small droplets of mercury were dispersed throughout this residue. Figure 1-111 shows metallic

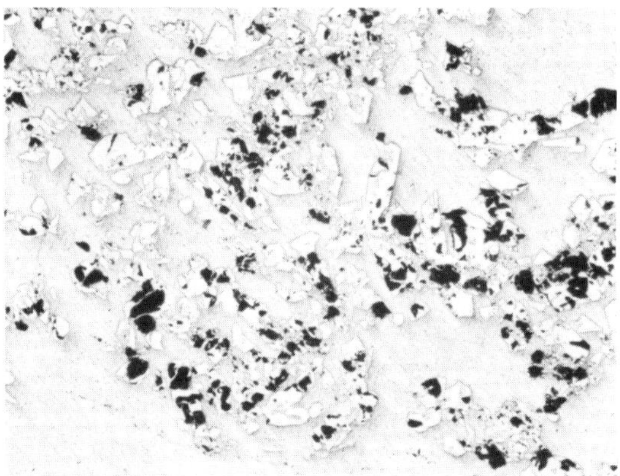

Fig. 1-109 Micrograph of light-colored phase found in the base metal (pipe) next to weld W-19. 30.5×

Table 1-25 Results of chemical analysis

	Composition, %							
Material	Mg	Mn	Cr	Zn	Si	Fe	Cu	Al
Pipe	4.59	0.80	0.24	0.11	0.13	0.16	0.05	bal
Elbow	4.12	0.75	0.07	0.01	0.06	0.13	...	bal
5083 specification (a)	4.0-4.9	0.4-1.0	0.05-0.25	0.25	0.40	0.40	0.10	bal
Weld metal	4.45	0.73	0.12	0.04	0.09	0.20	0.02	bal
5183 specification (a) (b)	4.3-5.2	0.5-1.0	0.05-0.25	0.25	0.40	0.40	0.10	bal

(a) Single values are maximums. Source: Ref 21. (b) Filler metal for gas metal arc welding

droplets on the interior surface of the pipe/elbow area. A mercury X-ray image of this area is shown in Fig. 1-112.

EDS analysis was performed to identify the light-colored phase shown in Fig. 1-110. This material was found in the pipe cross section next to the excavated weld. The EDS spectrum of these light areas confirmed the presence of chromium and manganese in amounts greater than normally present in the 5083 aluminum alloy.

Scanning Auger microscopy was used to examine the fracture surface of the through-wall crack. Wide-energy-range spectra were taken at various points on the surface. All spectra were similar, showing a surface that was essentially pure aluminum oxide, with small amounts of carbon and magnesium. Depth-profile studies of the fracture surface, using ion etching, revealed a uniform composition as a function of depth, with no additional elements detected as the etching progressed. A specific search for mercury was conducted, both during etching and after removing about 1 μm from the surface; none was detected.

Hardness. Vickers hardness values (200 g load) for the base metal and weld metal in the area of the through-wall crack were as follows: elbow base metal, 79 HV; pipe base metal, 85 HV; weld metal (root pass), 83 HV. Hardness values for the phases shown in Fig. 1-110 were as follows: gray-colored matrix, 92 HV (35 HRB); light-colored chromium-manganese-rich phase, 411 HV(42 HRC). It was suspected that the high-hardness light-colored phase was an aluminide intermetallic compound.

Tensile Properties. Tensile tests were performed at 25 °C (78 °F). Tensile specimens were machined from the elbow/pipe weld (W-3) opposite the through-wall crack. Two new welds were made in the laboratory from 5083 piping for additional tests. The filler metal was 5183. These new welds were designated C and D. Tensile specimens were machined along the pipe axis to include the weld and base metal. Results of tensile testing are given in Table 1-26. Table 1-27 lists typical tensile properties for wrought

Fig. 1-110 SEM micrograph showing the polished transverse section containing the segregation in the base metal next to weld W-19. 310×

aluminum alloy 5083 and as-welded for 5083 with 5183 filler metal. The tensile properties of the two specimens obtained from the elbow/pipe weld W-3 met or exceeded the typical results for welded 5083.

Two specimens from each of the new laboratory welds were tested in liquid mercury at 25 °C (78 °F). After immersion in the mercury bath and prior to testing, the oxide coating on three of the tensile specimens was broken with a diamond scribe to facilitate wetting of the aluminum by mercury. Table 1-26 shows that the mercury environment resulted in severe embrittlement of the welded aluminum alloy, as measured by the significant reduction in elongation and reduction of area. The oxide coating on one of

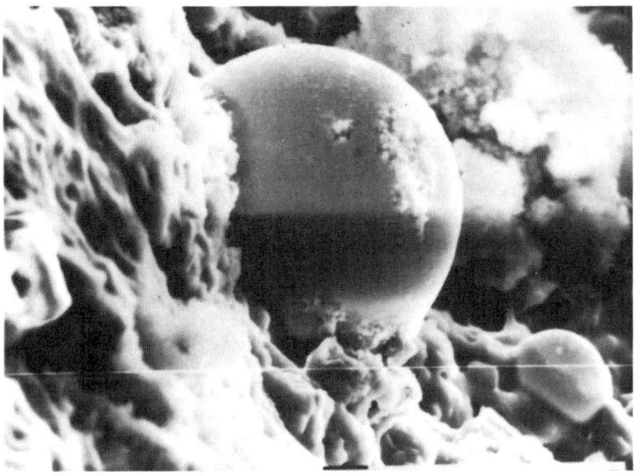

Fig. 1-111 SEM micrograph showing deposits on the interior surface of the weld

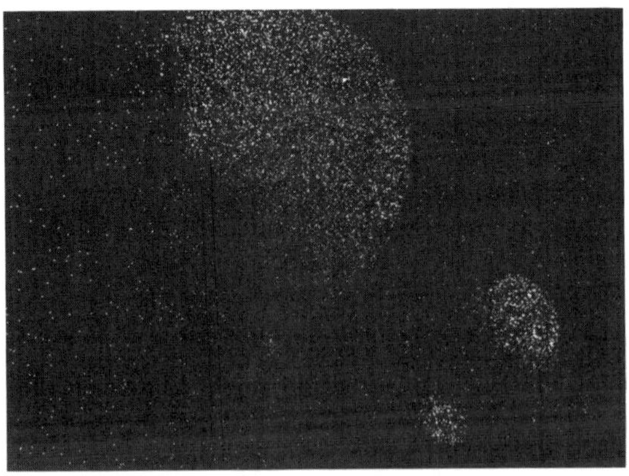

Fig. 1-112 Mercury X-image of the deposit on the inside surface of the pipe

the laboratory weld specimens was left intact. This specimen also embrittled in mercury, but had a higher ultimate tensile strength than the scribed specimens. This behavior suggested that plastic deformation of the aluminum substrate will cause rupture of the protective oxide and result in wetting of the aluminum.

The tensile specimens tested in mercury had different fracture paths (Table 1-26). Furthermore, specimen D-3 contained a considerable amount of exfoliation cracking in the base metal (pipe) that initiated at the tensile fracture surface and propagated in a direction parallel to the tensile axis. A metallographic section prepared from the facture surface of this specimen indicated that the crack initiated in the weld heat-affected zone (HAZ), with the crack propagating in a mixed intergranular-transgranular mode. After a short distance, the crack traveled into the base metal, propagating a transgranular mode with little or no associated plastic deformation.

Discussion. Because mercury had been detected in the process just upstream from the piping failure, mercury liquid metal embrittlement was the suspected mode of failure. The distinctive visual effects on the weld cap of the through-wall crack, such as the Branch cracking and the gray staining, supported this hypothesis. Fortunately, this mode of failure was suspected when the through-wall fracture was found. Safety

Table 1-26 Results of tensile testing

Specimen No.	Origin	Environment	Strain rate	Ultimate tensile strength MPa	ksi	Elongation in 25 mm (1 in.), %	Reduction of area, %	Failure location (c)
W-3-1	Production weld W-3	Air	50%/min	308.2	44.7	21.9	51.6	WM/HAZ
W-3-2	Production weld W-3	Air	50%/min	302.6	43.9	21.9	44.1	WM/HAZ
D-3	Laboratory weld D	Hg	0.5%/min	146.8	21.3	3.0	8.9	HAZ/BM
D-1	Laboratory weld D	Hg	50%/min	146.1	21.2	0	8.0	WM
C-1	Laboratory weld C	Hg	0.5%/min	169.6	24.6	3.0	8.5	WM
C-3	Laboratory weld C	Hg (a)	0.5%/min	218.5	31.7	(b)	6.0	BM

(a) Oxide coating on specimen not broken with diamond scribe prior to testing. (b) Failure occurred outside gage length. (c) WM metal; HAZ, heat-affected zone; BM, base metal

Table 1-27 Typical tensile properties of wrought and as-welded aluminum alloy 5083

Material	Average MPa	Average ksi	Average elongation in 50 mm (2 in.), %
Wrought 5083-O (a)	289.5	42.0	22.0
As-welded 5083 with 5183 filler metal (b)	296.5	43.0	16.0

(a) Source: Ref 21. (b) Source: Ref 22

precautions were taken to avoid the inhalation of mercury fumes during removal of the samples.

Intergranular cracking was the preferred mode of failure in the weldment and was also reflected in the exfoliation type of attack that ran parallel to the pipe walls. This type of behavior is similar to that observed in sensitized alloys and suggests a rapid reaction between the liquid mercury and the grain boundaries of the alloy. The specific nature of this interaction is not presently known. However, the interaction apparently involves only a few atomic layers of the fracture surface, as no mercury was observed in the secondary cracks to magnification of 1000×, nor could mercury be identified on fractured aluminum surfaces by surface analytical methods, such as Auger spectroscopy.

The presence of mercury in the secondary cracks, on fracture surfaces, and on free surfaces was unequivocal, however. Exposure of these cracks or surfaces to aqueous environments, including humid air, produced voluminous flocculent oxides containing spheroids of mercury.

Severe pitting on the fracture surface of the through-wall crack at weld W-3 indicated a corrosion reaction of mercury droplets with the aluminum. This suggested that the mercury was in contact with the fractured surface for an extended period of time. This pitting corrosion on the fractured surface was more severe near the inside diameter of the fracture, which also suggested that the crack did not progress as a single event from the inside to the outside diameter. The history of weld W-3 supported the suggestion that the crack did not propagate as a single event. Weld W-3 was normally at a temperature below the melting point of mercury and was exposed to liquid mercury for only brief periods before the failure occurred.

Mercury degradation of weld W-19, detected during ultrasonic inspection, appeared to have been aggravated by segregation of the hard chromium-manganese-rich inclusions. These inclusions could have acted as crack initiation sites. Also, the particles appeared to be more reactive to mercury than the matrix. Another factor that contributed to the degradation of weld W-19 was its backing ring. This crevicelike geometry helped to collect mercury next to the root of the weld.

The tensile tests performed in liquid mercury on the laboratory-welded specimens demonstrated that 5083-O aluminum alloy welded with 5183 filler metal was highly susceptible to mercury liquid metal embrittlement. The fracture modes in these specimens simulated the cracking found in the plant failure, including exfoliation in the base metal.

Detection of the mercury liquid metal embrittlement by nondestructive methods was difficult. Ultrasonic examination was effective in detecting mercury degradation when branch cracking and the exfoliation type of attack were present. However, because of the large number of fabrication-related weld defects, there was no assurance that all mercury-degraded welds had been removed from the production equipment. Replacement of the mercury-contaminated aluminum equipment thus became a high priority.

The presence of mercury contamination can produce very porous welds. Repair welding of the pipe segments that were removed required special techniques by experienced welders employed by a company familiar with making repairs on mercury-contaminated equipment. The welders wore full facemasks with bottle-supplied air in order to avoid breathing the poisonous mercury vapor during welding.

Conclusion and Recommendations. It was concluded that the through-wall failure in the pipe weld resulted from mercury liquid metal embrittlement. The most probable cause of the embrittlement was the use of unrefined feedstock in the ethylene plant. This feedstock was found to contain mercury at levels as high as 40 ppb. In a plant operation where 2 billion pounds of feedstock are processed per year, mercury impurity levels in the parts per billion range present a serious threat to the integrity of the aluminum equipment.

Prior to restarting the plant, all of the aluminum alloy piping was pneumatically tested with dry nitrogen and was checked for leaks. A mercury-removal system consisting of a sulfur-impregnated activated-charcoal bed was subsequently installed upstream of the aluminum equipment. The plant operated for 1 year in this manner until a new aluminum alloy cold box could be constructed and installed. This installation replaced the cold box that had been in service prior to the installation of the mercury-removal system.

Weld quality control during the construction of the new cold box consisted of 100% radiographic examination of all butt welds. All radiographs were filed for future reference. The use of weld backup rings was minimized in the construction. When backup rings were used, the upstream side of the ring was seal welded to minimize mercury entrapment.

A complete chemical analysis of trace impurities in the new feedstock for the ethylene plant could have prevented the mercury-related failure of the aluminum equipment. Upon detection of the mercury in the feedstock, a mercury-removal system could have been installed upstream of the aluminum equipment before the feedstock was processed.

Stress-Corrosion Cracking in a Downcomer Expansion Joint*

A type 321 stainless steel downcomer expansion joint (Fig. 1-113) was found to be leaking approximately 2 to 3 weeks after installation. The expansion joint handled process gases at 1790 kPa (260 psi) pressure and temperatures of 810 to 850 °C (1490 to 1560 °F).

The expansion joint was the second such coupling placed in the plant after failure of the original bellows. An independent analysis of the failure of the original expansion joint revealed that the coupling had failed from stress-corrosion cracking (SCC) of the bellows.

The as-received joint was disassembled (Fig 1-114) and sectioned (Fig 1-115 and 1-116) to remove the bellows and inner liner. The outer surface of the bellows was examined for cracks using a stereoscopic microscope and dye penetrant. Upon disassembly, two areas of corrosion were found and were submitted for energy-dispersive

---

*Adapted from R.D. Bowman, in *Handbook of Case Histories in Failure Analysis,* Vol 2, K.A. Esaklul, Ed., ASM International, 1993, p 222–224

X-ray analysis (EDXA). The cracked portions were removed for microscopic examination.

Visual Examination. After the assembly was sectioned, no cracks were visible or detected in the bellows. However, the outer surface of the bellows contained corrosion in the form of red dust and white powdery byproduct (Fig. 1-117). Removal of the inner liner revealed a cracked portion (Fig. 1-118).

Liquid Penetrant Testing. The bellows was dye penetrant inspected for possible cracks using the fluorescent dye method. Thorough inspection did not reveal any through cracks in the main bellows assembly.

Macrofractography. Stereoscopic examination of the outer surface of the bellows revealed no cracks, but did locate several dents on the tops of the convolutes. The dents represented possible stress-concentration sites.

Sectioned and mounted areas from the cracked portion of the inner liner and a portion of the sealed weld of the outer bellows were examined microscopically. The cracks were examined after etching and were determined to be branch cracks, a pattern typical of SCC. The crack initiated on the outer surfaces of the liner and propagated inward.

Fig. 1-113 As-received failed expansion joint

Microscopic examination of the sectioned bellows and braiding seal welds showed misalignment of the bellows with respect to the pipe. This resulted in the application of a weld bead on top of the convolution, thus introducing welding and restraint loading into a highly stressed area of the bellows. Further examination of the fractured bellows revealed stress-corrosion branch cracking that originated on the inner surface of the bellows and propagated outward.

Crack Origins/Paths. The branch cracking through the inner liner initiated on the outer surface and propagated inward. The cracking in the bellows, however, originated on the inner surface and propagated outward.

Corrosion or Wear Deposits. EDXA analysis of the corrosion by-product from the outer bellows surface revealed significant peaks for chlorine and phosphorus (Fig. 1-119). The strong chlorine peak indicated that the bellows failure was caused by chloride SCC.

Discussion. Microscopic examination of specimens revealed that SCC of the assembly had initiated between the outer bellows and the inner liner and had propagated in both directions. EDXA analysis confirmed the assumption of SCC. The significant presence of chlorine would have caused corrosive attack and helped initiate SCC.

Conclusion and Recommendations. The failure was caused by SCC. The chloride that was present became trapped between the braid and bellows after the bellows fractured and contaminant leakage occurred. The contaminants were trapped inside the bellows in a concentration high enough to initiate the crack.

The design of the expansion joint allowed hydrostatic test water or other fluids to become trapped between the bellows and the liner. This fluid remained in the assembly until boiled out when processing temperatures became sufficiently high. Hydrostatic test water is the most probable source of the contaminants.

Welding of the pipe, bellows, and braiding also resulted in a high stress concentration in the bellows due to residual stresses and restraint problems in a critical area. The

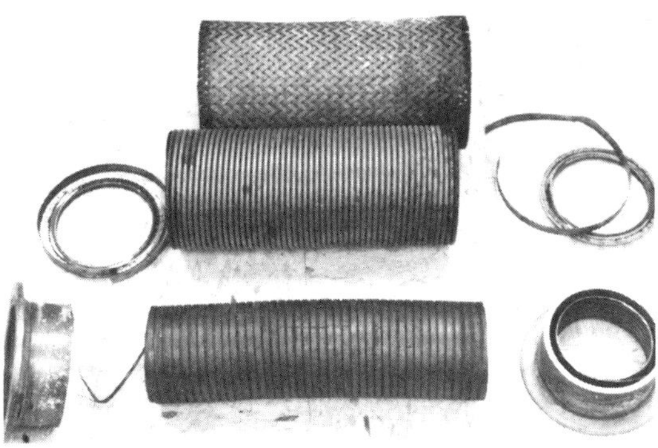

Fig. 1-114 Disassembled expansion joint. Note outer braided sleeve, bellows, and inner guide tube

dents found on the outer surfaces probably caused stress concentrations in the bellows as well.

Better manufacturing controls should be implemented. Steps to remove any fluids from the assembly after hydrostatic testing should be taken in conjunction with checking the test water for chloride content.

Better care in the use of the assembly to prevent dents and stress concentrations would reduce the stress that is capable of forming cracks as well. Prevention of misalignment of the bellows with respect to the welded pipe might prohibit welding and restraint loading of an already highly stressed area.

Thermal Fatigue Failure of Alloy UNS NO8800 Steam Superheating Tubes*

Several tubes of the steam superheating (SSH) coils of two hydrocracker (HCR) charge heater units in a refinery failed prematurely in service. The tubes were made of Alloy UNS NO8800 (Alloy 800). Unit 1 suffered five successive tube failures over a period of about three months within two years after it was commissioned. All failures were in the finned tubes located near the intermediate support towards the inlet side and were similar in nature, having circumferential cracks. Two of the tubes were completely fractured into two pieces. It was also noticed that the fins at the intermediate support were damaged, indicating restriction in movement. Unit 1 was decommis-

Fig. 1-115 Sectioned tube at flange-to-bellows joint (bellows at left)

*Adapted from M. Islam, in *Handbook of Case Histories in Failure Analysis*, Vol 2, K.A. Esaklul, Ed., ASM International, 1993, p 235–239

sioned after retubing with the same alloy having double the original wall thickness. Slight modifications were also made to the tube supports.

One of the finned tubes in Unit 2 failed in an identical manner three years after start up. In addition, hydro-testing of the unit during a shutdown shortly afterwards revealed leaks in another finned tube that was located between the intermediate support and the outlet.

The tubes had a thickness of 2.75 mm (0.108 in.) and a diameter of 88.9 mm (3.50 in.). Some of the tubes were circumferentially finned on the outside with type 304 stainless

Fig. 1-116 Sectioned bellows for microscopic examination

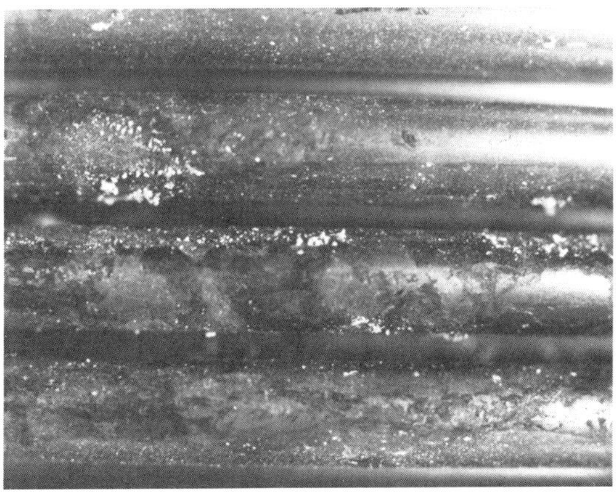

Fig. 1-117 Corrosion particles on surface of bellows (white powder)

steel while others were bare. The fins (thickness 1.25 mm or 0.0492 in., width 12.8 mm or 0.504 in.) were spaced approximately 7 mm (0.28 in.) apart and were joined to the tube surface by electric resistance welding. The SSH coils had superheated steam inside and hot flue gas on the outside. The design temperature of the unit was 767 °C (1412 °F), metal temperature, but the normal operating temperature was 325 °C (616 °F). The design pressure was 1.42 MPa (206 psig). Figure 1-120 shows the arrangement of SSH coil in the HCR units and the locations of the failed tubes.

Failure analysis was carried out on a tube section from Unit 1, approximately 200 mm (8 in.) long (which was completely fractured in two), and a tube section from Unit 2, about 350 mm (14 in.) long which had 40 fins.

Visual Examination. The examined tube section of Unit 1 was completely fractured into two pieces and had 26 welded fins. Some of the fins showed a lack of bonding. Visually, the position of the fracture appeared to coincide with the base of one of the welded fins. No secondary cracks were visible. Figure 1-121(a) shows a photograph of the failed tube with the matching fracture faces with one of the fins removed. Figure 1-121(b) is a photograph of the fracture surface. The arrow in Fig. 1-121(a) corresponds to the area marked "X" in Fig. 1-121(b).

The tube from Unit 2 was cut in half and a section 350 mm (14 in.) long with 40 fins was examined. Some of the fins showed a lack of bonding. No cracks were readily visible on the outer side. Four circumferential cracks 2 to 6 mm (0.08 to 0.24 in.) in length were visible on the inside surface, two of which are shown in Fig. 1-122(a). The cracks were all located in the 9 o'clock position in the direction of the steam flow. Examination of the outer surface after removal of the fins revealed that the four cracks observed on the inner surface had completely penetrated through the tube wall. Sixteen other circumferential cracks of various dimensions were also found on the outer surface. Fig. 1-122(b) shows some of these cracks. It was noticed that all the cracks were located adjacent and to the

Fig. 1-118 Inner surface of bellows, showing circumferential crack

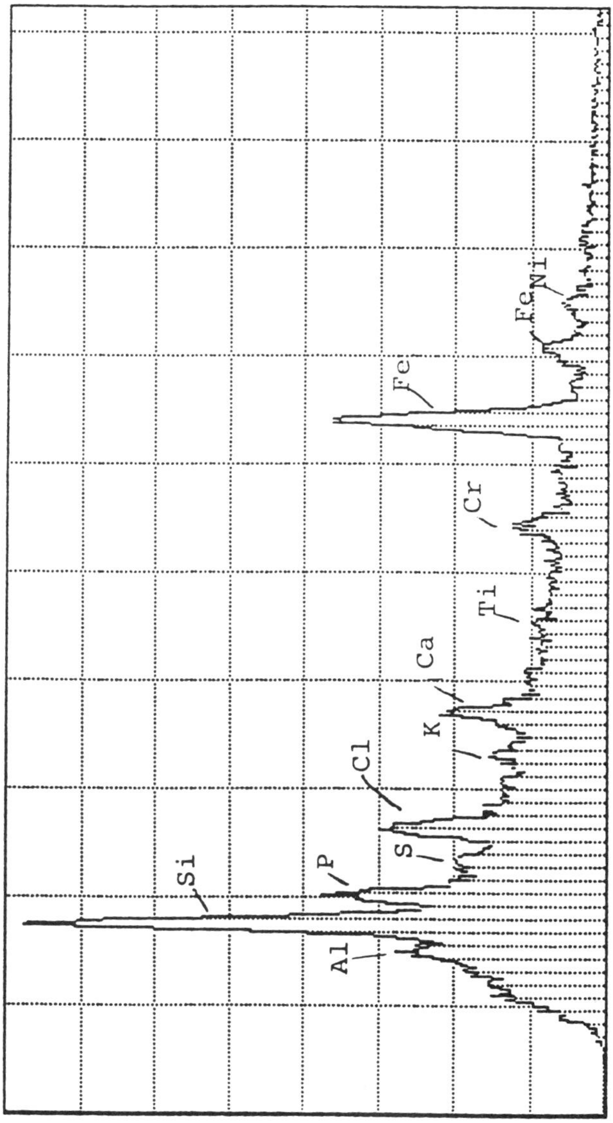

Fig.1-119 EDXA chart of contaminants on bellows surface, showing significant peaks for chlorine and phosphorus

right side of the welded zones. None of the cracks originated in areas where there was lack of welding.

SEM Examination. The fracture surface of the tube in Unit 1 was examined with the scanning electron microscope (SEM). The fracture surface was not treated for oxide removal prior to the SEM examination. Figure 1-125(a) shows a low magnification view of the area marked "X" in Fig. 1-121, while Fig. 1-125(b) shows a higher magnification micrograph from the same area. Fatigue striations are clearly visible in Fig. 1-125(b). The relatively wide spacing of the striations suggests the occurrence of low cycle fatigue. The direction of striations indicates that the fracture path is from the steam side to the flue gas side. However, the direction of other fatigue striations found in another area of the fracture surface suggests a fracture path in an opposite direction. The fracture surface was mostly transgranular in nature, except for some intergranular cracking near the steam side.

One of the cracks shown in Fig. 1-122 (the tube of Unit 2) was carefully broken open to expose the fracture surface for examination with the SEM. Some intergranular and transgranular fracture features were seen in addition to fatigue striations. Again, as in the case of Unit 1, the fatigue cracks were found to propagate from both the steam side and the gas side.

Metallography. Cross-sectional and longitudinal sections from the Unit 1 tube were prepared from near the fracture area and from an area slightly away from the location of fracture. Microstructural analysis indicated that the grain size of the alloy was not uniform and varied between ASTM grain sizes No. 2 and 3. Intergranular and intragranular carbides were observed in the microstructure of the material.

Examination of the metallographic cross sections revealed the presence of intergranular fissuring in addition to oxidation with intergranular and transgranular penetration on both sides of the tube (Fig. 1-123a, 1-123b). The intergranular fissuring and oxidation were more pronounced on the inner surface (steam side) than on the outer

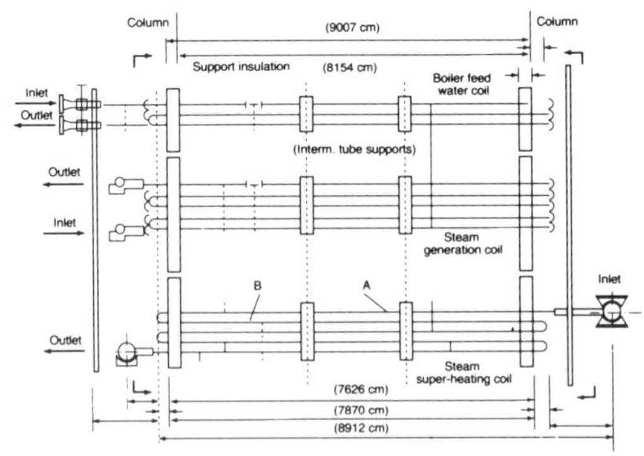

Fig. 1-120 Schematic arrangement of the steam superheating coil in the hydrocracker charge heater units showing locations of tube failures in (a) Unit 1 and (b) Unit 2

surface (flue gas side). A metallographic cross section through another crack revealed that the crack was transgranular without branching (Fig. 1-123c) and had progressed about one-fourth of the wall thickness.

Examination of metallographic cross sections of the Unit 2 tube showed that cracks started near the base of the weld and were transgranular in nature without crack branching (Fig. 1-124a). As in the case of the tube in Unit 1, intergranular fissuring was observed on both the steam and the flue gas sides and again was more prominent on the steam side (Fig. 1-124b, 1-124c). However, unlike the tube of Unit 1, there was no oxidation layer in the tube of Unit 2, and the grain boundary oxidation was much less than in the tube of Unit 1. The grain size of the Unit 2 tube was found to be between ASTM No. 4 and 5.

Mechanical Properties. Rockwell B hardness tests were carried out on both the inside and the outside surfaces of the Unit 1 tube. Tensile tests were conducted on longitudinal specimens machines from the tube material. None of these specimens had surface cracks at or near the welded areas. Table 1-28 shows the tensile and hardness results in addition to the values provided by the manufacturer and the ASTM specifications. The values given in the table show that the tube material has yield strength, ultimate tensile strength, and grain size higher than those specified by the manufacturer or the ASTM specification B 407-88 (ASME SB-163). Therefore, these results suggests that the material had age-hardened during service.

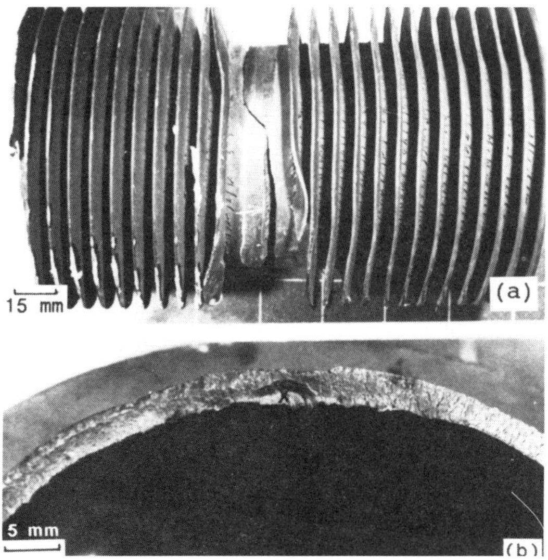

Fig. 1-121 Photographs showing: (a) fractured finned tube from Unit 1 (arrow indicates location of area marked "X" in photograph); and (b) fracture surface of the failed tube

Because of the extensive cracking of the tube in Unit 2, tensile tests could not be performed. However, hardness tests were done and the value was found to be 82 on the Rockwell B scale.

Discussion. The metallurgy of Alloy UNS NO8800 (Alloy 800) is rather complex. Minor constituents, in particular Ti, Al, and C, together with the annealing heat treatment, play a critical role in determining its service performance characteristics. It has been established that the maximum Ti+Al level should be 0.6% and the Ti to C ratio should be a minimum of 12 to maximum the stress-corrosion cracking resistance, the creep rupture strength, and rupture ductility properties (Ref 23, 24). Variations of the above limits create instability in the microstructure.

Table 1-29 gives the chemical composition of the two heats of UNS NO8800 used in the HCR units as provided by the manufacturer, in addition to ASTM B 407-88 specifications. It can be seen that the Ti+Al level should be 0.6% and the Ti to C ratio should be a minimum of 12 to maximize the stress-corrosion cracking resistance, the creep rupture strength, and rupture ductility properties (Ref 23, 24). Variations of the above limits create instability in the microstructure.

Table 1-29 gives the chemical composition of the two heats of UNS NO8800 used the HCR units as provided by the manufacturer, in addition to ASTM B 407-88 specifications. It can be seen tat the Ti+Al content of the failed alloy is between 0.6 to 0.72% and the Ti/C ratio is around 6. Hence some microstructural instability may be expected during service. However, the ASTM specification does not specify a Ti and Al level for

Table 1-28 Tensile hardness properties of UNS NO8800

Date	Yield strength, MPa	Ultimate tensile strength, MPa	Elongation, %	Hardness, HRB	Grain size ASTM No.
Laboratory	337	670	34	85	2-3
Manufacturer	238	556	55	69	4-5
ASTM B 407-88	172 (min)	450 (min)	30 (min)	95 (max)	5

Table 1-29 Chemical composition of UNS NO8800

Element	Composition,%		
	Heat 1	Heat 2	ASTM B 407-88
Ni	31.02	31.35	30.0-35.0
Cr	20.20	20.39	19.0-23.0
Fe	46.00	45.00	39.5 min
C	0.06	0.06	0.05-0.10
Mn	0.72	0.89	1.50 max
Si	0.46	0.49	1.0 max
Cu	0.14	0.14	0.75 max
Ti	0.36	0.34	0.15-0.6
Al	0.36	0.26	0.15-0.6
S	0.005	0.002	0.015 max

Alloy 800, but specifies that the level should be 0.85 to 1.20 in the higher temperature grades. The specification also does not specify a Ti/C ratio limit.

The service performance of UNS 8800 in various applications is generally good. It is not surprising, therefore, that not many instances of failures have been reported for this alloy. The one most relevant to the present failure case is that reported by Lippold (Ref 25), which describes the creep-fatigue failure of Alloy 800 in superheated steam.

In order to identify the most probable cause of failure of the SSH coils, it is necessary to consider the findings of the investigation in parallel with the operating condition of the HCR. According to information gathered from the refinery, it was found that Unit 1 was, on occasion, operated at the design temperature of 767 °C (1412 °F), and the SSH coil was also run dry for some time. The normal operating temperature should be 325 °C

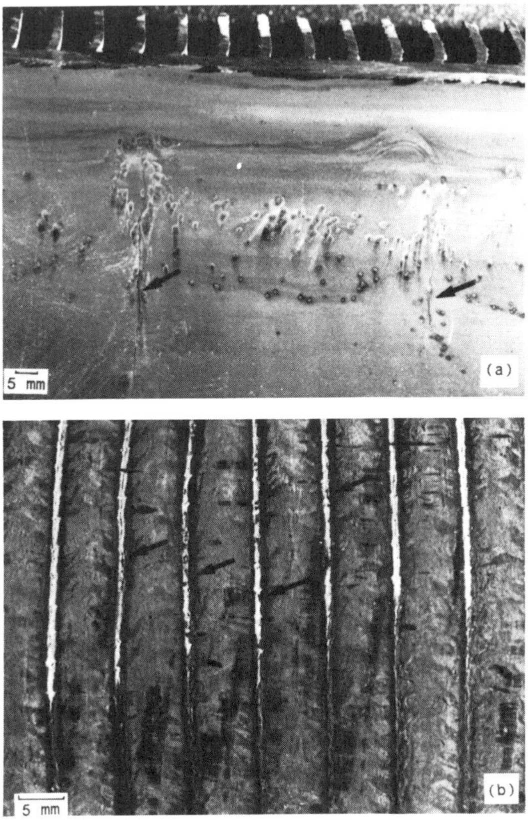

Fig. 1-122 Photographs showing: (a) two of the four visible cracks observed on the inner surface of the tube of Unit 2; and (b) the outer surface of the tube after removal of the fins. Arrows indicate locations of cracks adjacent to the welds.

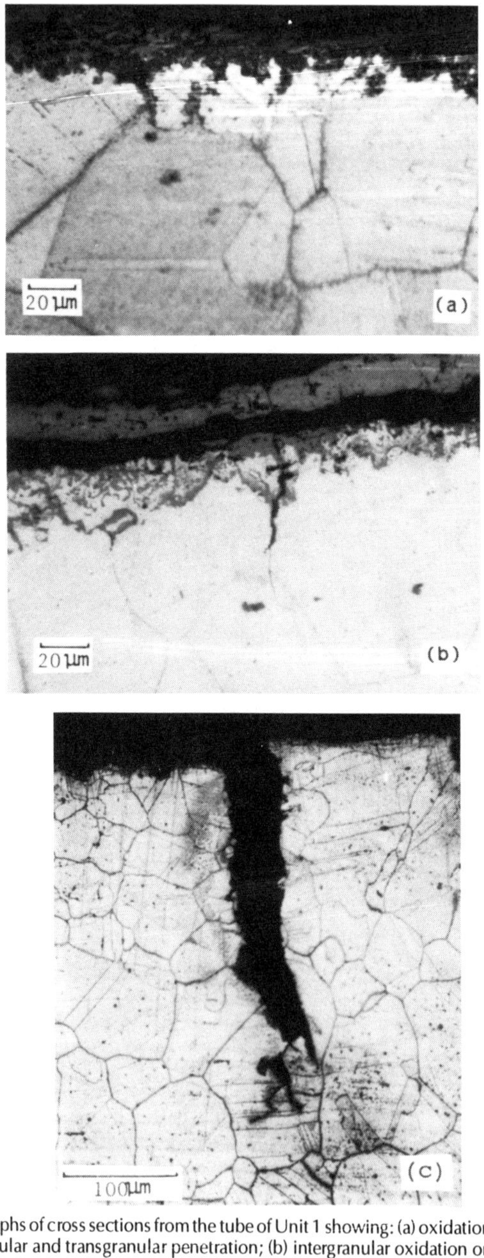

Fig. 1-123 Micrographs of cross sections from the tube of Unit 1 showing: (a) oxidation of the flue gas side with intergranular and transgranular penetration; (b) intergranular oxidation on the steam side; and (c) transgranular crack originating from the gas

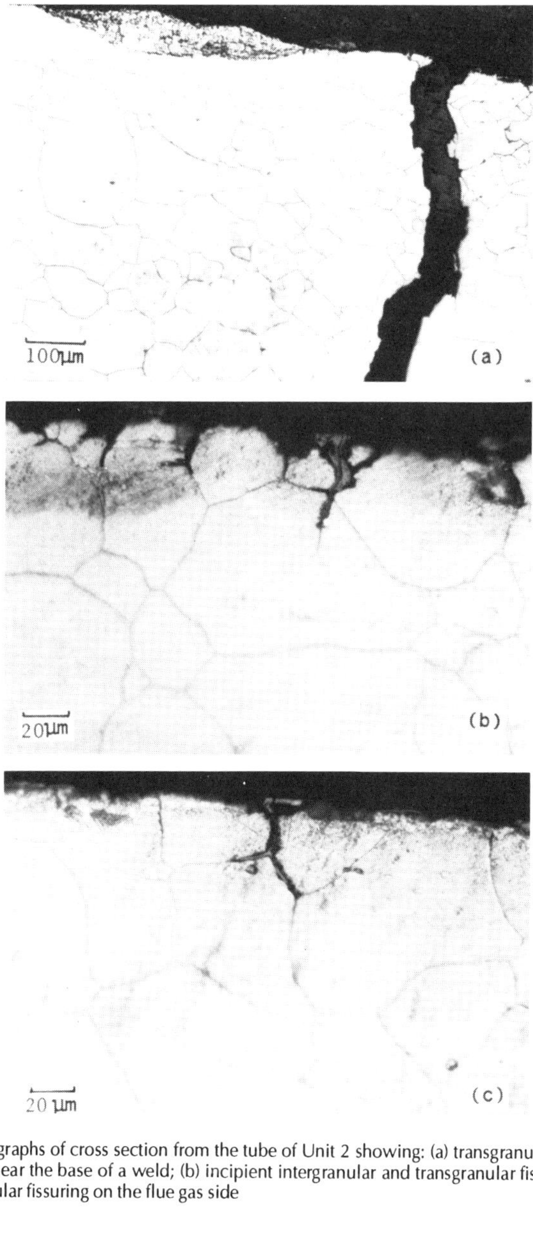

Fig. 1-124 Micrographs of cross section from the tube of Unit 2 showing: (a) transgranular crack originating from near the base of a weld; (b) incipient intergranular and transgranular fissuring on the steam side; and (c) intergranular fissuring on the flue gas side

(616 °F). ASTM B 407-88 specifies 595 °C (1100 °F) as the maximum service temperature for Alloy UNS NO8800. The time-temperature-precipitation diagram for Alloy 800 presented in Lippold's paper (Ref 3) indicates that in the temperature range above 540 °C (1004 °F), carbides start to precipitate within the first week (~100 h) of service if there is carbon available in solution. Tubes that operate in the temperature range below 480 °C (896 °F) should not form carbides in several years. High temperatures would also induce age hardening and grain coarsening. The total effect would be to deplete the chromium content of the alloy, particularly at the grain boundaries and the surface.

The tube material of Unit 1 suffered from oxidation and intergranular attack from both the steam side and the gas side (Fig. 1-122a, b). The microstructure was quite coarse (ASTM grain size No. 2 and 3), and there was carbide precipitation at the grain boundaries. All these features are consistent with high-temperature operating conditions. Moreover the presence of striations on the fracture surface (Fig. 1-123b) clearly indicates that fatigue was involved, and that cracking was from both sides.

The occurrence of fatigue failure in the tube of Unit 1 is not readily explainable because in general the fatigue life of Alloy 800 is quite good (Ref 25). However, the situation appears logical when one considers the information available in the literature on the low cycle fatigue of Alloy 800. The fatigue life of tubular samples was found (Ref 26) to be only 20% of that of solid bar samples when tested in static steam at 650 °C (1202 °F) with a total strain range of 0.5%. A more dramatic decrease in fatigue life was observed when a tensile and/or compressive hold period was incorporated into the fatigue test (Ref 26, 27). Nearly a 50-fold decrease in the fatigue life was observed because of specimen geometry (solid versus tubular) and fatigue loading conditions (hold versus no hold time) (Ref 25). Tensile or compressive hold time during fatigue can result in creep damage to the material during the hold period. As a result, the material damage is termed creep-fatigue behavior rather than low cycle fatigue.

It is well known that steam equipment may fail by fatigue if mechanical service stresses are fluctuating or vibrating, or if thermal cycles or thermal gradients impose sufficiently high peak stresses. Some form of mechanical constraint and temperature fluctuation give rise to thermal fatigue (Ref 28). Constraint may be external (for example, imposed by rigid mountings) or it may be internal, in which case it is set up by a temperature gradient within the part. It has also been established that a reduction in fatigue life takes place due to the occurrence of intergranular cracking and due to the effect of oxidation (Ref 29). Grain boundary oxidation may produce a notch effect that may limit the fatigue life. In the present case, creep damage was not observed in the alloy as apparent from the absence of any creep voids or intergranular fissures within the alloy matrix.

It therefore appears that thermal fatigue was responsible for the failure of the tubes in Unit 1. Operation of the unit at high temperatures resulted in aging, grain coarsening, and carbide precipitation, all of which led to enhanced oxidation of the alloy and grain boundary attack, thus reducing the fatigue life of the alloy. Moreover, the presence of grain boundary oxidation acted as stress raisers that, together with hold times during fatigue, led to the premature failure of the tubes of Unit 1.

Unit 2 was found to have been operated under tighter control (i.e., the unit was not run dry and there were no excursions to high temperature). The observed microstructure (finer grain size, less carbide precipitation) and the less severe oxidation of the alloy in comparison to that of Unit 1 appears to be consistent with the operating conditions of Unit 2. The metallographic and fractographic evidence indicate that the failure of the tube of Unit 2 was similar to that of Unit 1. However, because of the longer time taken to form incipient intergranular fissures, the fatigue initiation time was longer in Unit 2, and hence the service life was longer (approximately 3 years compared to 2 years for Unit 1).

In the case of Unit 2, the fatigue cracks were mostly from the flue gas side. A few cracks were found propagating from the steam side. Crack initiation and propagation from the outer side of the tube is quite logical because the highest tensile stresses are concentrated on the outer diameter of the tube. It can be hypothesized that when the fins are properly welded to the tube, complex stress patterns are set up in the weld zone due to thermal gradients as well as geometric restrictions leading to more rapid cracking.

Conclusion. The failure of the Alloy 800 tubes in the HCR units was due to thermal fatigue. The source of fatigue was possibly due to temperature fluctuations in addition to restriction of movement of the coils.

The initiation and propagation of fatigue cracks were mostly from the outside (flue gas side). However, some cracks were found to propagate from the steam side.

In the case of Unit 1, uncontrolled operating conditions (running dry and excursions to high temperatures) resulted in age hardening of the material and enhanced general and grain boundary oxidation with incipient intergranular cracks, leading to early initiation of fatigue cracks. In the case of Unit 2, with better controlled operating conditions, the grain boundary oxidation and incipient intergranular cracking were less severe, which manifested in a longer fatigue life.

Fatigue Failure at Fillet-Welded Nozzle Joints in a Type 316L Stainless Steel Tank*

Upon arrival at the erection site, an AISI type of 316L stainless steel tank exhibited cracks on its shell at two of four shell/nozzle fillet-welded joint regions. The tank is used for storage of liquid sodium, the coolant in fast breeder test reactor.

After fabrication, the tank was transported to the erection site by road, a distance of about 800 km (500 mi). During transport, the nozzles were kept at an angle of 45° to the vertical because of low clearance heights in road tunnels. When the tank arrived at the erection site, cracks were observed on its shell at two shell/nozzle fillet-welded joints (Fig 1-126 and 1-127).

The tank was fabricated in accordance with the section of ASME Pressure Vessel Code that applies to nuclear components. Every stage of the fabrication was stringently

*Adapted from D.K. Bhattacharya, S.K. Ray and P.R. Rodriguez, in *Handbook of Case Histories in Failure Analysis*, Vol 2, K.A. Esaklul, Ed., ASM International, 1993, p 256–258

controlled, as were the choice of parent metal and filler wires. The root pass of the fillet weld was made by the tungsten inert gas welding, followed by manual metal arc welding with basic coated electrodes. After each pass, the weldments were inspected visually and by using dye penetrants. After fabrication was completed the tank passed a hydraulic test.

Small interconnecting holes were drilled around the two nozzles to remove the cracked regions. The nozzles were then separated from the cut-out disks by machining

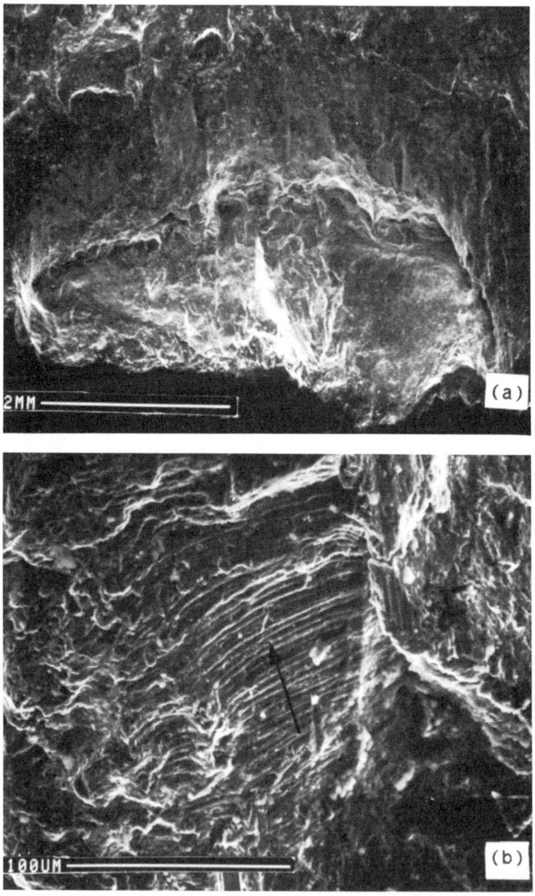

Fig. 1-125 SEM micrographs from the tube from Unit 1 showing: (a) low magnification view of the area marked "X" on the fracture surface shown in Fig. 1-121(b); and (b) fatigue striations observed in same area in Fig. 1-125(a) marked "Y." Photo is oriented in same direction as Fig. 1-121(b). Direction of crack propagation is from the steam side (bottom of micrographs) to the flue gas side of the tube.

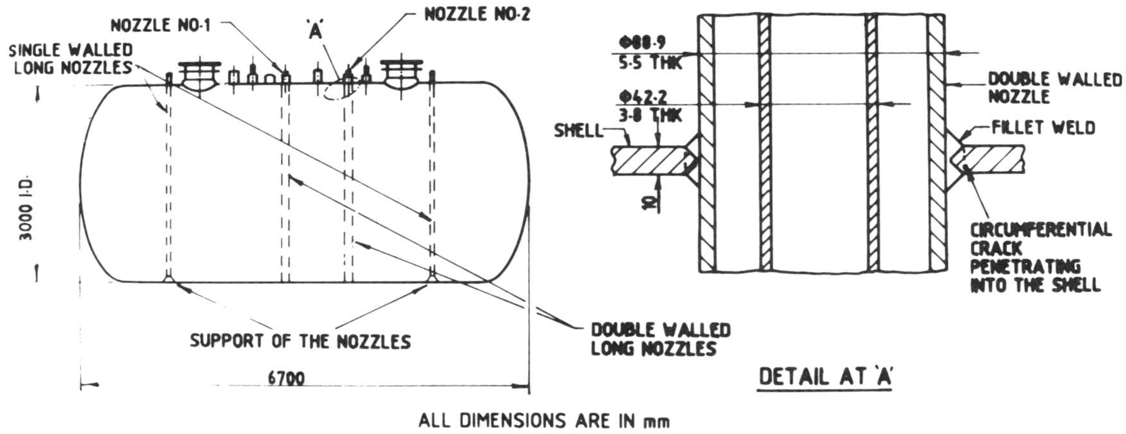

Fig. 1-126 Diagram of the cylindrical tank. Double-wall nozzles 1 and 2 were not supported at their ends inside the vessel. The other two nozzles, which were single walled, were supported as shown.

the fillet weldments, taking care that sufficient weldment material remained with the disks (Fig. 1-128). The two disks contained all the radial cracks and the circumferential cracks intact.

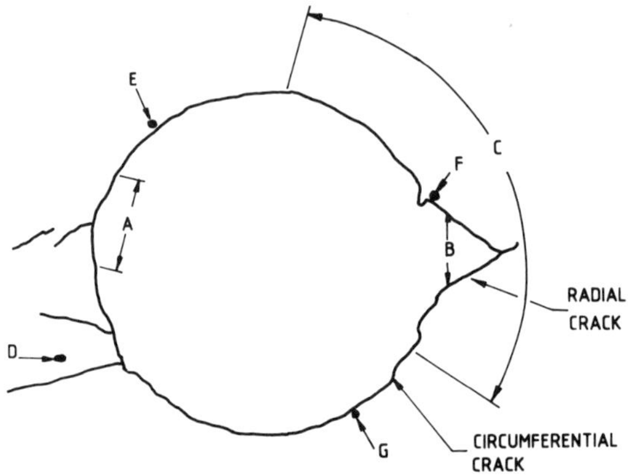

Fig. 1-127 Diagram of the circumferential and radial cracks observed at the fillet weld/shell boundary and in the shell of nozzle 2

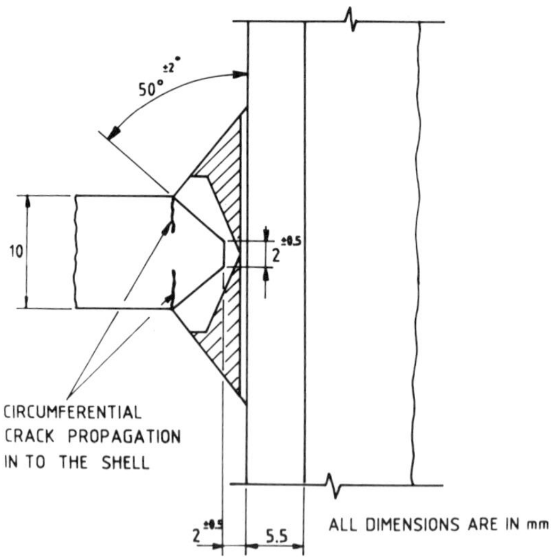

Fig. 1-128 Fillet weld joint between the shell and double-wall nozzles. The hatched portion in the weld was removed by lathe machining.

The circumferencial cracks were found to have started both from the outside and the inside surfaces of the shell and propagated toward the center of the shell thickness. The disks were cut radially to reveal the circumferential crack surfaces.

Visual Examination. Both the outside and the inside surfaces of the tank exhibited severe grinding marks. Most of the radial cracks followed the course of the grinding marks, at least in the initial stage of crack propagation (Fig 1-129).

Figure 1-129 shows the outside surface of the shell at position A of Fig. 1-127. Interconnecting drill marks are visible at the top, with weldment at the bottom. The lower crack (marked "1") followed the course of the grinding marks after initiation. Crack "u" was connected to crack "1" by a faint wavy crack ("w").

Surface Examination. Figure 1-130 shows a typical opened crack. Cracks propagated from the outside as well as the inside surface of tank shell and met somewhere near the center of the shell thickness. Ratchet marks ("R") were visible at the edges of the fracture

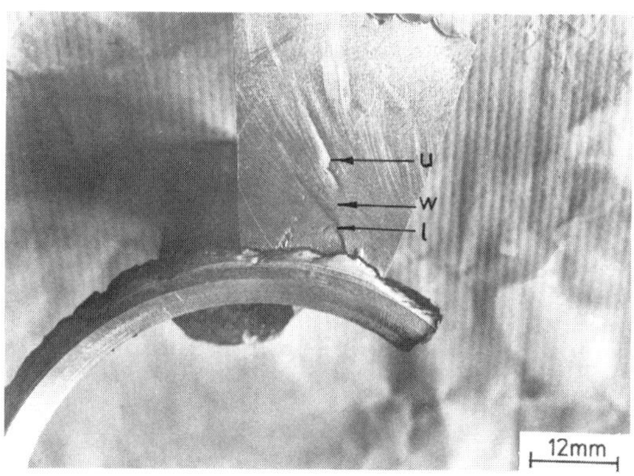

Fig. 1-129 Outside surface of the shell at position A of Fig. 1-126

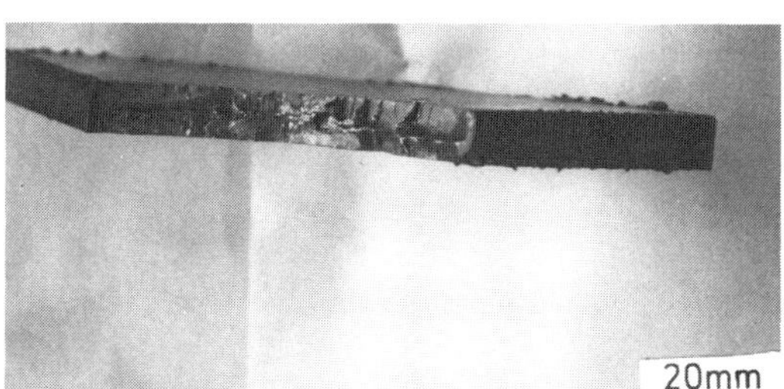

Fig. 1-130 Surface of a portion of an opened circumferential crack. P, plasma-cut surface (created by removal of the sample by sawing); R, ratchet mark

surface. These ratchet marks indicated that the loading was a rotating bending type, and crack propagation started at numerous points on the surface.

Electron Microscopy/Fractography. The fracture surfaces were examined at various locations using the two-stage replica method. A cellulose nitrate replica was shadowed by chromium, and a thin carbon layer was deposited on its surface. Next, the cellulose nitrate was dissolved in alcohol. The carbon replicas with chromium shadowing material were examined in a transmission electron microscope (TEM).

Figure 1-131 shows a TEM fractograph of a typical radial crack. Striation marks characteristic of fatigue fracture were visible, caused by discontinuous or repetitive fatigue crack propagation. Each striation corresponded to the crack advance during one cycle of loading. Examination of the striations showed that the cracks had advanced on

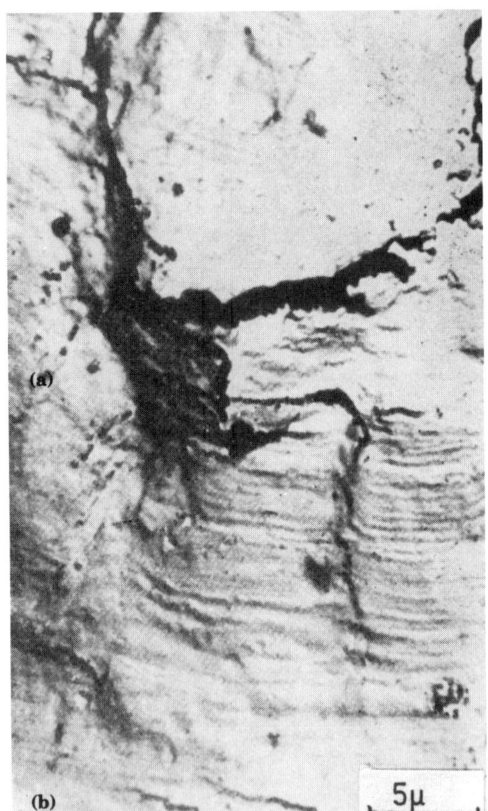

Fig. 1-131 TEM fractograph of a typical radial crack

many "mini-fronts." This behavior is expected when the loading is nonuniform, as in rotating bending.

Microstructural Analysis. Figure 1-132 is a typical optical micrograph of a radial crack. Because of the large width of the radial cracks, it was not possible to determine whether the cracks were transgranular or intergranular. Many finer branched cracks originated from the main radial cracks (Fig. 1-132). These cracks were transgranular. A few fine cracks parallel to the main cracks were also observed. This type of crack formation is characteristic of fatigue crack propagation.

Crack Origin/Paths. Crack fronts originated at numerous points on both the inside and the outside surfaces of the tank shell. The cracks propagated across the shell wall because of rotating bending fatigue loading and met at the approximate midpoint of the shell thickness.

Discussion. Evidence indicated that the nozzle joints failed by rotating bending fatigue. The following features pointed to fatigue failure:

- Initiation of cracks at a number of points
- Ratchet marks on the fracture surfaces, characteristic of fatigue failure under rotating bending loading
- Periodic branched cracks inclined to the main cracks and occasional cracks parallel to the main crack fronts
- Appearance of fatigue striations in the TEM fractographs

Conclusion and Recommendations. Severe grinding marks were also visible, and the cracks often followed these marks. Utmost care must be taken not to introduce grinding marks or other stress raisers near weld joints.

Fatigue failure was consistent with the type of loading present. The nozzles, positioned at angles of 45° to the vertical during transport of the tank, acted as cantilevers.

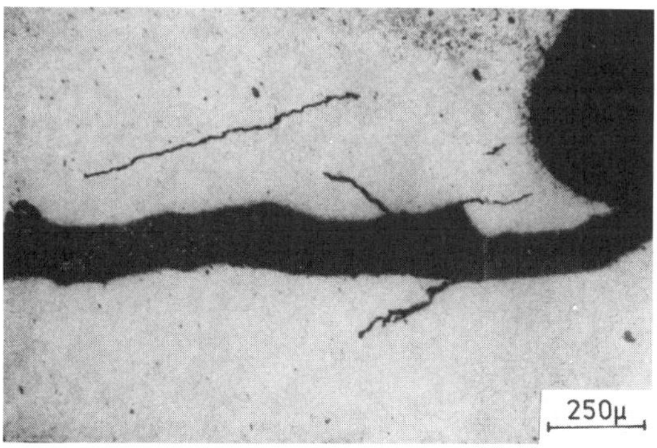

Fig. 1-132 Micrograph showing branched and parallel cracks associated with a wide radial crack

Cyclic loading resulted from the usual jerks and vibrations associated with road journeys. Such loading could have been avoided had the two nozzles been supported at their ends inside the vessel, similar to two other nozzles that did not show failure. In addition to avoiding transport problems, this design modification would reduce fatigue loading that occurs in service due to vibration of the nozzles during filling and draining of the tank.

Metallurgical Analysis of Steam Turbine Rotor Disc*

The steam turbine drives an air compressor in a nitric acid plant. The turbine is started using saturated steam, but is normally operated using superheated steam.

The steam turbine disc was fabricated from an A470 Class 4 alloy steel forging, which is a common steam turbine disc material. Approximately one-third of the disc was available for examination and sectioning.

Visual Examination. The steam turbine rotor disc segment, which is shown as received for analysis in Fig. 1-133, was visually examined. Multiple cracks were observed over much of the surface of the damaged disc.

Nondestructive Evaluation. The disc surfaces were examined using the dry magnetic particle technique. Multiple cracks, oriented in the radial direction on the sides of the disc, were observed. Figure 1-134 shows a closeup view of this damage.

Surface Examination. The turbine disc surfaces were examined using a stereomicroscope at magnifications from 7 to 67.5×. Cracking was observed on the sides of the turbine disc.

Microstructural Analysis. A section was removed form an area of the steam turbine disc segment which did not exhibit visible damage. This sample was mounted, ground, polished, and etched for metallographic examination and photographic documentation. The photomicrographs in Fig. 1-135 show the typical turbine disc microstructure at two different magnifications. The microstructure consists of tempered martensite and bainite.

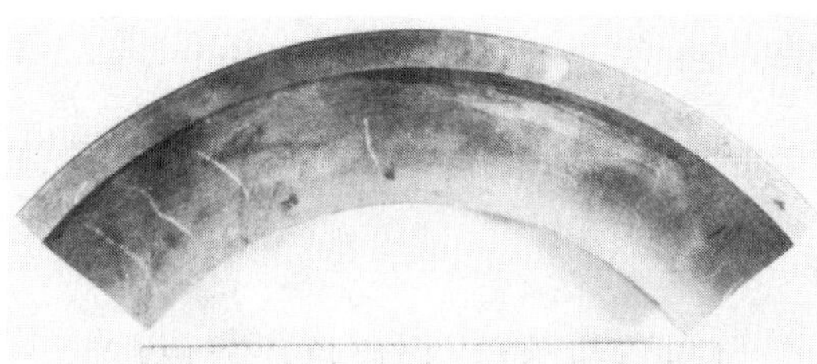

Fig. 1-133 The rotor disc segment is shown as received for analysis.

---

*Adapted from L.J. Hodas, in *Handbook of Case Histories in Failure Analysis*, Vol 2, K.A. Esaklul, Ed., ASM International, 1993, p 307–310

Crack Origins/Paths. Samples representing the typical damage were sectioned from the turbine disc segment. These samples were mounted, ground, polished, and etched for metallographic examination and photographic documentation. The photomicrographs in Fig. 1-136 through 1-138 show typical examples of the appearance of the damage. Multiple cracking was observed on the turbine disc sides and in the blade attachment areas. The photomicrographs in Fig. 1-136 show that some of the cracking has initiated from the bottoms of corrosion pits. The cracking morphology is mixed mode, that is, both intergranular (around grain) and transgranular (through grain). The upper photomicrograph in Fig. 1-138 shows two immediately adjacent cracks exhibiting different crack morphologies. The disc surfaces and primary crack paths show considerable evidence of on-going corrosion, in the form of pitting on the disc surfaces and build-up of corrosion deposits within the large cracks.

Table 1-30 Chemical composition of steam turbine rotor disc (weight percent)

Element	Disc	Chemical requirements per ASTM A470 Class 4
Carbon	0.27	0.28 max
Manganese	0.51	0.20-0.60
Sulfur	0.011	0.018 max
Phosphorus	0.010	0.015 max
Silicon	0.28	0.15-0.30
Nickel	3.55	2.50 min
Chromium	0.41	0.75 max
Molybdenum	0.41	0.25 min
Vanadium	0.09	0.03 min

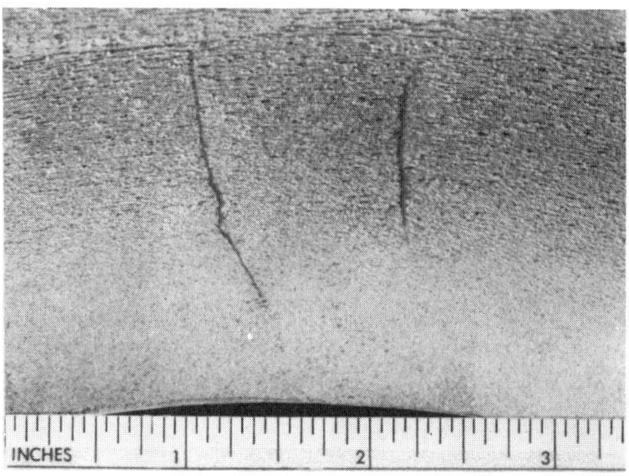

Fig. 1-134 This photomacrograph shows a close-up view of the typical MT indications observed on the disc side surfaces.

Chemical Analysis/Identification. A sample of the turbine disc segment was analyzed for chemical composition. The results of the compositional analysis of the steam turbine disc material are presented in Table 1-30. The compositional requirements for A470 Class 4 are listed in the table for comparison. The disc material meets the chemical requirements of the indicated specification.

Associated Environments. The deposits present at the termini of laboratory opened cracks were analyzed in situ using energy dispersive X-ray spectroscopy (EDS) The typical results of an in situ EDS analysis of the crack surface deposits are presented in Fig. 1-139. The deposit contains major amounts of iron, silicon, calcium, potassium and

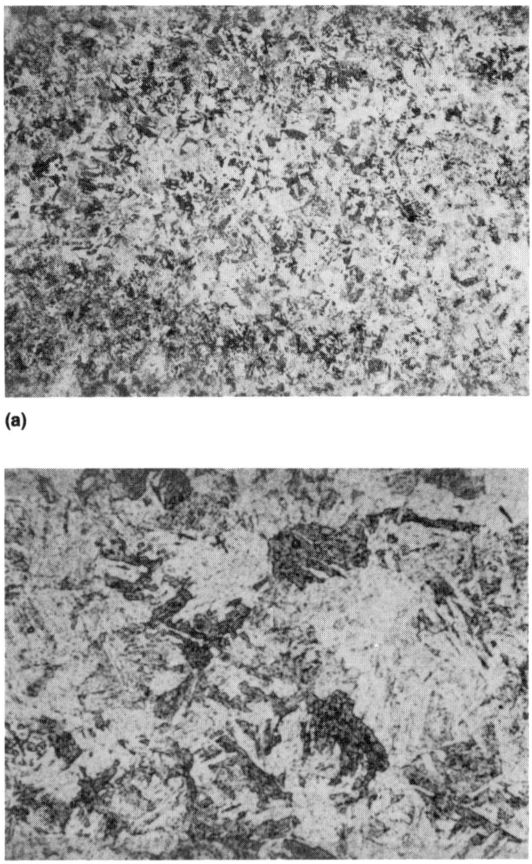

(a)

(b)

Fig. 1-135 The typical bulk microstructure observed in the disc is shown at two different magnifications. The microstructure consists of tempered martensite and bainite.

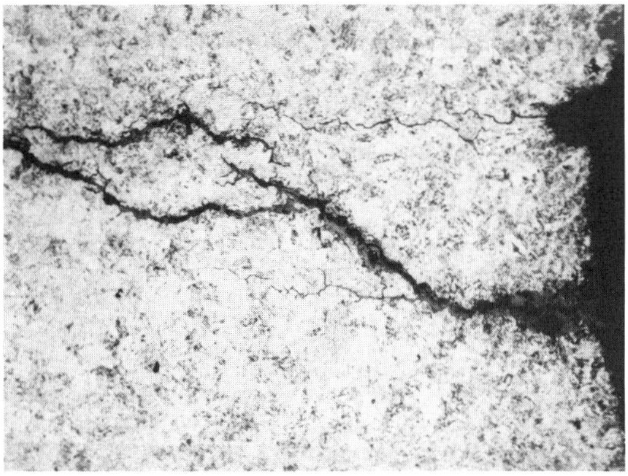

(a)

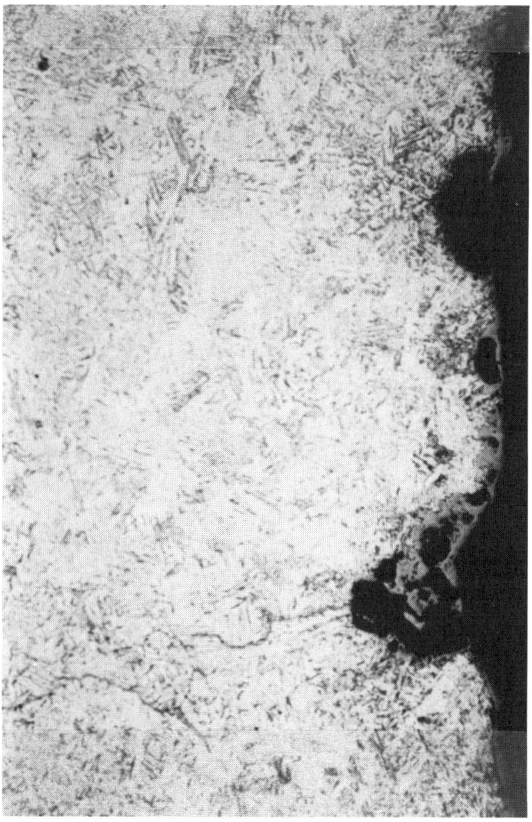

(b)

Fig. 1-136 These photomicrographs show typical examples of cracking which was observed in the blade attachment area. The cracks have propagated from pit bottoms.

chlorine plus minor to trace amounts of aluminum, sulfur, sodium, magnesium, phosphorus, titanium, chromium, manganese, nickel, and zinc. The iron, nickel, magnesium, and chromium are base metal constituents. The other deposit constituents likely come from feedwater carryover.

Mechanical Properties. The bulk metal hardness of the steam turbine disc was measured using the Rockwell test. The bulk metal hardness readings averaged Rockwell C (HRC) 20.2. This hardness value corresponds to a tensile strength of about 758 MPa (110 ksi) per the ASTM A370 approximate tensile strength conversation. This value meets the specification requirements.

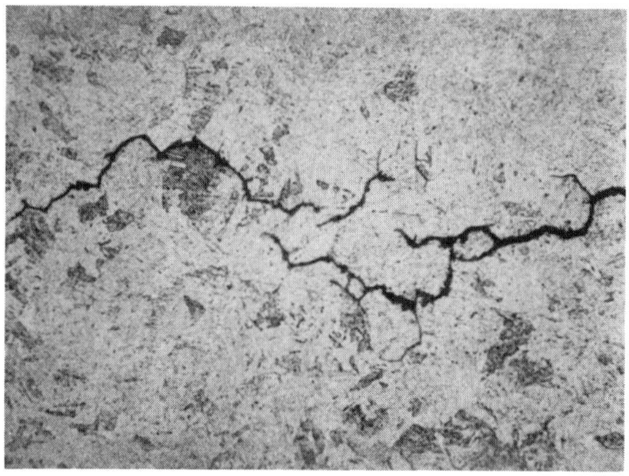

(a)

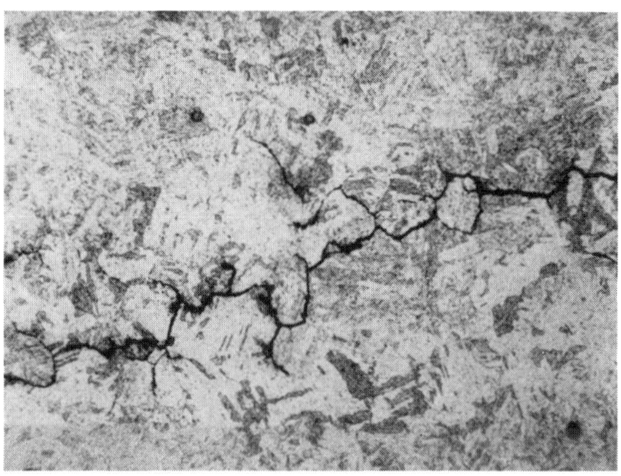

(b)

Fig. 1-137 These photomicrographs show typical examples of cracking which is primarily intergranular in nature.

Discussion. The multiple, radially oriented mixed mode cracking observed in the disc is indicative of caustic-induced SCC. Both sodium and potassium were identified at the fracture terminus areas. The source of the caustic can be a result of carryover or the introduction of feedwater into the turbine or severe steam quality problems. Chlorine (as chlorides) was also detected in the fracture surface deposits. The chlorides are likely the primary cause for the pitting observed on the disc surfaces and within the turbine blade attachment area. Both pitting and SCC occur in wet environments.

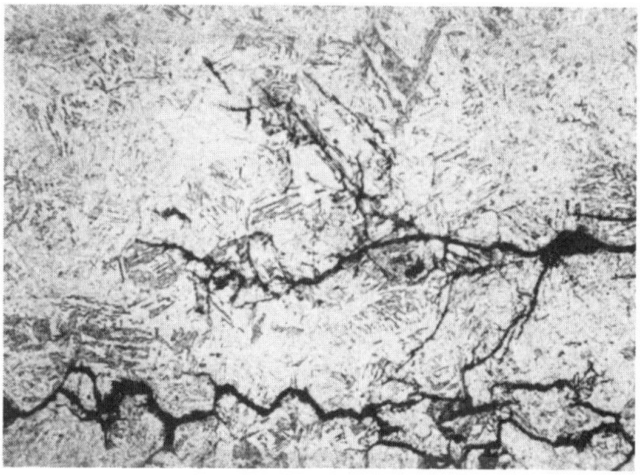

(a)

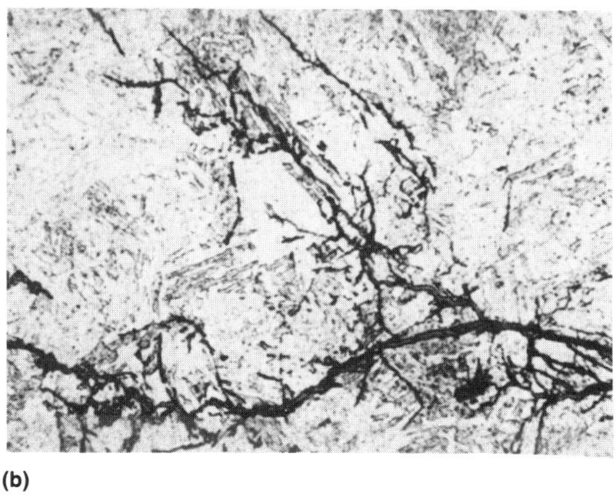

(b)

Fig. 1-138 The top photomicrograph shows two immediately adjacent cracks. The lower crack is primarily intergranular while the upper crack is primarily transgranular. The bottom photomicrograph shows a higher magnification view of the branched, transgranular cracking.

The disc composition, microstructure, and hardness appeared to be normal. No material defects or anomalies were observed that could have contributed to the cracking.

Conclusions and Recommendations. The cause of the numerous cracks observed was caustic-induced SCC.

An evaluation of several factors in the operation of the plant's steam system will be required to identify the source of the caustic. The operation of the steam drum/steam separators and the effectiveness of the plant water treatment program should be addressed. Additionally, the procedure of using saturated steam during start-up should be assessed with regard to the potential for inducting feedwater into the steam turbine.

Corrosion Fatigue and Subsequent Rupture of a Yankee Dryer Roll on a Modified Paper Machine*

A steam-pressurized shell in a paper-drying machine ruptured during normal operation. Paper machines producing soft types of paper (tissue and toweling) or similar products utilize a single large Yankee dryer for final drying of the paper sheet. This Yankee dryer was internally heated with steam at 450 kPa (65 psig) and was operated at 9 rev/min.

The machine was operating normally, 2½ h after a warm startup from a brief shutdown period. Rupture occurred without warning and resulted in 30% of the shell material separating from the dryer, as shown in Fig. 1-140.

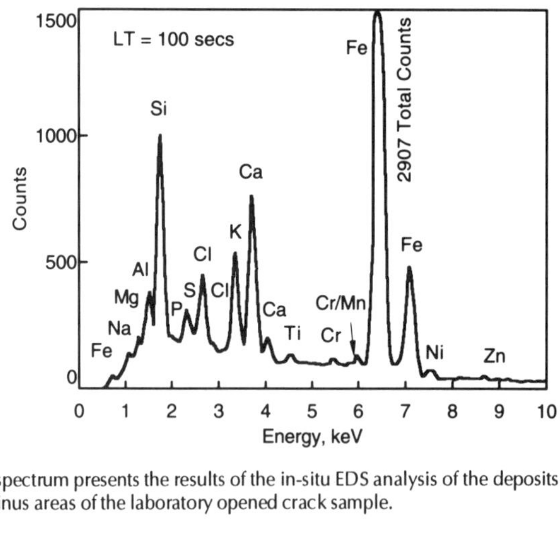

Fig. 1-139 The spectrum presents the results of the in-situ EDS analysis of the deposits present in the fracture terminus areas of the laboratory opened crack sample.

*Adapted from M.H. Gilkey, in *Handbook of Case Histories in Failure Analysis,* Vol 1, K.A. Esaklul, Ed., ASM International, 1993, p 132–135

The component was designed according to the provisions of ASME SA-278, "Specification for Gray Iron Castings for Pressure Containing Parts for Temperatures up to 650 °F," Class 60.

Yankee dryers are typically used for many years (often more than 50), until the shell thickness is reduced below a minimum acceptable value by the periodic regrinding of the working surface. Although shell or head fracture is unusual, there has been at least one reported instance of cracking at a differently designed head-to-shell joint because of the wedging effect of corrosion products.

The fracture appeared to have initiated at a circumferential notch along the drive edge of the shell. The notch is shown in Fig. 1-141. A sample was removed from the edge of the largest shell section, adjacent to the fracture. This sample was selected because of a large indication detected during ultrasonic examination.

Visual Examination. All fracture surfaces had the appearance of brittle fracture. The entire exposed edge of the shell exhibited general and pitting corrosion. It was also covered with a thin, uniform layer of corrosion products. The gap between the ID of the shell and the edge of the head was completely filled with a hard, densely packed material, presumably a mixture of corrosion products and foreign deposits.

The OD surface (face) of the shell was smooth and free of corrosion products. This was the working surface in contact with the product, and it was well maintained.

Ultrasonic Examination. Within days of the failure, sections of the ruptured Yankee dryer were examined using ultrasonic flaw detection methods. The examination focused on the edges of the shell and the circumferential notch between the shell overhang and the bolting flange.

At the drive, or "untended," edge of the shell, crack indications were discovered around 270° of the circumference. A diagram of the shell edge, featuring the indication and the ultrasonic transducer, is shown in Fig. 1-142.

Surface Examination. The fracture surfaces observed at the mill indicated a brittle fracture. The fracture pattern and the fracture surface morphology indicated the initia-

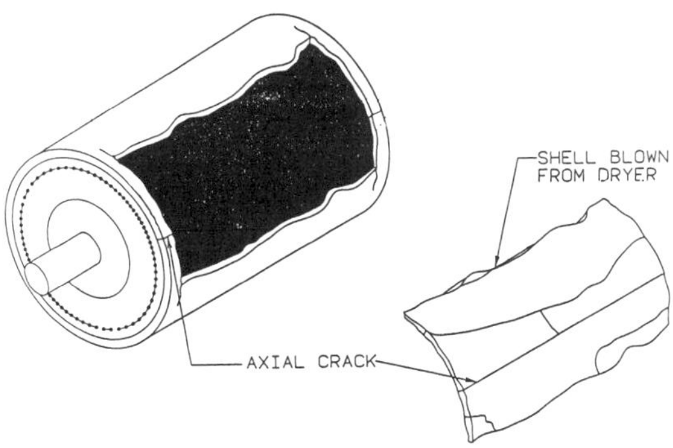

Fig. 1-140 Diagram of Yankee dryer, showing section of the shell that ruptured in service

tion point to be at the edge of the shell on the drive side of the machine. A sample was subsequently removed from this location.

Corrosion Patterns. The as-received sample is shown in Fig. 1-143 and 1-144. The thick scale of corrosion products plus deposits where the shell overhang formed a narrow gap with the edge of the head is visible in Fig. 1-144 (arrow). The maximum scale thickness was determined to be 3.7 mm (0.145 in.). The corner notch between the overhang and bolting flange was covered with corrosion products. There was no visual evidence of a crack.

Microstructural Analysis. The unetched microstructure of the sample presented in Fig. 1-145 consisted of graphite flakes in an iron matrix, typical of gray cast iron. Etching in 2% nital revealed a bainite matrix with scattered massive carbides (Fig. 1-146).

Crack Origins/Paths. The location and size of the crack are highlighted in the cross section of the submitted sample shown in Fig. 1-147. The location was exactly as predicted by ultrasonic examination. A sample containing the entire crack was later removed and cold mounted in epoxy. The mounted and polished sample is shown in Fig. 1-148. It is evident that, although the propagation path is jagged on a local scale, it is generally linear. There is no evidence of crack branching, but there is a secondary surface crack that joins the primary crack. The primary crack is significantly wider at the surface than near the tip.

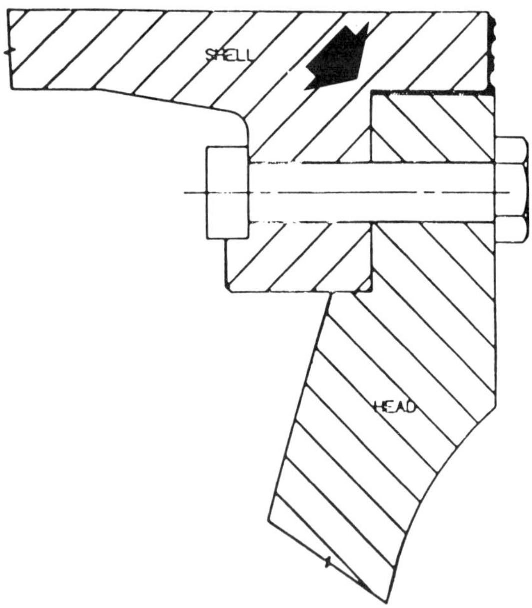

Fig. 1-141 Diagram of cross section through head-to-shell joint. Arrow indicates suspected fracture origin.

Corrosion Morphology. The two cracks are shown in Fig. 1-149. The primary crack was partially filled with corrosion products. Near the crack tip, the crack was filled with corrosion products and clearly propagated along some of the graphite flakes (Fig. 1-150). This tendency to follow the weak graphite flakes was the likely cause of the jagged propagation path seen in Fig. 1-148. The presence of corrosion products indicated that the crack had existed for a significant period of time and was not formed during the rupture incident.

Chemical Analysis. The shell material was analyzed for its chemical composition, which was found to be Fe-2.76C-1.7Si-2.72Ni-1.23Mo-0.94Mn-0.7Cu-0.04Cr-0.07P-0.11S. The cast iron was alloyed with nickel and molybdenum to make it hardenable to a tensile strength near 410 MPa (60 ksi).

Corrosion or Wear Deposits. The thick scale of corrosion products from the headshell gap was analyzed using qualitative emission spectroscopy. It was determined to have a composition similar to the shell chemistry presented above, but with minor amounts of (in decreasing order) lead, sodium, calcium, magnesium, aluminum, and titanium. The lead was probably from a sealing compound used during assembly of the dryer. The other elements presumably originated in the product, the process water, or the sealing compound.

Associated Environments. A sample of the product was analyzed and shown to contain iron, sodium, and titanium.

Mechanical Properties. Machined tensile specimens were also tested. Four tests produced a range of ultimate tensile strength from 350 to 400 MPa (51 to 58 ksi). The

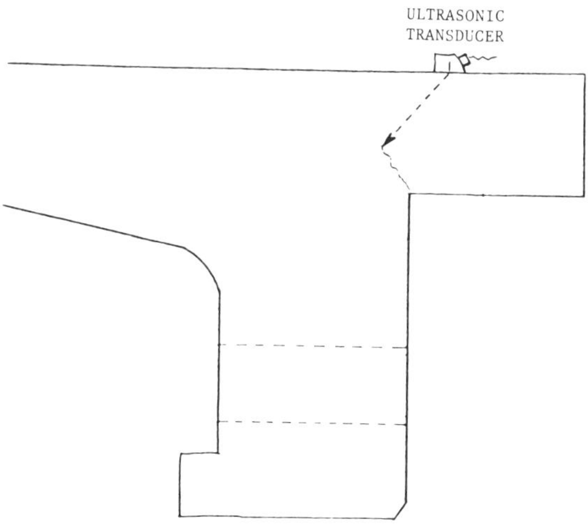

Fig. 1-142 Diagram depicting ultrasonic inspection of the shell and location of the indication

average was 375 MPa (54 ksi), which was below the minimum specified value of 410 MPa (60 ksi).

Discussion. The fracture pattern and the appearance of the fracture surfaces indicated that the fracture had initiated at the drive edge of the shell. Ultrasonic examination of this area revealed a crack indication around most of the circumference. The indication was located at a sharp corner (notch) at the junction of the bolting flange and the overhang.

A cross section through the shell edge confirmed the presence of a crack at the corner. The overall linear propagation, free of branching, and the presence of corrosion products throughout the crack indicated that it initiated and propagated by a corrosion fatigue mechanism.

Conclusion and Recommendations. The Yankee dryer ruptured when a circumferential corrosion fatigue crack reached a critical size. Cracking occurred because of the combination of the corrosive environment inherent in the manufacturing process and the cyclic stresses of normal operation, such as dryer rotation and periodic startups. Significant static stress also was applied to the crack initiation site by the pressure of oxides forming at the surfaces of the already filled headshell gap. The iron to hydrated iron oxide transformation involves a significant increase in volume, which resulted in a buildup of static stress in the 90° corner of the shell.

Fig. 1-143 As-received shell section removed from the Yankee dryer

Ultrasonic examination of the shell, especially at the headshell joint area, could have detected the crack before it reached a critical size. It is likely that the crack propagated for at least several months prior to the rupture incident. Annual or semiannual inspection may have detected the cracking.

The visibly corroded condition of the external surfaces adjacent to the headshell joint also should have alerted the operators and/or inspectors of the dryer to the need for further inspection, including ultrasonic examination. It is known in the pulp and paper

Fig. 1-144 Pitted edge of the shell and the surfaces that were in contact with the head. Arrow indicates the thick layer of corrosion products in the head-shell gap.

industry that such corrosion can indicate corrosion product buildup in the joint, which can lead to shell and head cracking.

Trends in Equipment Failure Rates

The trends in equipment failure rates are clear from the preceding data. In aviation failure rates are decreasing whereas in hydrocarbon processing failure rates are increasing.

Dominant Failure Mechanisms

In aviation the dominant failure mechanisms appear to be fatigue (46%), corrosion (27%), stress corrosion (16%), and corrosion fatigue (11%), Fig. 1-4. In hydrocarbon processing, the distribution is as in Fig 1-62. In mechanical failures (which constitute 50% of the total), fatigue is 15%, mechanical overload 10%, and poor fabrication (welding) 4%. In corrosive failures (the remaining 50%), general corrosion 15%, stress corrosion cracking 13%, pitting 8%, intergranular corrosion 6%, and preferential weld corrosion (off-shore) 6%.

Examination of the preceding service failure analyses reveals similar trends in the aviation and hydrocarbon processing examples given, with almost all the aviation failures involving fatigue and some corrosion mechanism. In the hydrocarbon processing failures given, there is a more even distribution of the mechanisms outlined.

Further distributions of industrial failures according to mechanism and cause were examined. The results of one such analysis, performed in 1992 (Ref 30) are shown in Fig. 1-151 and 1-152, respectively. It can be seen that the dominant mechanism is fatigue, at

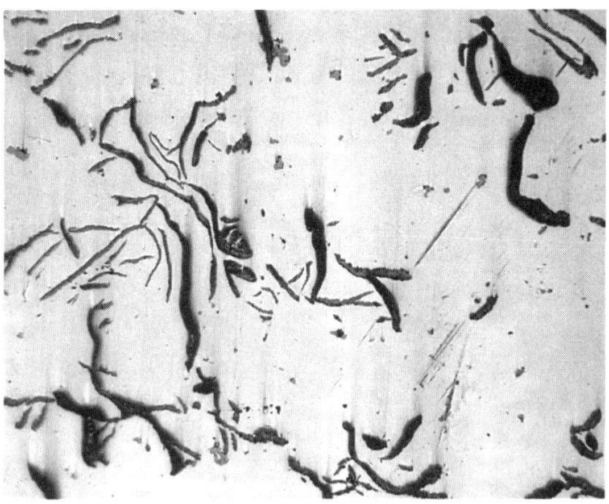

Fig. 1-145 Micrograph showing graphite flakes characteristic of gray cast iron. Unetched. 64×

77%; although this is made up of 58% simple fatigue, 11% corrosion/thermal fatigue and 8% low cycle fatigue. The "other mechanisms" are assessed at 23%. The "other mechanisms" are made up of static fracture (13%), corrosion (3%), and stress corrosion cracking (5%). This grouping can be more appropriately arranged into fatigue (58%), corrosion (19%) [= corrosion 3% + SCC 5% + corrosion/thermal fatigue 11%], and static fracture (21%) [= static fracture 13% + low cycle fatigue 8%].

In this 1992 analysis (Ref 30), the dominant causes are ascribed to the fabricator (design, manufacturing, installation) at 78%. This figure is broken down: underestimation of external force (33%), inappropriate structure (e.g. shape) (24%), misselection of material (19%), and assembling and setting (2%). The view expressed in this work is that this grouping would have been better housed under the heading, "design."

The cause ascribed to the user (operation, maintenance, modification) at 22% is also broken down: inappropriate replacement (16%) and inappropriate repairing (6%). Similarly, this would have been more meaningful if the heading "maintenance and inspection" would have been used.

The results of a 1985 survey (Ref 31) are shown in Table 1-31, which contains mechanisms and Table 1-32, which lists causes. The dominant mechanisms of failure in "engineering components" according to this survey indicate corrosion (29%), fatigue (25%), brittle fracture (16%), mechanical overload (11%), high-temperature corrosion (7%), SCC/corrosion fatigue/HE (6%), creep (3%), and wear/abrasion/erosion (3%). This grouping can be more appropriately arranged into corrosion (41%) [= corrosion 29% + high-temperature corrosion 7% + SCC/corrosion fatigue/HE 6%], fatigue (25%), and static fracture (27%) [= brittle fracture 16% + mechanical overload 11%].

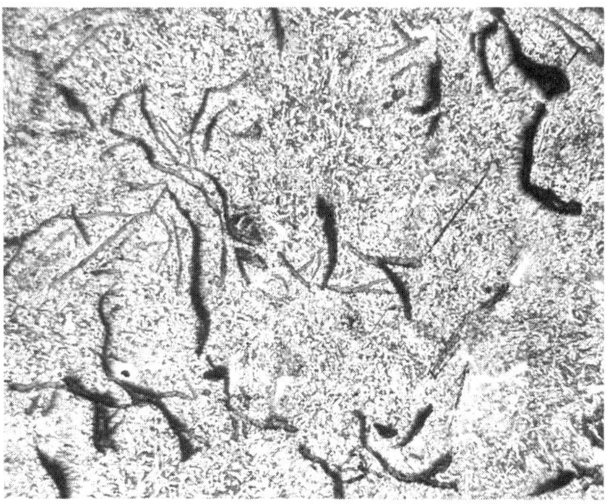

Fig. 1-146 Micrograph of the same area shown in Fig. 1-145 after etching in nital. Matrix microstructure is bainite with scattered carbides. 64×

In the same study, failure mechanisms in aircraft components are also listed: fatigue (61%), mechanical overload (18%), corrosion (3%), high-temperature corrosion (2%), SCC/corrosion fatigue/HE (8%), stress rupture (1%), and wear/abrasion/erosion (7%). As before, these can be re-grouped to give fatigue (61%), corrosion (11%), [= corrosion 3% + SCC/corrosion fatigue/HE 8%], and static fracture 18%.

The causes of the failures in engineering components were improper materials selection (38%), fabrication imperfections (15%), faulty heat treatment (15%), design errors (11%), unanticipated service conditions (8%), uncontrolled environmental conditions (6%), material mix (2%), and defective material (7%). This grouping could be restructured to give "design" 57% [= improper materials selection 38% + design errors 11% + unanticipated service conditions 8%]and "manufacture" 39% [= fabrication imperfections 15% + faulty heat treatment 15% + material mix 2% + defective material

Table 1-31 Frequency of failure mechanisms

	Percentage of failures	
Mechanism	Engineering components	Aircraft components
Corrosion	29	3
Fatigue	25	61
Brittle fracture	16	...
Overload	11	18
High-temperature corrosion	7	2
SCC/corrosion fatigue/HE	6	8
Creep	3	...
Stress rupture	...	1
Wear, abrasion/erosion	3	7

SCC = stress-corrosion cracking; HE = hydrogen embrittlement. Source: Ref 31

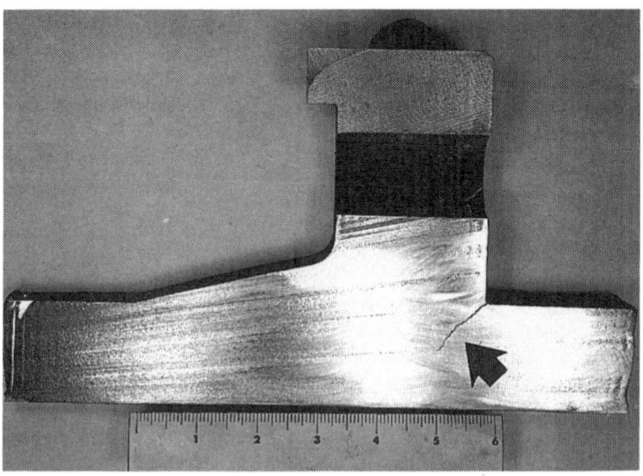

Fig. 1-147 Section through sample. Arrow points to crack at 90° corner of shell.

7%]. For aircraft components, the split was improper maintenance (44%), fabrication imperfections (17%), design errors (16%), unanticipated service conditions (10%), and defective material (7%). Similarly, "maintenance" 44%, "design" 26% [= design errors 16% + unanticipated service conditions 10%] and "manufacture" 24% [= fabrication imperfections 17% + defective material 7%] is a more appropriate grouping.

Further, in the ASM International (Ref 32) distribution of failure mechanisms (Table 1-33), which excludes "corrosion," the dominant fracture mechanism is fatigue [+ corrosion fatigue] at 38.5%, "miscellaneous" (quasi-cleavage, mixed mode fractures, overload, material defects, manufacturing defects) at 26.5%, "decohesive rupture" (creep, stress corrosion cracking, hydrogen embrittlement, intergranular fracture, thermally induced embrittlement) at 14%, dimple rupture at 13% and cleavage at 8%. Also, as Table 1-33 shows, failures occurred primarily in low alloy steels, carbon steels, and wrought aluminum alloys.

Clearly, in aviation failures, fatigue (~60%) and corrosion dominate the failure mechanisms. With causes of failure, it appears that poor maintenance (~44%), at least in

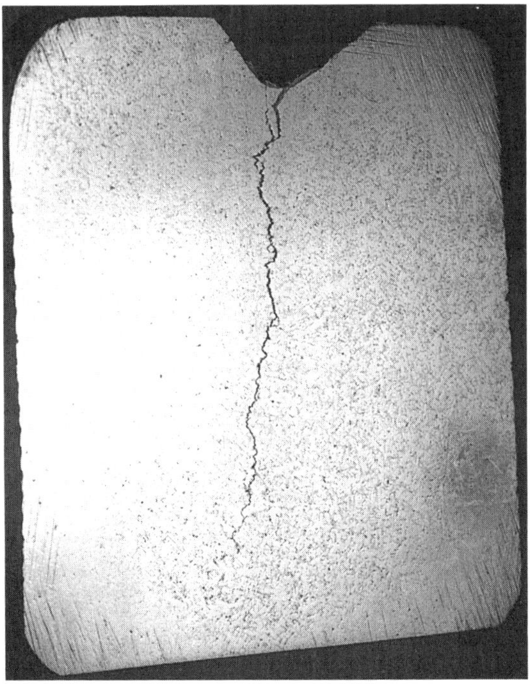

Fig. 1-148 Mounted and polished sample containing the crack

the 1985 (Ref 31) survey, design conditions (~26%) and manufacturing defects (~24%) dominate.

In hydrocarbon processing failures, as with the results of (Ref 30, 31, and 32), there appears to be less dominance of fatigue (~40%), with corrosion (~40%) and static fracture (~20%) having more predominant effects. Causes of failures are apparently attributable to design (~60%), manufacture (~20%), and maintenance and inspection (~20%).

Predicting possible causes of failure at the design stage or at a proposed change of design stage was initially addressed by the Chemical Process Industry (Ref 33) by the introduction of a team approach to identify, evaluate, and recommend ways to control hazards associated with the design, operation, and maintenance of process facilities. The team members required a working knowledge of the process and a demonstrated expertise in some aspect of the operation to be studied. Team members performed their assignment by studying one or more aspects of the process either individually or collectively, depending on the method used.

Fig. 1-149 Micrograph showing two parallel cracks at the surface. Corrosion products are evident in the enlarged primary crack. 28.4×

The investigations required the following:

- Audits of operator practices and job knowledge in the field
- Studies of the suitability of equipment and materials of construction
- Reviews of the chemistry involved in the process and the process control systems
- Audits of operating and maintenance records.

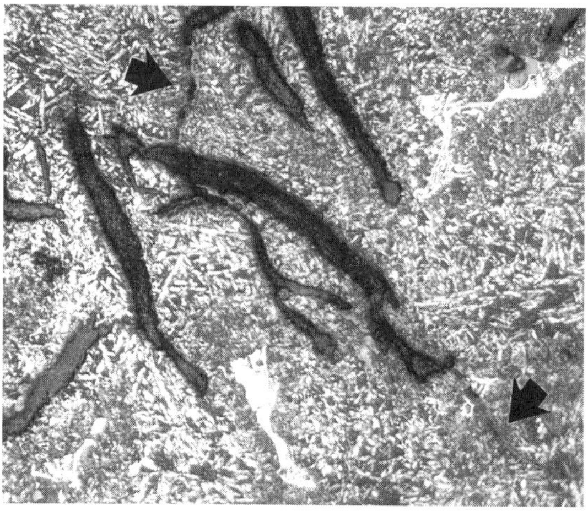

Fig. 1-150 Photomicrograph of the crack near the tip. The crack propagated along a graphite flake, which caused it to change direction. Nital etch. 172.5×

Table 1-32 Frequency of causes for failure

Cause	Percentage of failures	
	Engineering components	Aircraft components
Improper material selection	38	...
Fabrication imperfections	15	17
Faulty heat treatment	15	...
Design errors	11	16
Unanticipated service conditions	8	10
Uncontrolled environmental conditions	6	...
Inadequate inspection/quality control	5	...
Material mix	2	...
Inadequate maintenance	...	44
Defective material	...	7
Unknown	...	6

Source: Ref 31

The team then identified and analyzed hazards using one or more of the methods summarized below.

The ""What if?"" Method

The "What if?" method, the most common one used, is the foundation method (Ref 34); all other methods were built upon this base. It represents a complete review for relatively uncomplicated processes. Review team members formulate and answer "What if?" questions at each handling or processing step to evaluate the effects of component failures or procedural errors on the process. The "What if?" method may be followed by a checklist evaluation to stimulate additional questioning.

Failure Modes and Effects Analysis (FMEA)

When a specific item of equipment, such as a reactor, is to be studied, review team members can use FMEA (Ref 35) to aid in assessing the effects of component failures and

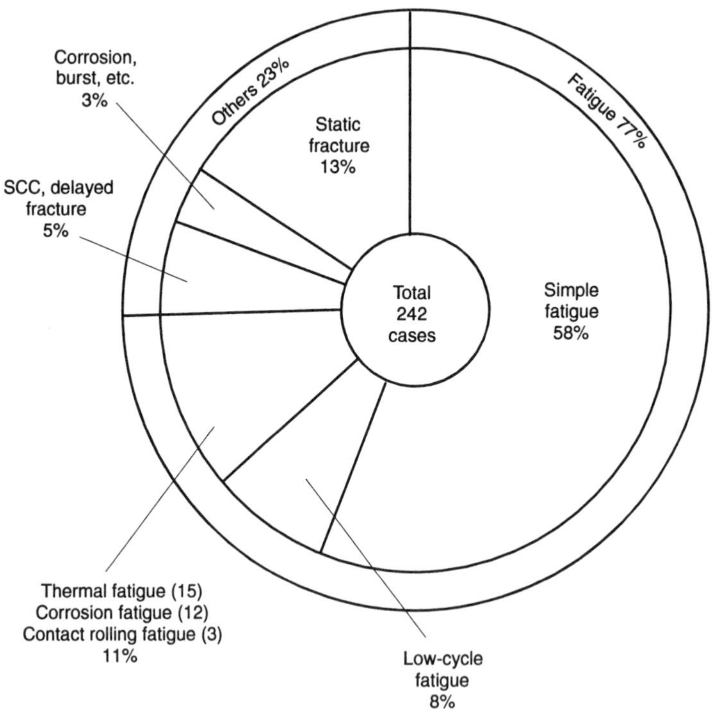

Fig. 1-151 Distribution of failures according to mechanism. Source: Ref 30

the occurrence of certain specified events or errors in the process. This semi-quantitative approach assists in prioritizing hazards.

Hazard and Operability Study (HAZOP)

The HAZOP examination procedure (Ref 36) systematically questions every part of a process to discover how deviations from the intention of the design can occur and to determine if the consequences of such deviations are hazardous.

Table 1-33 Causes of fractures for various ferrous and nonferrous alloys

	Fracture sources			Fracture mode				
Material	Parts	Test specimens	Total	Dimple rupture	Cleavage	Fatigue(a)	Decohesive rupture(b)	Miscella-neous(c)
Irons	...	19	19	5	6	1	3	4
Gray irons	1	1	2	...	2	1	...	...
Ductile irons	4	45	49	8	17	11	...	13
Malleable iron/white irons	1	4	5	1	...	...	...	4
Low-carbon steels	18	11	29	1	2	11	7	8
Medium carbon steels	40	1	41	2	...	34	...	5
High-carbon steels	34	...	34	...	1	21	3	9
AISI/SAE alloy steels	66	66	132	15	4	46	15	52
ASTM/ASME alloy steels	4	10	14	...	...	6	5	3
Austentistic stainless steels	18	11	29	3	1	8	12	5
Martensitic steels	3	...	3	...	...	2	1	...
Precipitation-hardening stainless steels	4	6	10	2	...	1	3	4
Tool steels	7	14	21	3	...	7	4	7
Managing steels	...	7	7	2	...	3	...	2
Superalloys(d)	3	18	21	2	...	11	6	2
Nickel alloys	...	3	3	...	...	...	...	3
Cobalt alloys	1	2	3	...	...	3	...	...
Copper alloys	7	8	15	3	...	5	2	5
Cast aluminum alloys	5	1	6	1	...	1	...	4
Wrought aluminum alloys	18	34	52	6	1	32	5	8
P/M aluminum alloys	...	4	4	...	...	3	1	...
Titanium alloys	3	36	39	15	4	8	8	4
Miscellaneous metals and alloys(e)	1	25	26	3	9	2	5	7
Total	**238**	**326**	**564**	**72**	**46**	**217**	**80**	**149**

(a) This category includes 21 parts and test specimens that failed because of corrosion fatigue. (b) This category includes creep rupture, stress-corrosion cracking, hydrogen embrittlement, thermally induced embrittlement and any other combination of applied stress and environment that results in an intergranular fracture path. Fractures resulting from the synergistic interaction of a corrosive environment and applied stress that are entirely transgranular are listed under "Cleavage." (c) This category includes quasi-cleavage; mixed-mode fractures; overload failures involving tensile, shear, or torsion stresses that do not result in an entirely dimpled rupture fracture surface; fractures due to such material defects as laps, seams, cold shuts, pre-existing cracks, inclusions, porosity, and grain-boundary discontinuities; photographs illustrating microcrack initiation and propagation fractures due to design deficiencies; and radiation-induced failures. (d) Superalloys include both iron-based and nickel-base alloys. (e) Miscellaneous metals and alloys include tungsten, indium, magnesium-base, iron-base, molybdenum-base, and tantalum-base materials.

The process relies on guide words, such as none, more of, less of, reverse, part of, as well as, and other than to identify the deviations. An example of part of a HAZOP on a rail car containing chlorine is given in Table 1-34 (Ref 37). Table 1-35 lists the recommendations for review.

These methods have the common appeal of assessing, what can go wrong and estimating some level of probability of occurrence. After each team member has the

Table 1-34 A part of a HAZOP on a rail car containing chlorine

Company: Safetech
Facility: Princeton
Revision: (0)
Session: (1) 7-26-96
Node: (1) Chlorine rail car
Drawings: CLC/01-07-66
Parameter: Pressure
Intention: Normal operation is 100 to 150 psig. Target pressure is 125 psig.

GW	Deviation	Causes	Consequences	Safeguards	S	L	R	Recommendations	By
More	1. Higher pressure	1.1 Fire exposure	1.1.1. Potential overpressurization of rail car resulting in release of chlorine	1.1.1.1. Rail cars provided with relief valve 1.1.1.2. Rail cars insulated	1	4	4	No recommendations	
			1.1.2. Potential rupture of the rail car if the rail car relief valve fails	1.1.2.1. Location of rail car minimize likelihood of exposure	1	5	5	No recommendations	
		1.2. High ambient temperature	1.2.1. Potential increase in pressure Not likely to approach rated pressure of rail car	1.2.1.1. Rail cars insulated 1.2.1.2. Location of rail car minimize likelihood of exposure 1.2.1.3.Pressure indicator, P1-1	5	1	5	No recommendations	
Less	2. Lower pressure	2.1. Leak in rail car or relief valve fails open	2.1.1. Release of chlorine to the atmosphere	2.1.1.1. Railcar emergency leak patch kit is available on site. 2. 2.1.1.2. Pressure indicator, P1-1	1	4	4	Consider conducting a failure modes and effects analysis (FMEA) of a typical pressure relief valve	PWP
			2.1.2. Potential exposure of personnel and potential offsite impact		1	4	4	No further recommendations	
		2.2 Empty rail car	2.2.1. Delay in treating	2.2.1.1. Rail car is weighed upon receipt	5	3	9	No further recommendations	
		2.3 Sudden change in ambient temperature	2.3.1. Potential for too low a flow to the treatment system		5	2	8	No further recommendations	

GW, guide word; S, severity; L, likelihood; R, risk. Source: Ref 37

opportunity to identify and analyze all the process hazards associated with the portion of the process, the team then meets to:

- define and evaluate these hazards
- make recommendations to remedy problems
- recommend priorities for correction of deficiencies

The team then issues a process hazards review report to document the hazards identified during the review, as well as the recommended actions to eliminate or control the hazard. This report aids management in ensuring that process safety needs are being met satisfactorily. Through the report, management can assess the scope and thoroughness of the review. Following the management review and acceptance of the process hazards report, the responsibility for corrective action is assigned to appropriate operating personnel.

The advantage of the preceding as a tool for preventing service failures is that the team-oriented method raises the awareness from viewpoints taken by varying disci-

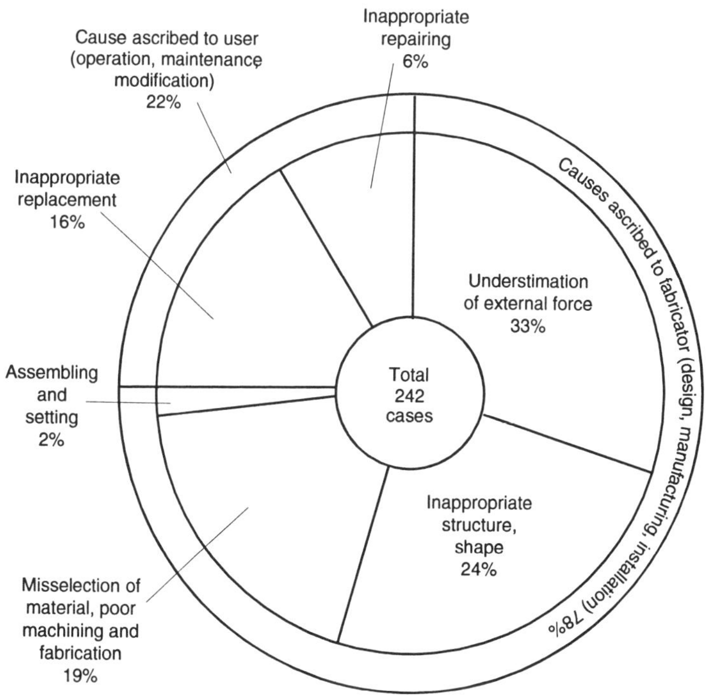

Fig. 1-152 Distribution of failures according to cause. Source: Ref 30

plines. Consider, for example, applications of the simple "What if?" and FMEA techniques to the situations giving rise to two historical catastrophic failures.

The Titanic

The *Titanic* left Southampton on her maiden voyage (Ref 38) on 10 April 1912. She took on more passengers at Cherbourg and, on the following day, at Queenstown. The ship left Ireland on 11 April with 2,228 passengers and crew. The U.K. Board of Trade regulations required that ships with a displacement of over 10,000 tons should carry 16 lifeboats. The *Titanic*, with a displacement of 46,382 tons (the average for 1910 was 1,400 tons), carried 16 lifeboats plus four collapsible boats, giving a total capacity of 1,178 persons.

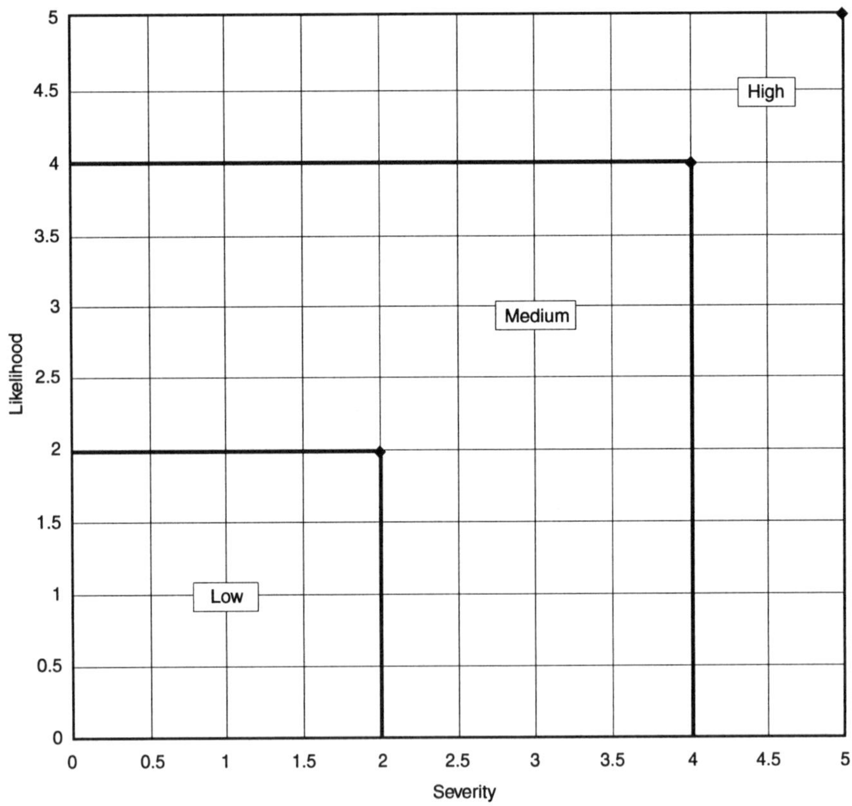

Fig. 1-153 Risk assessment matrix

The maximum complement of the Titanic was 2,300. At the time, the requirement for cargo ships was that they should carry lifeboats with a capacity of twice the ship's complement.

The Titanic was one of three sister ships, the other two being the *Olympic* and the *Britannic*. All three were built with the latest available technology, employing riveted and caulked construction of the double-bottomed hull. The hull was divided into watertight compartments. The ships were designed to remain afloat with two compartments flooded.

On the night of April 14th the *Titanic* was steaming at full speed, (~22 knots). The weather was still and frosty, cloudless with no moon, and the sea dead calm. The impact occurred at 11:40 P.M. The iceberg was sighted when it was only a quarter of a mile ahead. The officer of the watch ordered full astern and put the helm hard over. The distance was too short, and the ship received a series of glancing blows over a length of ~250 ft. These caused plating to buckle and rivets to shear or fail in tension, cracking the caulking. Water then flooded in five or six of the forward compartments. The *Titanic* sank at 2:20 A.M. on 15 April with the loss of more than 1,500 lives. Soon after the inquiry, the *Olympia* went into dock and was fitted with a double bottom which extended up the sides of the hull and rows of lifeboats along the decks. The Board of Trade regulations were amended, and lifeboat drill became a regular feature of transatlantic and other crossings.

The outcome of the application of the "What if?" technique to the *Titanic* is given in Table 1-36. There are no real surprises in Table 1-36. If this approach had been used at the time of the ship's design, the consequences of the hazards would have been realized. What is lacking in this simple "What if?" approach is that no risk is assigned to the scenarios. Because risk = likelihood (or "What if?" or probability) × consequences (or severity or cost), it is possible to assess the risks associated with the "What if?" scenarios. The risk ranking may be shown on a 5 × 5, likelihood × severity matrix or grid, as shown in Fig. 1-153. The task is then reduced to assigning appropriate values to the scenarios.

Table 1-35 User Report. Likely hazards from the exercise

Company: Safetech
Facility: Princeton
Revision: (0)
Session: (1) 7-26-96
Node: (1) Chlorine rail car
Drawings: CLC/01-07-66
Parameter: Pressure
Intention: Normal operation is 100 to 150 psig. Target pressure is 125 psig.

Ref #	Deviation	Recommendations	By	Comments
1	2. Lower pressure	Consider conducting a failure modes and effects analysis (FMEA) of a typical pressure relief valve	PWP	

Source: Ref 37

For all the scenarios shown in Table 1-12. the likelihood is very low, a ranking of unity. The severity or consequences or cost is very high, a ranking of five . The risk ranking in all cases then becomes $1 \times 5 = 5$ which is high risk on the grid in Fig. 1-153.

Because the issue is the loss of human life, then the conclusion is that the recommendations shown in Table 1-36 should have been in place prior to the maiden voyage of the *Titanic*. This view is echoed by the retrofitting of the sister ship, *Olympia*.

The dominant failure mechanisms in the *Titanic* catastrophe were plate buckling and fracture of the rivets. Examination of the wreck revealed that the stern section was severely damaged by implosion, but it did not break up. The forward portion was also bent but not broken. These observations provide clear evidence of ductile behavior.

The brittle fracture problem arose 30 years after the *Titanic*. With a monolithic hull, a running crack could split a ship in two, as was the case with the Liberty ships of WWII, as shown in Figure 1-154 (Ref 39). In a riveted ship, however, such a crack would propagate only as far as the edge of the plate in which it initiated. The failure rate in shipping has decreased since records were established by Lloyd's of London (Ref 40), as shown in Fig. 1-155 (Ref 40).

Fig. 1-154 Brittle fracture of a Liberty Ship. Source: Ref 39

Flixborough

On 1 June 1974, there was a devastating explosion at the Nypro Works, Flixborough, South Humberside, U.K. (Ref 41). The vapor cloud explosion killed 28 men on site and injured 89 people, including some members of the community. The explosion severely damaged the entire plant, destroying a number of major buildings and hundreds of houses and other buildings in the community. Buildings destroyed on site included the control building in which 18 persons died, and the main office block. The capital loss due to the explosion and fire is estimated at $161 million in 1992 U.S. dollars.

A Court of Inquiry set up to determine the causes and circumstances of the explosion found that it resulted from the failure of a bridging assembly. The assembly had been inserted as a temporary measure in the gap left when one reactor in a train of six developed a crack and had to be removed for repairs Fig. 1-156. The bridging assembly consisted of a specially fabricated piece of pipe attached at either end to an existing expansion bellows Fig. 1-156. This created a mechanically unstable system, because a difference in level between the outlet of one reactor and the inlet of the other required that the bridging piece be cranked (Ref 41). Therefore, each bellows was subjected to a nonaxial stress. No calculations were made, however to determine the stresses in this temporary system and whether the bellows could sustain the stress. The system was devised and installed, after consultation with senior management, by the acting chief mechanical engineer, who possessed neither the qualifications nor the experience appropriate to such a position. The system was not pressure-tested to an appropriate standard, and when the system was brought online and pressurized to the normal operating pressure, it was subjected to a mechanical couple. After a period of time, the couple caused both bellows to squirm and burst and the pipe to jack-knife. The purpose of the reactor train was the liquid-phase oxidation of cyclohexane. It contained 120 tons of cyclohexane as a liquified vapor which, when the pressure was released, flashed off 30

Table 1-36 What if?

Hazard	What if?	Consequences	Existing protection	Recommendations
Iceberg	Titanic hits iceberg?	Hull fractures and water floods in.	None	Insufficient protection to hull. Investigate double bottom up side of hull.
	Titanic hits iceberg?	Hull dented. Rivets fractured. Caulking cracks. Water seeps in.	None	Insufficient protection to hull. Investigate double bottom up side of hull.
Sinking	Titanic sinking?	Possible loss of life. Abandon ship.	16 lifeboats	Insufficient number of lifeboats
Fire	Titanic ablaze?	Abandon ship. Possible loss of life	16 lifeboats	Insufficient number of lifeboats
Fire	Abandon ship?	Possible loss of life	16 lifeboats	Insufficient number of lifeboats
Unknown	Abandon ship?	Possible loss of life	16 lifeboats	Insufficient number of lifeboats

to 50 tons of vapor through the open branches. A vapor cloud ignited, leading to the vapor-cloud explosion.

Consider now the application of an FMEA to this situation. The results contained in Table 1-37 present no surprises at all. If this approach had been used at the time of the design alteration, the effects of the possible failure modes would have been realized. At the very least, the outcome would have been to recommend an engineering review of the situation, as in Table 1-37. Assigning risk would also have been straightforward. There would have been a high probability of failure and a high cost or consequence of that failure. The overall risk would have been high.

Such failures as Flixborough, where a major plant was destroyed by a vapor cloud explosion following a full-bore rupture, have become the dominant cause of hydrocarbon plant catastrophes, as described earlier Table 1-30. The Flixborough Court of Inquiry (Ref 41) went on to indicate that, in making a change to a part of the plant that required careful consideration, on-site personnel were being asked to do work outside those areas in which they were trained and competent. This led the Court of Inquiry to recommend that the training of engineers should be broadly based. Although it may well be that the occasion to use such knowledge will not arise in an acute form until an engineer has to take executive responsibilities, it is impossible at the training stage to know who will achieve such a position. All engineers should, therefore, learn at least the elements of

Table 1-37 Failure modes and effects of analysis (FMEA) of Flixborough

Description	Function	Failure modes	Causes	Effects	Recommendations
Bellows	Allow for expansion	Squirm due to lack of constraint	No reactor to contain squirming	Vapor cloud	Mechanical engineering review

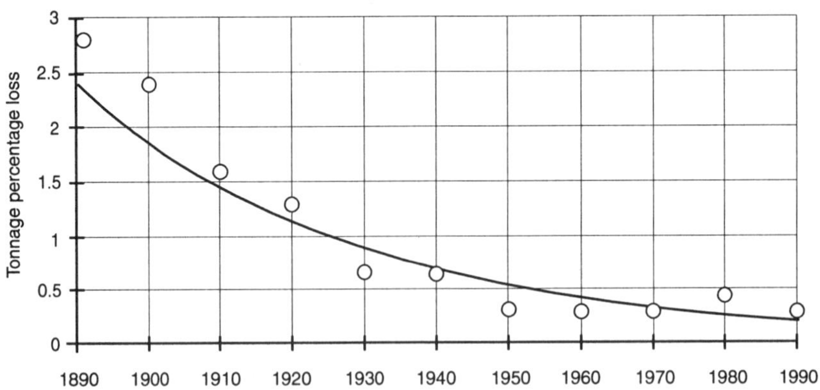

Fig. 1-155 Annual percentage of tonnage lost. Shipping worldwide 1891–1990. Annual decrement = 2.5%. Source: Ref 40

branches of engineering other than their own in both academic and practical training. Also, the Court of Inquiry determined that it is essential (a) persons given certain responsibilities be competent to carry out those responsibilities, (b) top management have a clear knowledge and understanding of individuals and the magnitude and type of demand made upon them, and (c) top management have a clear knowledge and understanding of the total workload placed on each individual in relation to his or her capacity. Even competent individuals have increased potential for errors of judgment when overworked. In times of crisis and extreme demand, it is tempting to overwork the willing, some of whom may not know their own limitations.

Management of Change (MOC)

A further consideration from the Flixborough disaster was that changes in the existing design were not managed at all. Following is a description of a typical, paper-driven management of change (MOC) system (Ref 42). It is intended to ensure that a logical, stepwise review is undertaken of all changes that may be proposed or required to be made to process and utility systems resulting from operational, maintenance, or design modifications. MOC is closely related to a HAZOP, but it embraces changes of a smaller magnitude that formally would not have initiated a HAZOP—a temporary spool-piece, for example. Adoption of MOC will ensure that no changes, however small, will escape appropriate review, the absence of which may result in an operational "incident." The flowchart for the MOC process is shown in Fig. 1-157 (Ref 42).

In the two catastrophic situations considered, it is clear that relatively simple approaches to the identification of hazards could possibly have averted the catastrophes. These tools have the common appeal of assessing "What can go wrong?" and estimating some level of probability of occurrence. Their usefulness is extended if some attempt is also made to assess risk. All of the preceding methods are extremely valuable in

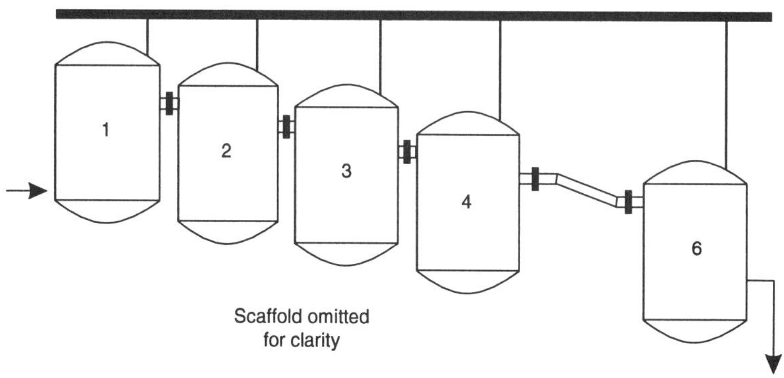

Fig. 1-156 Reactor train at Flixborough showing cracked bridging piece between the expansion bellows connected to reactors 4 and 6. Source: Ref 41

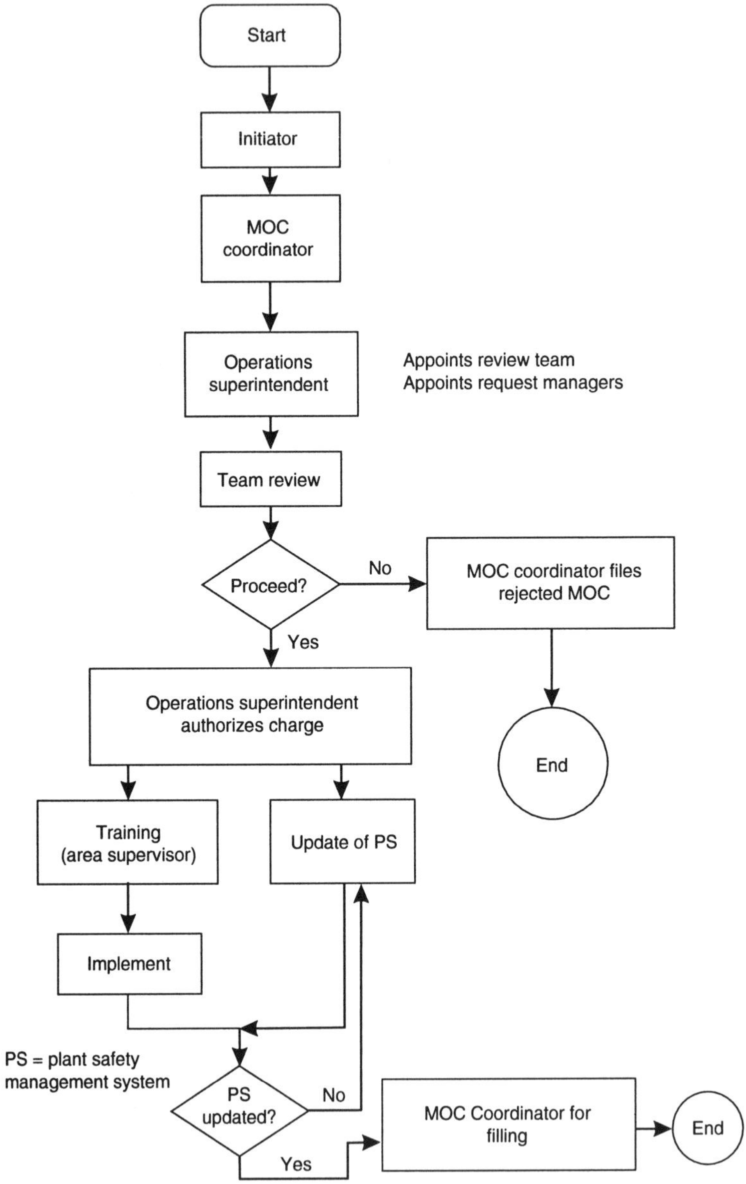

Fig. 1-157 Management of changes (MOC) system. Source: Ref 42

reducing equipment failure rates and providing platforms of guidance in the prevention of equipment service failures.

REFERENCES

1. "Statistical Summary of Commercial Jet Aircraft Accidents, Worldwide Operations 1959–1992", Boeing Commercial Airplane Group, Seattle, WA, 1993
2. C.J. Peel and A. Jones, Analysis of Failures in Aircraft Structures, *Met. Mater.*, Vol 6, 1990, p 496–502
3. A.K. Koul, R. Thamburaj, M.D. Raizenne, W. Wallace, and M.C. Malherbe, Practical Experience With Damage Tolerance Based Life Extension of Turbine Engine Components, *Proceedings of International Conference of Advisory Group for Aerospace Research and Development (AGARD) of NATO*, AGARD-CP393, paper 23, April 1985
4. N.C. Bellinger, P. Stoute, G. Gould, A. Fahr, and A.K. Koul, "The Nondestructive and Destructive Data for Compressor Discs for Detecting Bolt Hole LCF Cracks," Report ST-490, National Research Council of Canada, Dec 1987
5. N.C. Bellinger, A. Fahr, A.K. Koul, G. Gould, and P. Stoute, "The Reliability and Sensitivity of NDI Techniques in Detecting LCF Cracks in Fastener Bolt Holes of Compressor Discs," Report LTR-ST-1651 National Research Council of Canada, Feb 1988
6. A.K. Koul, N.C. Bellinger, and A. Fahr, Damage Tolerance Based Life Prediction of Aeroengine Compressor Discs: A Deterministic Fracture Mechanics Approach, *Int. J. Fatigue*, Vol 12 (No.5), 1990, p 379
7. A.K. Koul, N.C. Bellinger, and G. Gould, Damage Tolerance Based Life Prediction of Aeroengine Compressor Discs: A Probablistic Fracture Mechanics Approach, *Int J. Fatigue*, Vol 12 (No. 5), 1990, p 388
8. A. Hull, R.V. Dainty, M.D. Raizenne, and D.W. Hoeppner, Fatigue Crack Growth Testing of Compressor Discs, *J. Can. Aerospace Inst.*, Vol 3, (No.4) 1986, p 321
9. D. Mahoney. *Large Property Damage Losses in the Hydrocarbon-Chemical Industries*, 15th ed. Marsh and McLennan, 1993
10. D. Mahoney, *Large Property Damage Losses*, 14th ed. and analysis, Marsh and McLennan, 1992
11. G. Drogaris, *Major Accident Reporting System*, Elsevier, Amsterdam, 1993
12. W.G. Garrison, Loss Prevention, *Hydrocarbon Processing*, Vol 1–2, Sept. 1988
13. J.A. Collins and M.L. Monack, Stress Corrosion Cracking in the Chemical Process Industry, *Mater, Protect, Perform*, Vol 12, 1973, p 11–15
14. P. Nelson and J.R. Still, Metallurgical Failures on Offshore Oil Production Installations, *Met. Mater*, Vol 4, 1988, p 559–564
15. Helmut Thielsch and Florence M. Cone, "Failures, Integrity and Repairs of Petrochemical Plant Reformers"
16. Marketing Brochure, Fired Heater Technology Corporation, Liberty Corner, NJ
17. "Incoloy Alloys 800 and 800HT," Inco Alloy International, Inc., 1986, p 17

18. "Superalloys," *Metals Handbook, Desk Edition,* American Society for Metals, Metals Park, OH, 1985, p 16-13
19. R. Davies, private communication, Det Norske Veritas Industry, Houston, 1987
20. J. Kapp, "Crack Growth in Mercury Embrittled Aluminum Alloys Under Cyclic and Static Loading Conditions," PhD. thesis, Rensselaer Polytechnic Institute, 1982
21. *Metals Handbook Desk Edition,* American Society for Metals, 1985
22. *Welding Handbook,* 7th ed., Vol 4, American Welding Society, 1982
23. J. Orr and J.B. Marriot, Proceedings of the CEC Conference on Alloy 800, Brussels, Sept., 1976
24. J. Orr, Proc. Petten International Conference Alloy 800, North Holland Publishing, 1978, p 25
25. J.C. Lippold, *Mater. Performance,* Vol 24 (No. 10), 1985, p 16
26. W.B. Jones and J.A. Van Den Avyle, "Fatigue and Creep-Fatigue Testing of Steam Filled Tubular Alloy 800 Specimens," Report SAND. 82-0856, Sandia National Laboratories, May 1982
27. H. Teranishi and A.J. McEvily, Proc. Petten International Conference on Alloy 800, North Holland Publishing, 1978, p 125
28. *Failure Analysis and Prevention, Metals Handbook,* 8th ed., Vol 10, American Society for Metals, 1975, p 122
29. C. Levaillant and A. Pineau, Low-Cycle Fatigue and Life Prediction—ASTM STP 770, 1982, p 169
30. S. Nishida, *Failure Analysis in Engineering Applications,* Butterworth- Heinemann, UK, 1992
31. G.J. Davies, Performance in Service, *Essential Metallurgy for Engineers,* E.J. Bradburg, Ed., Van Nostrand Reinhold, London, 1985, p 126–155
32. *ASM Handbook,* Vol 12, Fractography, ASM International, 1992, p 217
33. F.P. Lees, *Loss Prevention in the Process Industries* Vol 1 and 2, London, Butterworth, 1980
34. C.H. Buschman, Ed., Loss Prevention and Safety Promotion in the Process Industries *Proc. of the First Int. Loss Prevention Symp.* Delft, The Netherlands, 1974
35. MIL-STD-1629A, U.S. Department of Navy, 1977
36. *A Guide to Hazard and Operability Studies,* Chemical Industries Association, Alembric House, London, 1977
37. PhaWorks 3 for Windows, Primatech Inc., Columbus, OH, 1997
38. R.D. Ballard, *The Discovery of the 'Titanic,'* Madison Press Books, Toronto, 1987
39. T. Jutla, Fatigue and Fracture Control of Weldments, *ASM Handbook,* Vol 19, *Fatigue and Fracture,* 1996, p 442
40. Lloyd's Casualty Returns, Annual Statistical Summary, Lloyd's Register of Shipping, London, First published 1891
41. The Flixborough Disaster, report of the Court of Inquiry, HM Stationery Office, London, 1975
42. P.F. Timmins, *Predictive Corrosion and Failure Control in Process Operations,* ASM International, 1996

CHAPTER 2

Solutions to Equipment Failures in Aviation

It was demonstrated in Chapter 1 that the equipment failure rate was decreasing in the aviation sector of industry and that some design problems could be solved by hazard identification methods. The purpose of the present chapter is to examine those factors which have contributed to the observed decrease in failure rate and attempt to develop a methodology for solving equipment failure problems which may be applied to other industry sectors.

During the 1950s, the loss of several Comet aircraft (British Overseas Aircraft Corp., Bristol, U.K.) within a short period of time prompted an investigation into the structural integrity of the fuselage of the Comet. A pressurization test on a sample aircraft revealed that the combination of fatigue and stress concentration at the corner of a window was the probable cause of the failures (Ref 1). These accidents and others prompted the development of fatigue methods for aircraft design. Some of these methods are outlined in the following paragraphs.

Safe-Life or Durability Method

The safe-life or durability method (Ref 2) is applied to structures with hard-to-detect cracks (e.g., in inaccessible locations) or when initial crack size limits are below inspectable limits. If no inspections can or will be done, a small initial crack (a_i) could be assumed to exist initially in the new structure. The time, t, for the crack to grow from a_i to a_p, the permissible size, is then the available safe life. In that case, the structure or component must be retired or replaced after, for example, $t/2$ hours. This is called the durability method. A durability analysis (based on *S-N* data) and/or tests are used to establish a "design life." To account for scatter or uncertainties in the operating conditions, retirement or replacement are then mandated at fractions (e.g., ⅕ to ⅓) of that design life. Crack growth and inspection intervals are not involved. The drawbacks of this approach are the critical dependence on the durability analysis and safety factor (which may be inadequate) and the high costs of retiring components without an obvious "cause." Nonetheless, some components (such as landing gear) are designed to safe-life criteria, because cracks are noninspectable (i.e., the materials have low toughness) and proof tests are not practical.

A problem with the durability approach is how the size of a_i is assumed. The assumption of an initial crack size may be difficult or unclear for high quality components, which may be essentially defect free. In welded structures, however, the assumption of an initial flaw is more realistic, since welds often contain defects such as porosity or lack of fusion.

Fail-Safe Method

The fail-safe method of aircraft fatigue design was developed during the 1960s and implemented in a number of commercial aircraft (Ref 3). The fail-safe philosophy is to design multiple load path structures, such that if an individual element should fail, the remaining elements would have sufficient structural integrity to carry the additional loads from the failed element until the damage is detected through scheduled maintenance inspections. In fail-safe designs, crack stoppers are also used which consist of materials with a high fracture toughness that supplement the residual strength of the surrounding structure and prevent cracks from propagating to failure. An example of a crack stopper is a stringer in a pressurized fuselage. The stringer reduces the amount of energy available for crack growth, slowing or stopping the advance of a crack that crosses it. An aircraft designed to fail-safe principles can sustain damage and remain airworthy until the damage is detected and repaired. Periodic inspection is therefore necessary to examine the structure to determine if the primary load-carrying member contains cracks. Safety is considered ensured when cracks are eliminated before reaching a_p. Therefore, cracks must be discovered before that point by means of periodic inspection. Whatever the inspection method, a certain size of crack, a_0, is not likely to be detected. This implies that discovery and repair must occur in between a_0 and a_p as shown in Fig. 2-1. For any inspection done before or at time t_1, the crack will be missed; should the next inspection be at t_2, the crack would already be too long (having reached a_p). Hence, the inspection interval, I, must be shorter than $t_{(inspection)} = t_2 - t_1$.

It is often taken as $I = t/2$, but a more rational procedure to determine inspection intervals can be employed. Once discovered, a crack must be repaired at the earliest opportunity. Since a_p is a permissible and not a critical crack, and because detection commonly occurs at sizes less than a_p, immediate repair may not be necessary, but any complacency will defy all analysis and inspection efforts.

Whether inspections must be performed every day (i.e., $I = 1$ and $t = 2$ days) or every year (i.e., $I = 1$ and $t = 2$ years), there would always be two inspections between a_0 and a_p. In daily inspections, the achieved safety is no different from daily or yearly inspections. If a crack is missed in daily inspection, a potential fracture will occur sooner, but if a crack is missed in yearly inspection, fracture will occur before the year is over. If short inspection intervals are not feasible, the option of selecting a more refined inspection procedure with a smaller a_0 is open. In this case, t, and hence the inspection intervals, will be longer.

The time at which crack initiation takes place, is not of concern, as shown in Fig. 2-2. Inspections scheduled at a $t/2$ interval will always give two opportunities for detection, provided that inspections are scheduled at $t/2$ intervals starting from hour zero. Similarly, if the interval chosen is $t/3$, there will always be three inspections between a_0 and a_p. The frequency of these inspections is typically assigned by the manufacturer based on service experience. This method does not consider the initiation and growth (linking up) of small cracks at fastener holes. As a result, the loss of several "fail-safe" aircraft in the mid 1970s emphasized the need to locate cracks and repair damage before failure occurred (Ref 4).

Damage Tolerance Method

In the early 1970s, the U.S. Air Force initiated a number of standards and specifications that led to a method of fatigue design called "damage tolerance" (Ref 5). The

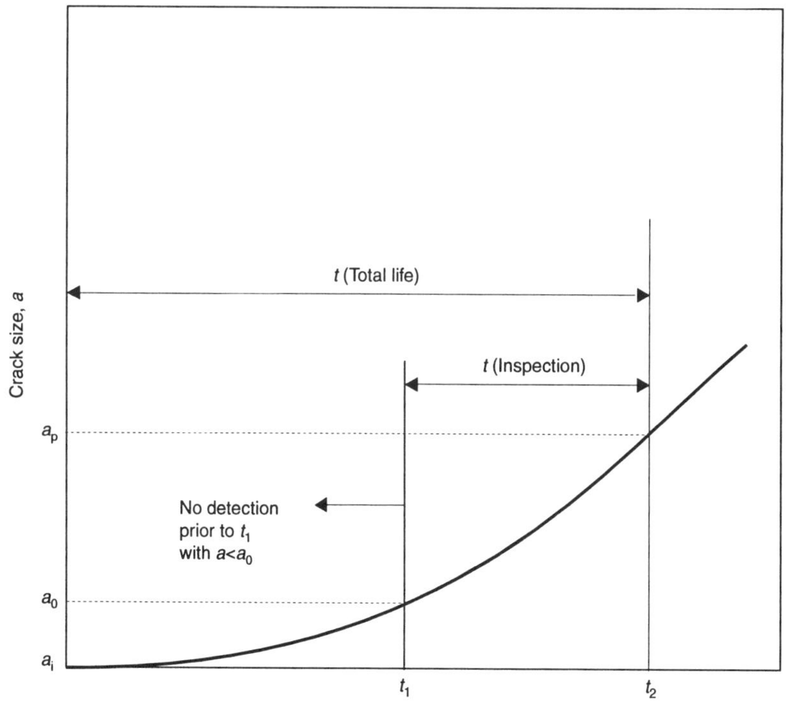

Fig. 2-1 Crack growth and time period when inspection can be performed

damage tolerance method is based on fracture mechanics. The objective of this approach is to detect cracks in principal structural elements (PSEs) before they propagate to failure. A PSE is any aircraft structure carrying flight, ground, or pressurization loads whose failure could result in the loss of the aircraft.

By establishing inspection intervals for these elements based on the time it takes a crack to grow from an initial detectable size to the critical crack length, the objective of the damage tolerance approach can be achieved.

Before inspection intervals can be established for an aircraft, its usage profile must be defined. This profile describes the various flight conditions such as taxi, climb, cruise, descent, and landing impact, and the amount of time spent at each gross weight, speed, and altitude. The usage profile is then used to create a load factor spectrum at the center of gravity of the aircraft, based on the frequency of occurrence of each increment of load factor, and to identify the PSEs of the aircraft.

An inspection interval must be established for PSE. This task is accomplished by converting the load factor spectrum into a stress spectrum for each location and incorporating the effects of the service environment. Using crack growth equations, such as the Forman equation (Chapter 3), the stress spectrum is combined with material

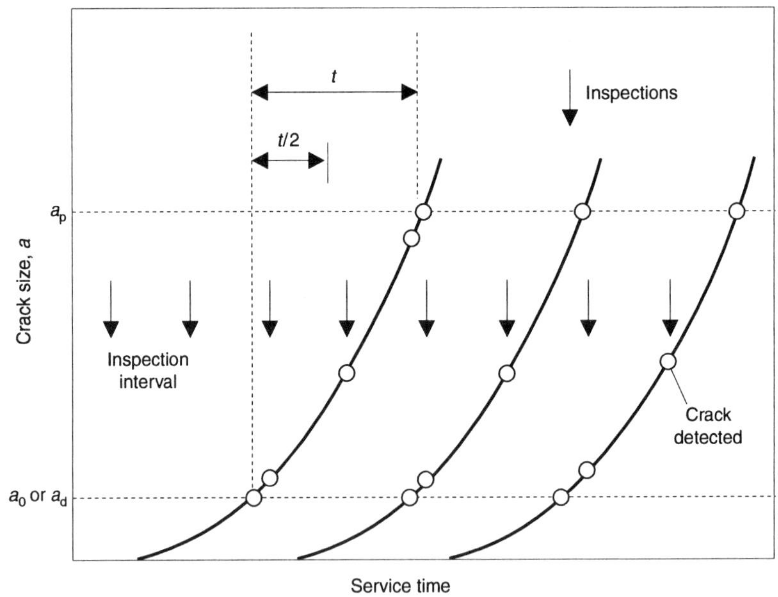

Fig. 2-2 Two possibilities for crack detection with $t/2$, regardless of when crack starts

properties data and stress intensity factor solutions applicable to each PSE to determine the number of cycles or flight hours to failure. This number is usually divided by a factor of two to arrive at the inspection interval. Dividing by a factor of two ensures that, should a PSE develop a crack, it will be inspected at least once before the crack propagates to failure. The determination of inspection intervals is further detailed in subsequent paragraphs.

Unlike the safe-life approach, where components are retired whether or not they are damaged, damage tolerant components are only replaced if a crack is found during an inspection. It is important to note that, although the detectable crack size dictates the time between inspections to a certain degree, any size crack found during an inspection mandates replacement of the damaged component. In addition to its economic advantages, the damage tolerance approach also allows for a reduction of safety factors in design. Since fracture mechanics provides a more precise characterization of crack behavior, the large scatter factors typical of *S-N* fatigue methods are not required.

Damage tolerant components are designed on the basis of residual strength. The residual strength of a cracked component decreases as the crack grows longer. After a certain amount of time, the residual strength of the component will fall below the maximum expected service stress, at which time failure of the component becomes probable. The inspection interval must be made short enough to locate the damage before a crack reaches a length likely to result in failure. The task of the designer is to produce components that have sufficient residual strength to withstand the maximum stresses expected, while at the same time maximizing the critical crack length. Some of the ways in which this task can be achieved are selecting a material with good fracture toughness, eliminating stress concentrations, and integrating crack stoppers into the design.

The preceding discussion has focused on the deterministic aspects of damage tolerance. Stochastic methods such as risk analysis also play a role, as shown in Chapter 1. Combining the data generated from deterministic analysis with stochastic factors such as manufacturing quality, then risk analysis attempts to assess the probability of fracture of principal structural elements. An initial flaw in a PSE is represented by an equivalent initial flaw size distribution (EIFSD), which is obtained from tests that determine the distribution of times it takes a crack of some initial size to reach a specified reference size (Ref 6). This distribution of times is referred to as the time to crack initiation (TTCI) distribution. This approach differs from a deterministic analysis, which assumes a single value for the initial flaw size. Once the TTCI distribution is known, the EIFSD may be obtained for any service time. Inspection intervals are then established based on the amount of time it takes the crack sizes at the lower end of the distribution to reach the critical crack length. These inspection intervals may later be modified to account for the quality of maintenance and inspection capability (the probability of detecting cracks).

Detection of cracks larger than the "detectable size" is not a certainty. It is affected by many factors: the skill of the inspector; the specificity of the assignment (e.g., one specific location, as opposed to a whole wing); the accessibility and viewing angles;

exposure (e.g., part of a crack may be hidden behind other structural elements); possible corrosion products inside the crack; and so on. In general, discrimination of flaws depends on detection response levels (or crack size) and the level of the application noise. Analysis of signal and signal plus noise are common in electronic devices, optics, and other discrimination processes. However, it is important to recognize that the dominant noise source in a nondestructive evaluation (NDE) process is not electronic noise that may be reduced by filtering, multiple sampling, and averaging techniques, but instead the noise is due to nonrelevant signals generated in applying the NDE procedure to a specific hardware element.

The outcome from an NDE measurement/decision process needs to be understood as a problem in conditional probability (Ref 7). When an NDE assessment is performed for the purpose of crack detection, the outcome is not a simple accept/reject (binary) process, as is frequently envisioned. It is actually the product of conditional acceptance due to the interdependence of the measurement and decision responses. In general, the four possible outcomes from an inspection process are:

- *True positive:* A crack exists and is detected.
- *False positive:* No crack exists but one is identified.
- *False negative:* A crack exists but is not detected.
- *True negative:* No crack exists and none is detected.

Typically, the probability of crack detection depends on crack size in the manner shown in Fig. 2-3. There is a certain crack size, a_0, below which detection is physically impossible. For example, for visual inspection this would be determined by the resolution of the eye, and for ultrasonic inspection by the wavelength. In reality, a_0 is larger than these physical limits. The probability of detection is never equal to 1, even for large cracks; any crack may be missed. It follows that the probability curve must have the general form shown in Fig. 2-3.

A crack is subject to inspection several times before it reaches the permissible size. At successive inspections, the crack will be longer, and the probability of detection higher, but there is still a chance that it will go undetected. Consider 100 cracks growing at equal rate (same population), all in the same stage of growth (same size). Let the probability of detection at a certain inspection be $p = 0.2$. The probability that a crack will be missed is then $q = 1 - p = 0.8$. That means that 80 cracks will go undetected. At the next inspection, the cracks are longer; let the probability of detection then be $p = 0.6$, so that $q = 0.4$. Thus of the remaining 80 cracks, $0.4 \times 80 = 32$ cracks will go undetected. Apparently the cumulative probability that a crack will be missed in successive inspection is $Q = q1 \cdot q2 \cdot \ldots \cdot qn$. In the above example, $Q = 0.8 \times 0.4 = 0.32$. Of the 100 cracks, 32 cracks remain undetected after two inspections. The cumulative probability of detection is $P = 1 - Q$. In the above example, $P = 0.68$. Of the 100 cracks, 68 were detected after two inspections, but 32 were missed.

Factors Influencing the Life of Aircraft

The life of an aircraft is dependent on a number of variables, such as flight profiles, usage rate, external environment, material selection, and design philosophy. The first two of these (flight profile and usage rate) affect the design life of the aircraft (i.e., the number of expected safe flight hours). The external environment affects corrosion, and therefore component life, while material selection (composite versus metal structure) and design philosophy effect individual components.

A study conducted by the National Transportation Safety Board, they found that, in the fifteen years between 1970 and 1985, corrosion or fatigue was involved in 77 out of 579 aircraft incidents (Ref 8). This number can be greatly reduced through improved

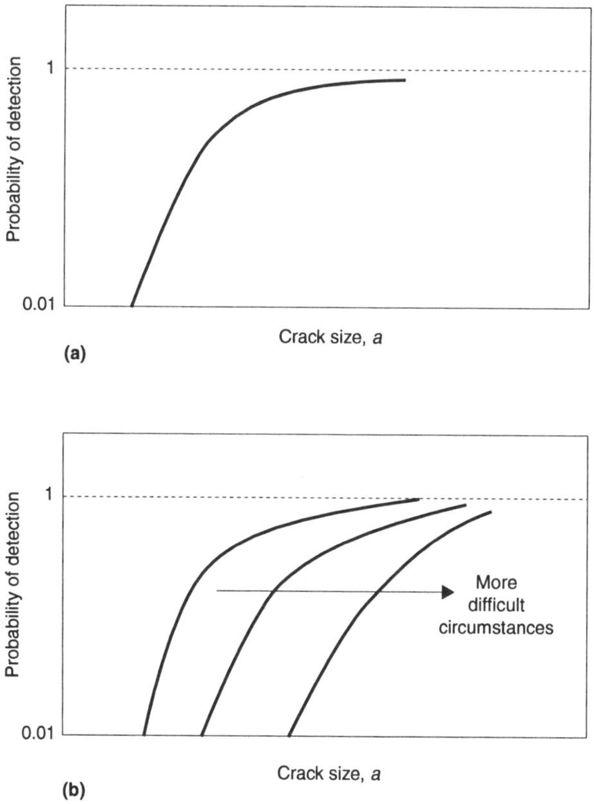

Fig. 2-3 Probability of crack detection in one inspection. (a) Basic curve. (b) Effect of accessibility, and specificity, or other difficulty factor

design methods and inspection procedures. While an aircraft is operational, the dynamic factors (fatigue, flutter, and vibration) are the responsible parameters for determining the life aspects of selected components, but while the aircraft is sitting on the ground, corrosion behavior is normally the dominant driver. There are factors, however, that exacerbate both these conditions. These are manufacturing and maintenance actions which allow for both the premature initiation and propagation of cracks and also enhance the onset and continuing damage of corrosion. For example, these may include dings, scratches, grinding marks, improper cleaning, and non-removal of drilled or reamed material.

Manufacturing Defects

Manufacturing defects may be subdivided into two broad categories: material defects and mechanical defects. Both types of defect generally exist in the aircraft at the time the product leaves the manufacturer. The material defect may manifest itself as a result of a metallurgical anomaly. This anomaly or metallurgical defect may occur as a result of faulty furnace control during heat treating, improper plating which may result in unwanted hydrogen in high strength steel, processing problems (forging laps, shuts, undesirable grain size), or welding defects.

For example, a material defect caused failure of an F-111 aircraft which used D6AC steel in the wing box. Material allowables were obtained through a testing program. Unfortunately, the test specimens to determine the toughness of the steel had a section thickness different from the material placed on the aircraft. The material located on the aircraft had a toughness substantially lower, (~50%) than the material used in the test program due to the difference in thermal behavior during heat treatment. Additionally, a forging lap was present in the material used for the joint. The crack was not observed during inspection, and the material was placed onto the aircraft with the forging lap present. The aircraft was flown safety for 107 flights, at which time catastrophic failure occurred, causing the destruction of the aircraft and the death of the pilot (Ref 9).

Subsequent inspection noted the presence of the flaw, and a rigorous test program indicated a significant difference in material toughness between the coupons used in the original test program and the forging used on the aircraft. As a result of this failure, means to periodically inspect the components were required. The method utilized was a proof test. Chambers were built that housed the F-111, and a system was designed and fabricated that would lower the temperature of the chamber below nil ductility transition (NDT) for the steel. The aircraft was placed into the chamber, the temperature stabilized, and fixtures loaded the wing box as required. If a crack equal to or greater than critical existed in the structure, failure would occur. If no failure occurred, the aircraft would be able to safely fly another flight interval. This inspection technique was repeated for each aircraft as appropriate.

There are a number of examples where failures of structural components initiated from manufacturing defects. For military aircraft, these include T-38 and F-5. Improper use of a router caused these aircraft to lose wings as a result of fatigue cracking (Ref 10).

Civilian aircraft include United Flight 232 which failed near Sioux City, Iowa, in July 1989 (Ref 11) and failures of titanium discs on an RB211 engine (Ref 12). The Sioux City mishap was due to a manufacturing problem in a titanium fan disc, and the RB211 failures were due to tensile residual stresses due to processing of the titanium. These residual stresses, when added to the normal operating stresses, caused an effective operating stress that had sufficient magnitude to lower the fatigue life of the structure. It can be seen that these manufacturing defects occur as a result of both manufacturing assembly and material processing.

Corrosion

Corrosion is a source of time-dependent metal damage that requires protection of structure according to FAA regulations (Ref 13). In addition, the FAA has issued an Advisory Circular (AC) on corrosion (Ref 14). This document discusses types of corrosion, areas on aircraft susceptible to corrosion, inspection techniques, agents which cause corrosion, and methods of corrosion removal and component rework.

In most cases, corrosion itself does not lead to structural failure, although it is the most expensive aspect of maintaining structural integrity. One operator of a fleet of aircraft indicated that 95% of parts removed on aircraft were a result of corrosion (Ref 15). The damage caused to an aircraft can manifest itself as corrosion and paint degradation, both of which are exacerbated by the presence of salt water, acid rain, and sunshine. As a result, the down time of an aircraft, whether it is kept in a hanger or on the flight line, and its primary geographical location all become important variables in the determination of inspection intervals and types of inspections that are required.

The effect of corrosion needs to be addressed by denoting types of corrosion and where they occur, inspection techniques to find the corrosion in situ, and repair or replacement methods to remove the offending material. Corrosion may be broken into two broad categories: enhanced fatigue crack growth due to corrosion and general corrosion. Boeing developed a corrosion task force to accumulate and review all corrosion problems data on its aircraft (Ref 16). Boeing developed a corrosion task force to accumulate and review all corrosion problem data on its aircraft (Ref 16). Boeing developed a program that identified, to a specific usage baseline, the location, type, and inspection intervals for its aircraft. If a particular aircraft was based in an area with a more severe environment, these intervals would be decreased. As a result of this task force, Boeing now has corrosion prevention and control programs on its aircraft and has developed a Corrosion Design Handbook. The feedback between aircraft users and manufacturers has led to designs which decrease corrosion.

Environmentally Enhanced Fatigue Crack Growth

Crack growth analysis is normally performed using material properties obtained in laboratory air. It is well known that the introduction of active environments (sump water, salt water, and fuel) increase crack growth rate (Ref 17, 18). Along with the environment, other important factors include the cyclic frequency and shape. It has been shown that the greater the time the material is exposed to a high stress in an aggressive environment, the greater the crack growth rate (Ref 19, 20). Analysts have developed many techniques to account for the enhanced crack growth rate in hostile environments but no universal method has yet been determined. Generally, the analyst uses the increased crack growth rate of the particular environment (cyclic, chemical, and temperature), and it has been found that the inspection intervals decrease. However, the alternating stresses that cause the crack to grow occur in flight, where the effect of salt water (a ground environment condition) is absent. Additionally, at high altitudes, temperatures are decreased, thereby slowing chemical reactions. In wet wings, the effect of the chemical environment is always present; however, the temperature decreases as altitude increases.

The complexity of accurately determining the advance of the crack is a substantive problem. An approach used by the United States Air Force to aid in solving this problem employs a corrosivity factor (Ref 21). This method documents the air bases to which a particular aircraft is assigned, and then each base is assigned a particular corrosivity factor. This empirical approach allows determination of crack growth rate, and thus inspection intervals, as a function of the particular location of the aircraft and its corrosive behavior as a function of that location.

Improper Maintenance

Manufacturers of aircraft systems present many documents as part of delivery. These include pilots handbooks, structural repair manuals, and corrosion manuals, and it is imperative that these manuals be followed in a rigid fashion. Other methods may be easier and less time consuming than those discussed in the manuals, but the manufacturer selected those particular procedures for good reasons. It is dangerous to deviate from these methods, for the deviation itself can place undue stress or cause latent damage to components other than those being examined. This high stress or latent damage can lead to premature failure of these other components, and that can result in undesirable consequences. Additionally, personnel must take special care to ensure that maintenance action itself does not result in damage to the aircraft. American Airlines Flight 191 crashed in Chicago in May 1979 as a result of improper maintenance actions (Ref 22). The crash of a U.S. Air Force T-38 trainer occurred as a result of a router mark on the lower skin of the wing. The mark was quite small, but as the stresses were significant, a crack initiated and grew until failure occurred. The fatigue crack at failure was only 0.097 in. deep. A similar mark also caused the crash of an F-5 aircraft. From these examples, it is evident that extreme care must be taken during maintenance and the inspection following maintenance to ensure that proper procedures and techniques are rigorously followed.

Multiple Site Damage

Since the accident involving Aloha Airlines Flight 243 in April 1988, interest in multiple site damage (MSD) has increased significantly (Ref 23). The U.S. Air Force, through their specifications, account for MSD by assuming that each fastener has a crack of a particular size, and the most critical location has a crack with a significantly larger size. As the aircraft is utilized, both types of cracks grow, albeit at different rates. Methods commonly used to retard MSD include fastener selection, metal working, and component geometry. Additionally, much analytical and test work has been completed to study the effect of many small cracks and their linking together.

The onset of MSD, especially in fuselage structure in the first row of rivets in joints, is a serious concern. In a riveted, load-carrying structure, the load is transferred from one panel to another through the rivets. If there are rows of rivets, (assuming a lap joint) the highest load is in the first row of rivets, the second highest load in the second row of rivets, and so on. The loads in the holes adjacent to these rivets are twofold. Both the far field stress (the pressure loads) and the local loading of the rivets on the skin (the bearing load) should be considered. It is the superposition of these separate loads that must be considered in this type of structure.

The localized or bearing loads are quite high, but the length over which they are dominant is quite small. However, the initiation of a crack by these loads is significant. Because a crack may be initiated in a short time period, the crack growth life of the structure must be carefully analyzed. If an existing crack is located on an airframe, or if the user wishes to make a change or addition which necessitates additional cutouts in the fuselage structure, a doubler must be added. The design of the doubler and the method of attachment to the existing structure must be ascertained. Because the loads from the skin are placed into the doubler through the fastener, and Newton's laws dictate displacement compatibility of the fastener/doubler/skin system, the thickness of the doubler greatly affects the load on the fastener. As the doubler thickness increases, the load induced on the fastener increases, with a resulting decrease in crack initiation time. This decrease in crack initiation time greatly affects the life of the overall structure. Other aspects of the fastener design also play a large role in structural life. These include fastener spacing and type and hole preparation.

Aircraft Life Extension Techniques

The damage tolerance portion of life extension depends primarily upon three independent variables. These are the material properties, the usage history, and the damage intrinsic to the material component itself. Knowledge of these variables and their interaction, if any, allows determination of component life. In other instances, fatigue initiation or crack growth can determine the life of a particular part or component. In these instances, knowledge of the geometry and location of the flaw and the loading and usage of the aircraft become increasingly important. To extend the life of the structure, it is necessary to quantify (either deterministically or statistically) the information obtained from field experience on either these or similar aircraft. Then analysis techniques that allow the user to determine inspection procedures and techniques and their appropriate timing would extend the life of the selected components.

When corrosion and its growth can be the driving factors behind the life expectancy of the particular structural member, the determination of component life is much less straightforward. There is limited knowledge of the methods for determining either corrosion initiation or propagation times, and this topic will be covered in Chapter 4. Periodic inspections to seek out the existence of corrosion is a viable method.

The FAA issued Advisory Circular 91-56, "Supplemental Inspection Program for Large Transport Category Airplanes" (6 May 1981), a supplemental inspection document (SID) (Ref 24), and Advisory Circular 91-60, "The Continued Airworthiness of Older Airplanes" (13 June 1983). The purpose of the latter is to expand the scope of the former to other airplanes (Ref 25). The SID informs the user of the changes required in maintenance actions as the aircraft accumulates service hours. The inspection program defined by the SID requires the user to continually re-evaluate inspection procedures and intervals based upon usage, service experience of the particular fleet of aircraft, and advances in technology (analyses, testing, and inspection procedures). The SID emphasizes obtaining this information on principal structural elements.

Corrosion Protection

Components must be inspected to ensure that corrosion has not initiated and grown to a condition where part clean-up or replacement is necessary. It is also important to design the particular components in a way that impedes corrosion. There are several approaches to counteract the onset and growth of corrosion. Many of these methods are applicable to only one type of component, but some lend themselves to several portions of the aircraft. The methods that the manufacturer may use to attain a longer corrosion initiation time can fall into several categories. These include appropriate selection of material, judicious choices for materials in contact with one another, design of the metallurgy of the material for the intended loading, and coating or surface treatment of the material itself.

Aluminum, as an alloy, has been used on aircraft for many years. However, the alloy itself and its heat treatment have changed substantially. Manufacturers used duralumin (2024) in the 1940s, and 7178, 7079, and 7075 in the 1960s. Two of these alloys, 7178 and 7079, were both of higher strength than the 2000 series alloys, but their stress-corrosion cracking susceptibility was substantially greater. As a result of the increased degradation in active environment, these alloys are no longer used in aircraft design. The alloy 7075 was first used in its high strength temper, T6, but was found that the alloy in this temper is susceptible to stress-corrosion cracking. Experiments determined that if this alloy is overaged and used in the T76 or T731 condition, resistance to stress-corrosion cracking is enhanced. In the overaged condition, however, the strength of the alloy is lessened.

Aluminum alloys are also susceptible to surface corrosion, pitting corrosion, and exfoliation. It was found that by cladding the alloy with pure aluminum, resistance to these types of corrosion is enhanced. For many aircraft usages, these alloys are now being used in the clad condition. It should be noted that the clad is removed at fastener holes due to the machining process. The fasteners are usually countersunk and as a result, the clad material is removed in their vicinity. Rivets for aircraft are generally manufactured from 2117 aluminum; therefore, no galvanic reaction occurs at the skin/rivet interface. When steel bolts are used, they are usually cadmium plated to minimize galvanic corrosion. Another surface treatment of aluminum, in addition to priming and painting a surface, is anodizing.

Another material commonly used in the manufacture of aircraft is titanium. This alloy is stiffer and stronger than aluminum, yet not as dense as steel. Its corrosion resistance is quite good, although it is susceptible to chloride attack. As a result, both sea water and methyl chloride can cause corrosion in titanium.

The remaining metal widely used on aircraft is steel. Because low-alloy steels are commonly used on aircraft, the effect of corrosion must be continually addressed. Steel is the preferred material in landing gear because of its high strength and stiffness. To minimize the amount of material used, the selected alloy is generally used in its high-strength condition. As the strength levels are increased, the material becomes chemically more active. For this reason, many of the steel alloys (4340, 4140, 300M, etc.) are coated or plated, usually with cadmium, to prevent exposure to the atmosphere.

Mechanical Techniques of Controlling Multiple Site Damage

A typical lap splice joint can attach a stringer to the outer skin of the aircraft and have a crack arrest strip integral to the design. Since these rivets are highly loaded due to both pressurization and fuselage bending loads, and they also have superimposed localized bearing loads, small cracks can initiate and propagate. As the Aloha Airlines accident has shown, a sudden failure in a panel can cause several additional panels to fail almost instantly. The multiple site damage (MSD) is the existence and linking of these small cracks into a large panel crack that can lead to catastrophic consequences.

The design of the doubler helps determine both the stress concentration at the doubler/skin interface and the inspectability of the structure itself. These two considerations are both important in evaluating MSD. The location of the skin/stringer or the skin/longer on interface attachment has a known load. The known load at the first rivet away from the attachment is substantially similar, but the thickness at that location is substantially smaller. The smaller cross-sectional thickness results in higher stresses. Because the length over which the stress increase occurs is limited, a stress concentration results. To avoid these stress concentrations, designs that alleviate the stress increase may be considered, including a tapered skin, a tapered doubler, or the tying in of the first rivet at different rows. To aid in inspecting the structure, the designer may include a doubler designed with tabs that allow inspection of the skin material in a straightforward manner.

The bearing stress on a rivet is dependent on its pitch (the distance between adjacent rivets). For an aircraft skin that is 0.040 in. thick and subject to a stress of 15 ksi, the shear now would be 600 lb/in. For a basic five-row doubler joint (0.040 in. skin and 0.050 in. doubler) with a rivet spacing of 1.00 in. and a rivet diameter of 0.125 in., a displacement compatibility analysis would indicate a maximum shear load of 176 lb/rivet on the outer row of rivets. The bearing stress from these rivets would be 35 ksi. As the allowable bearing stress (for yielding conditions) on 2024T3 alloy is 88 ksi, these stresses are acceptable. If the rivet spacing were increased to 2 in. (50 mm), then the bearing stress on the same rivets would become 55 ksi (380 MPa). Even though the static margin is high, the change in stress due to the fastener spacing is significant.

Another important factor in improving component life is hole preparation. If a hole is drilled through the skin and doubler, the resulting stress concentration factor is three. However, if the hole is cold worked by expanding the material through the use of a mandrel and split sleeve, then a residual compressive stress is placed into the material. This residual compressive stress greatly reduces the effect of the bearing stress at the edge of the hole, and crack initiation time is greatly lengthened (Ref 26). Furthermore, small cracks or discontinuities near the edge of the hole are closed, and component life is greatly enhanced.

Inspection

The inspection programs for each particular aircraft are described in its supplemental inspection document (SID). The rationale for these inspections, on the many principal structural elements (PSEs), can include both a deterministic and a stochastic approach. As the aircraft ages, the number of inspection locations based on initial flaw assumptions using equivalent initial flaw size distribution (EIFSD) usually grows. As a result, the aging of the aircraft causes an increase in the inspection time. The determination of when and where to inspect becomes increasingly complicated. Using a lead-the-fleet aircraft or a full-scale test article gives a good indication of the growing flaw population. The continuous analysis of the data from the lead-the-fleet aircraft and all other aircraft inspected can be used to tune the inspection intervals. A goal of this iterative testing and analysis procedure is to determine the economic life of the aircraft itself. It is at this juncture that the aircraft is retired because further usage can no longer be economically justified.

Documentation necessary to accomplish this complex undertaking is lengthy. At McDonnell Douglas, three volumes of information are prepared (Ref 27). These are:

- Volume I: General Information, PSE Description, Inspection Requirements
- Volume II: Inspection Methods
- Volume III: Inspection Planning and Status Report

McDonnell Douglas plans to annually update and republish Volume III by a computer program specifically written for this purpose (Ref 27). Volume III will contain information on fleet status, PSE population, projected supplemental inspection start dates and intervals for each PSE, inspection planning data, and a record of past inspection (Ref 27).

Inspection Procedures for Corrosion

Inspection procedures used on aircraft include visual, radiography, ultrasonic, acoustic emission, eddy current, magnetic particle, and dye penetrant techniques (Ref 28). The type of inspection procedure used is dependent on the location of the component tested, the material, and the type of corrosion. In many instances, it is necessary to gain access to the area of interest. The preparation of a corrosion prevention and control plan, as previously described, is paramount. The user must know where to inspect, how to inspect, and how often to inspect. These inspection documents are normally quite detailed. A typical mechanism is the definition and use of PSEs. These are structural components that can be grouped together in terms of corrosion susceptibility or inspectability. For each PSE (each of which can consist of many components), the user is informed of all aspects of corrosion susceptibility and inspection procedures. This general plan will call out many other specifications (ASTM, SAE, and Military), which discuss particular methods and procedures of accomplishing the necessary inspection. In many instances, it is necessary to remove plating or paint and to repaint or re-plate the affected area after the inspection.

Repair Limits for Corrosion Damaged Structure

The aircraft user receives details on maintainability or repair of the aircraft from the manufacturer. The user is given information on methods of first identifying and then removing corrosion from the aircraft. In some instances (e.g., pitting corrosion), the removal method can be as straightforward as sanding or using chemicals to remove the offending material. With a more severe corrosion such as exfoliation, segments of material must be removed. If the removal of the material is within the prescribed limits of the manufacturer, the manufacturer's repair manual is sufficient. However, if the repair falls outside the limits set by the manufacturer, approval to exceed these limits must be obtained from the FAA.

Additionally, corrosion is generally insidious. The user does not know of the existence of corrosion until it is seen, and then the type and destruction it causes cannot be deterministically evaluated. The repair of corrosion is generally an iterative process of clean-up and evaluation. For example, stress-corrosion cracking follows grain boundaries in aluminum alloys, and its depth and severity cannot be accurately determined by nondestructive examination. After the corroded area has been identified and the corrosion product removed, the repaired area should have the same surface treatment as the unaffected area adjacent to it. This can necessitate painting, plating, or using some other surface treatment on the material.

Derived Methodology for the Prevention of Equipment Failure

The following factors represent contributions to an overall methodology which has resulted in decreasing failure rates in the aviation sector of industry.

- the adoption of the fracture mechanics based defect tolerant design
- the concept of principal structural elements
- the development of alloys for corrosion and fatigue resistance
- the assumptions made in the retardation of multiple site damage
- the control of manufacturing procedures
- the need for careful maintenance and procedures
- the requirement of a corrosion prevention and control plan
- the need for detailed inspection procedures and precise inspection intervals

Attempts will now be made to address some of the service failures in Chapter 1 in terms of the preceding methodology, in order to determine how effective this methodology would have been in preventing these service failures.

Fatigue Fracture of Aircraft Engine Compressor Disks*

Several compressor disks in military fighter and trainer aircraft gas turbine engines cracked prematurely in the bolt hole regions. The disks were made of precipitation-hardened AM355 martensitic stainless steel. Experimental and analytical work was performed on specimens from the fifth-stage compressor disk (judged to be the most crack-prone disk in the compressor) to determine the cause of the failures. Failure was attributed to high-strain low-cycle fatigue during service. It was also determined that the cyclic engine usage assumed in the original life calculations had been underestimated, which led to low-cycle fatigue cracking earlier than expected. Fracture mechanics analysis of the disks was carried out to assess their damage tolerance and to predict safe inspection intervals.

Conclusion and Recommendations. The aircraft under investigation were not equipped with airborne data recorders for monitoring the cyclic usage of the engines. It was thus necessary to convert the cyclic SII into flying hours. The SII values obtained through different fracture mechanics calculations are compared in Table 2-1. These calculations indicate that the SII for service-induced cracks is of the order of 900 to 2200 flying hours, depending on the assumed cyclic usage of the engines and the method followed to calculate the SII value.

In order to accurately predict the SII in terms of flying hours, it was recommended that airborne data recorders be used on a number of aircraft. The intent was to monitor different engine usage parameters over a few hundred flights to accurately define an average mission profile for the engines.

It was also recommended that each aircraft be fitted with an airborne data recording system so that the engines could be inspected after a certain amount of cyclic usage, rather than relying on an average mission profile for converting the cyclic usage into

---

*Adapted from A.K. Koul and R.V. Dainty, in *Handbook of Case Histories of Failure Analysis,* Vol 1, K.A. Esaklul, Ed., ASM International, 1992, p 241–250

Table 2-1 Comparison of safe inspection interval values predicted by different test methods

	Safe inspection interval in hours assuming:	
Basis of prediction	3 cycles/ mission hour	5 cycles/ mission hour
Worst-case crack growth rate	1530	918
Quantitative fractography of service cracks	1818	1091
Quantitative fractography of spin-pit cracks	2167	1300

Note: The SII value predicted by the engine manufacturer was 555 h.

flying hours. In this manner, damage-tolerance-based maintenance methodology can be effectively used to retire the cracked disks with little danger of catastrophic failures.

Analysis of Wing Main Spar Cracking*

Following the crash of a Mirage III-0 aircraft (apparently caused by engine failure), a small crack was detected in a bolt hole in the wing main spar (AU4SG aluminum alloy). Because this area was considered to be critical to aircraft safety and similar cracking was found in other spars in service, the Royal Australian Air Force requested that the crack growth rate during service be determined. The loading history of the aircraft was made available in the form of flight-by-flight records of the counts from the vertical accelerometer sensors fitted to the airframe and a series of "overstress' events recorded during the life of the aircraft. The bolt hole was examined by eddy current testing, visual examination, high-powered light microscopy, and scanning electron microscopy. Simulation tests were also conducted. The use of simulation specimens permitted actual crack growth rate data to be determined for the configuration of interest.

Conclusion and Recommendations. Many spars in the fleet were found to be cracked. The results of the investigation revealed the reasons for crack growth rates in service. This in turn allowed a program of hole refurbishment to be safely implemented. Refurbishment was based on reaming the holes and using interference-fit bushings.

Failed Mixer Pivot Support of An Army Attack Helicopter**

A forged, cadmium-plated electroslag remelt (ESR) 4340 steel mixer pivot support of the rotor support assembly located on an Army attack helicopter was found to be broken in two pieces during an inspection. Visual inspection of the failed part revealed

*Adapted from N.T. Goldsmith and G. Clark, in *Handbook of Case Histories of Failure Analysis,* Vol 1, K.A. Esaklul, Ed., ASM International, 1992, p 9–11

**Adapted from V.K. Champagne, G. Wechsler, and M. Pepi, in *Handbook of Case Histories of Failure Analysis,* Vol 2, K.A. Esaklul, Ed., ASM International, 1993, p 388–391

significant wear on surfaces that contacted the bushing and areas at the machined radius where the cadmium coating had been damaged, which allowed corrosion pitting to occur. Optical microscopy showed that the crack origin was located at the radius within a region that was severely pitted. Electron microscopy revealed that most of the fracture surface failed in an intergranular fashion. Energy dispersive spectroscopy determined that deposits of sand, corrosion and salts were found within the pits.

Conclusion and Recommendations. The cracking of the mixer pivot support initiated at the machined radius within a severely pitted region. The fracture did not originate where fretting was most severe. The failure began due to hydrogen charging as a result of corrosion. This condition was aggravated by the stress concentration effects of pitting at the radius and the high notch sensitivity of the material. Failure was hydrogen assisted, and was most likely a combination of stress-corrosion cracking and corrosion fatigue.

It was recommended that the components be removed from service at the first indication of corrosion. Visual inspection with the aid of a magnifying lens could be used to detect corrosion in the field. The component could continue to be utilized when hardened to 54 to 57 HRC as long as there is no visual evidence of corrosion. In this way, the ballistic properties could be maintained. However, a more conservative approach (which would sacrifice some of the ballistic properties of the material) would be to heat treat the component to a softer condition. A softer material would be less notch-sensitive, and the inspection intervals could then be longer since the critical crack size would be increased.

Failure of a Transport Aircraft Relay Valve Guide*

A 52000 bearing steel valve guide component operating in the fuel supply system of a transport aircraft broke into two pieces after 26 h of flight. The valve guide fractured through a set of elongated holes that had been electrodisharge machined into the component. Analysis indicated that the part failed by low-cycle fatigue. The fracture was brittle in nature and had originated at a severely eroded zone of craters in a hard, deep white layer that was the result of remelting during electrodischarge matching.

Conclusion and Recommendations. The brittle cracking of the valve guide was caused by low-cycle fatigue, which was attributed to the presence of a preexisting crack in the outer corner of the bore. This crack resulted from an unduly overheated structure (white layer) formed during machining.

Stereoscopic macroexamination as well as borescopic inspection of the electrodischarge machined bore surfaces should be performed before such components are accepted from the exporter.

---

*Adapted from A.K. Das, in *Handbook of Case Histories of Failure Analysis*, Vol 2, K.A. Esaklul, Ed., ASM International, 1993, p 15–18

Fatigue Fracture of a Helicopter Tail Rotor Blade due to Field-Induced Corrosion*

A helicopter tail rotor blade spar failed in fatigue, allowing the blade to separate during flight. The 2014-T652 aluminum alloy blade had a hollow spar shank filled with lead wool ballast and a thermoset polymeric seal. A corrosion pit was present at the origin of the fatigue zone and numerous trails of corrosion pits were located on the spar cavity's inner surfaces. The corrosion pitting resulted from the failure of the thermoset seal in the spar shank cavity. The seal failure allowed moisture to enter into the cavity.

Conclusion and Recommendations. The tail rotor blade failed by fatigue which initiated from a trough-shaped corrosion pit located at the inner wall of the spar shank cavity. The pitting was a result of galvanic attack caused by entry of moisture into the spar cavity. The moisture served as an electrolyte for galvanic corrosion between the lead wool ballast and the aluminum spar material. Corrosion pits were formed along the contact points between the lead wool fibers and the spar cavity surface. The moisture entry was due to the failure of a thermoset polymeric seal inside the spar cavity.

The failure could have been prevented by ensuring the integrity of the thermoset polymeric seal, and by barrier coating the wall of the spar cavity with an insulating material. Alternate methods of encapsulating the lead wool would also have prevented the galvanic corrosion, corrosion pitting, and resulting fatigue failure.

Processing-Induced Fatigue Fracture of a Helicopter Tail Rotor Blade**

A helicopter tail rotor blade spar failed in fatigue, allowing the outer section of the blade to separate in flight. The 7075-T7351 aluminum alloy blade had fiberglass pockets. The blade spar was a hollow "D" shape, and corrosion pits were present on the inner surface of the hollow spar. A single corrosion pit, 0.38 mm (0.015 in.) deep, led to a fatigue failure of the spar. The failure initiated on the pylon side of the blade. Dimensional analysis of the spar near the failure revealed measurements within engineering drawing tolerances. Though corrosion pitting was present, there was an absence of significant amounts of corrosion product and all of the pits were filled with corrosion-preventative primer. This indicated that the pitting occurred during spar manufacture, prior to the application of the primer. The pitting resulted from multiple nickel plating and defective plating removal by acid etching. Post-plating baking operations subsequently reduced the fatigue strength of the spar.

Conclusions and Recommendations. The tail rotor blade failed by fatigue which initiated from a 0.38 mm (0.015 in.) deep acid etch pit located at the inner wall of the spar cavity. The pitting was a result of a multiple plating and plating removal etch process during manufacturing. An unauthorized series of baking operations reduced the mechanical properties of the spar. The reduced mechanical properties further contributed to the fatigue process.

---

*Adapted from R.H. McSwain, in *Handbook of Case Histories of Failure Analysis,* Vol 2, K.A. Esaklul, Ed., ASM International, 1993, p 30–32

**Adapted from R.H. McSwain, in *Handbook of Case Histories of Failure Analysis,* Vol 2, K.A. Esaklul, Ed., ASM International, 1993, p 33–35

Better processing controls during the etching and plating procedures would have lessened the possibility of defective plating and subsequent stripping. The post-plating baking which reduced the mechanical properties should not have been performed.

Fatigue Fracture of a Transport Aircraft Crankshaft during Flight*

A 4340 steel piston engine crankshaft in a transport aircraft failed catastrophically during flight. The fracture occurred in the pin radius zone. Fractographic studies established the mode of failure as fatigue under a complex combination of bending and torsional stresses. SEM examination revealed that the fracture origin was a subsurface defect—a hard refractory (Al_2O_3) inclusion—in the zone close to the pin radius. Chemical analysis showed the crankshaft material to be of inferior quality.

Conclusions and Recommendations. The catastrophic fracture was attributed to the presence of a subsurface defect in the form of a lone hard refractory inclusion in the critically loaded pin radius zone, which became the initiation site for a hidden fatigue crack. The uncleanliness of the material further reduced its fatigue resistance.

It was recommended that the supplier exercise utmost care to maintain product quality, including the use of the dc method of magnetic particle inspection.

It appears that the use of the outlined derived methodology would have prevented the aviation service failures described here and in Chapter 1. The following question now arises as to whether the methodology would have also prevented the following hydrocarbon processing service failures.

Hydrotest Failure of a Carbon Steel Pressure Vessel**

A carbon steel (ASTM A515 grade 70) pressure vessel failed by brittle fracture while being hydrotested in the fabricating shop. The fracture origin was a small crack at a welding arc strike associated with the toe of a nozzle weld. A fracture mechanics calculation indicated that this imperfection, although small, initiated fracture because of the local geometry and stress conditions and the low toughness of the steel. It was recommended that (1) the probability of flaws be reduced by welding over or grinding out arc strikes, (2) the local stresses be lowered by postweld stress relief and improved weld toe geometry, and (3) toughness be improved by specifying fine-grain steel and/or by normalizing.

Conclusions and Recommendations. The hydrotest failure of this pressure vessel was the result of a preexisting crack in an arc strike at a nozzle weld on the inside of the vessel. Although the crack was only 1.5 mm (0.06 in.) deep and 15.2 mm (0.60 in.) long, it was located in a highly stressed area, and the material was of insufficient toughness to resist rapid crack propagation.

---

*Adapted from A.K. Das, in *Handbook of Case Histories of Failure Analysis,* Vol 2, K.A. Esaklul, Ed., ASM International, 1993, p 36–38

**Adapted from W.R. Warke, in *Handbook of Case Histories of Failure Analysis,* Vol 1, K.A. Esaklul, Ed., ASM International, 1992, p 143–146

Failure could have been prevented by either eliminating the flaw, lowering the stresses, or increasing the toughness of the steel. Arc strikes are common in welded fabrication and are a source of problems on occasion. Because of the low heat input and the quenching action of the cold steel plate, high hardness and microcracking are not uncommon. Welders often weld back over their arc strikes, or subsequently grind them out. In the confined quarters of the inside of this pressure vessel, however, the arc strike could have been easily overlooked.

Postweld heat treatment would have lowered the residual stresses, and a less steep weld toe would have reduced the stress concentration at the toe. The toughness of the steel could have been increased by normalizing, but a better course of action would be to use A516 grade 70 steel. Specification A516 requires that the steel be produced by a fine grain practice, which is normally met by aluminum deoxidation. Normalized A516 steel usually has excellent toughness.

Stress-Corrosion Cracking of C-Mn Steel in a CO_2 Absorber in a Chemical Plant*

The source of cracking in the circumferential weld seam in a JIS-SM50B carbon-manganese steel pipe used in a CO_2 absorber was investigated; the absorber had been in service for 18 years. The seam had been weld-repaired twice, and the repair welds had been locally stress relieved. Longitudinal seams in the same vessel, which had been stress relieved in a furnace, showed no tendency toward cracking. The solution passing through the vessel contained CO_2-CO-H_2O, KHCO, and Cl^- ions. Nondestructive testing revealed that the cracks originated in the heat-affected zone and propagated into the base metal and weld. Severe branching of the cracks characteristic of stress-corrosion cracking was observed. Microexamination revealed that crack propagation was transgranular, further supporting the possibility of stress-corrosion cracking. Simulation tests carried out in the vessel confirmed this mode of cracking.

Conclusions and Recommendations. The basic cause of failure was stress corrosion caused by locked-in residual stresses resulting from improper and inadequate stress relieving.

To avoid SCC, the welding seams should be furnace heat treated at a temperature of 600 to 640 °C (1110 to 1180 °F) for a minimum of 1 h per inch of section thickness.

Cracking of Inconel 800H in a Steam Methane Reformer Furnace**

During 5.7 years of service, dye penetrant inspection of Inconel 800H pigtail connections regularly showed cracks at weld toes. Weld repairs were not able to prevent reoccurrence but often aggravated the condition. Samples containing small, but detectable, reducer-to-pigtail cracks showed intergranular cracks originating at weld toes and filled with oxidation product, which precluded determination of the cracking mecha-

---

*Adapted from S.K. Chauhan, in *Handbook of Case Histories of Failure Analysis,* Vol 1, K.A. Esaklul, Ed., ASM International, 1992, p 191–193

**Adapted from S.R. Gertler, in *Handbook of Case Histories of Failure Analysis,* Vol 2, K.A. Esaklul, Ed., ASM International, 1993, p 168–171

nism. All weldments exhibited high degrees of secondary precipitates, with original fabrication welds exhibiting higher apparent levels than repair welds. SEM/EDS analysis showed base metal grain boundary precipitates to be primarily chromium carbides, but some titanium carbides were also observed.

Conclusions and Recommendations. The damage mechanism is assumed to be synergistic. Thermally-driven tube distortion is the biggest contributor, as it imposes cyclic, moment stresses which are highest at the subject connection. Welding and high-temperature operation result in: increased creep and intergranular oxidation rates; precipitation of brittle chromium and titanium carbides at grain boundaries; introduction of high residual stresses; changes in mechanical properties; and stress risers at weld defects and weld transitions. All of these factors contribute to accelerated deterioration and reduced repair weldability.

It was recommended that the moment stresses be reduced and/or materials and components be changed. Refinements in welding procedures and implementation of preweld/postweld heat treatments were also recommended.

Failure of Carbon Steel Superheater Tubes*

Two superheater tubes from a 6.2 MPa (900 psig) boiler failed in service because of creep rupture. One tube was carbon steel and the other was carbon steel welded to ASTM A213 Grade T22 (2.25Cr-1.0Mo) tubing. The failure in the welded tube occurred in the carbon steel section. Portions of the superheater were retubed five years previously with Grade T22 material. The failures indicated that tubes were exposed to long-term overheating conditions. While the carbon steel tube did not experience temperatures above the lower transformation temperature 727 °C (1340 °F), the welded tube did experience a temperature peak in excess of 727 °C (1340 °F).

Conclusions and Recommendations. The tube 1 failure was caused by creep mechanism via long-term overheating. Microstructural analysis indicated that the tube 1 peak metal temperatures were great than 454 °C (850 °F) but less than 727 °C (1340 °F). Secondary creep cracking indicates long-term overheating conditions were present prior to failure of tube 2. However, microstructural analysis indicated that tube 2 experienced a brief high-temperature excursion greater than 727 °C (1340 °F) at the time of failure. The long-term overheating of tubes 1 and 2 was likely caused by an excessive heat flux and/or insufficient steam distribution. The high-temperature event coinciding with the violent rupture of tube 2 most likely resulted from a sudden loss in steam distribution through this tube circuit. Although SEM-EDS analysis identified some carryover deposits on the internal surfaces, only a small amount of transported deposit was present. Internal deposits were not a primary cause of the tube failures.

A review of steam flow in the superheater should be conducted to ensure adequate steam distribution in the circuit. The entire superheater should be upgraded to ASTM A213 Grade T22 tubing.

---

*Adapted from J.P. Ribble, in *Handbook of Case Histories of Failure Analysis*, Vol 2, K.A. Esaklul, Ed., ASM International, 1993, p 177–180

Alloy analyses of the submitted tube samples indicates the overheating failures occurred in low-carbon steel sections. The portion of tube 2 that did not contain the failure tested as being consistent with ASTM A213 Grade T22 boiler tubing specification. The similarity of the chemical composition of the failed sections suggests these sections were installed at the same time. When the superheater was retubed, the entire superheater assembly should have been upgraded to Grade T22 material.

Failure of a Pipe Slip-on Flange*

A cracked 356 mm (14 in.) diam slip-on flange (Ni-Cr-Mo-V steel) was submitted for failure analysis. Reported results and observations indicated that the flange was not an integral forging or a casting, as specified. It had been fabricated by welding and machining a ring insert within a flange with a larger internal diameter. The flange cracked because the welds between the flange and the insert were inadequate to withstand the bolting pressures.

Conclusion. It was concluded that the flange failed at a weak welded joint. The flange did not conform to the ANSI specification, which called for an integral component.

The defective flange was one of a large shipment of similar components imported by the customer. A warning was issued to the end users of the flanges, which were then inspected nondestructively for conformance to specifications

Mercury Liquid Embrittlement Failure of 5083-O Aluminum Alloy Piping**

The failure mode of through-wall cracking of a butt weld in a 5083-O aluminum alloy piping system in an ethylene plant was identified as mercury liquid metal embrittlement. As a result of this finding, 226 of the more than 400 butt welds in the system were ultrasonically inspected for cracking. One additional weld was found that had been degraded by mercury. A welding team experienced in repairing mercury-contaminated piping was recruited to make the repairs.

Conclusions and Recommendations. It was concluded that the through-wall failure in the pipe weld resulted from mercury liquid metal embrittlement. The most probable cause of the embrittlement was the use of unrefined feedstock in the ethylene plant. This feedstock was found to contain mercury at levels as high as 40 ppb. In a plant operation where 2 billion pounds of feedstock are processed per year, mercury impurity levels in the parts per billion range present a serious threat to the integrity of the aluminum equipment.

Prior to restarting the plant, all of the aluminum alloy piping was pneumatically tested with dry nitrogen and was checked for leaks. A mercury-removal system consisting of a sulfur-impregnated activated-charcoal bed was subsequently installed upstream of the aluminum equipment. The plant operated for 1 year in this manner until

---

*Adapted from M.D. Chaudhari, in *Handbook of Case Histories of Failure Analysis*, Vol 2, K.A. Esaklul, Ed., ASM International, 1993, p 185–186

**Adapted from J.J. English and D.J. Duquette, in *Handbook of Case Histories of Failure Analysis*, Vol 2, K.A. Esaklul, Ed., ASM International, 1993, p 207–213

a new aluminum alloy cold box could be constructed and installed. This installation replaced the cold box that had been in service prior to the installation of the mercury-removal system.

Weld quality control during the construction of the new cold box consisted of 100% radiographic examination of all butt welds. All radiographs were filed for future reference. The use of weld backup rings was minimized in the construction. When backup rings were used, the upstream side of the ring was seal welded to minimize mercury entrapment.

A complete chemical analysis of trace impurities in the new feedstock for the ethylene plant could have prevented the mercury-related failure of the aluminum equipment. Upon detection of the mercury in the feedstock, a mercury-removal system could have been installed upstream of the aluminum equipment before the feedstock was processed.

Stress-Corrosion Cracking in a Downcomer Expansion Joint*

A type 321 stainless steel downcomer expansion joint that handled process gases was found to be leaking approximately 2 to 3 weeks after installation. The expansion joint was the second such coupling placed in the plant after failure of the original bellows. The failed joint was disassembled and examined to determine the cause of failure. Energy-dispersive X-ray analysis revealed significant peaks for chlorine and phosphorus, indicating failure by chloride stress-corrosion cracking (SCC). Cracks in the liner and bellows exhibited a branched pattern also typical of SCC. Cracks through the inner liner initiated on the outer surface of the liner and propagated inward, whereas cracks in the bellows originated on the inner surface and propagated outward. Stress-corrosion cracking of the assembly was caused by chloride contaminants trapped inside the bellows following hydrostatic testing.

Conclusions and Recommendations. The failure was caused by SCC. The chloride that was present became trapped between the braid and bellows after the bellows fractured and contaminant leakage occurred. The contaminants were trapped inside the bellows in a concentration high enough to initiate the crack.

The design of the expansion joint allowed hydrostatic test water or other fluids to become trapped between the bellows and the liner. This fluid remained in the assembly until boiled out when processing temperatures became sufficiently high. Hydrostatic test water is the most probable source of the contaminants.

Welding of the pipe, bellows, and braiding also resulted in a high stress concentration in the bellows due to residual stresses and restraint problems in a critical area. The dents found on the outer surfaces probably caused stress concentrations in the bellows as well.

Better manufacturing controls should be implemented. Steps to remove any fluids from the assembly after hydrostatic testing should be taken in conjunction with checking the test water for chloride content.

---

*Adapted from R.D. Bowman, in *Handbook of Case Histories of Failure Analysis*, Vol 2, K.A. Esaklul, Ed., ASM International, 1993, p 222–224

Better care in the use of the assembly to prevent dents and stress concentrations would reduce the stress that is capable of forming cracks as well. Prevention of misalignment of the bellows with respect to the welded pipe might prohibit welding and restraint loading of an already highly stressed area.

Thermal Fatigue Failure of Alloy UNS NO8800 Steam Superheating Tubes*

Alloy UNS NO8800 (Alloy 800) tubes of the steam superheating coils of two hydrocracker charge heaters in a refinery failed prematurely in service. Failure analysis of the tubes indicated that the failures could be attributed to thermal fatigue as a result of temperature fluctuations as well as restriction to movement. Fatigue cracks initiated intergranularly from both the flue gas and steam sides. Enhanced general and grain boundary oxidation coupled with age hardening of the alloy led to the formation of incipient intergranular cracks that acted as sites for the initiation of the fatigue cracks.

Conclusion. The failure of the Alloy 800 tubes in the HCR units was due to thermal fatigue. The source of fatigue was possibly due to temperature fluctuations in addition to restriction of movement of the coils.

The initiation and propagation of fatigue cracks were mostly from the outside (flue gas side). However, some cracks were found to propagate from the steam side.

In the case of Unit 1, uncontrolled operating conditions (running dry and excursions to high temperatures) resulted in age hardening of the material and enhanced general and grain boundary oxidation with incipient intergranular cracks, leading to early initiation of fatigue cracks. In the case of Unit 2, with better controlled operating conditions, the grain boundary oxidation and incipient intergranular cracking were less severe, which manifested in a longer fatigue life.

Metallurgical Analysis of Steam Turbine Rotor Disc**

Numerous cracks observed on the surface of a forged A470 Class 4 alloy steel steam turbine rotor disk from an air compressor in a nitric acid plant were found to be the result of caustic-induced stress-corrosion cracking (SCC). No material defects or anomalies were observed in the disc sample which could have contributed to crack initiation or propagation or secondary crack propagation. Chlorides detected in the fracture surface deposits were likely the primary cause for the pitting observed on the disc surfaces and within the turbine blade attachment area. It was recommended that the potential for water carryover or feedwater induction into the turbine be addressed via an engineering evaluation of the plant's water treatment procedures, steam separation equipment, and start-up procedures.

Conclusions and Recommendations. The cause of the numerous cracks observed was caustic-induced SCC. An evaluation of several factors in the operation of the plant's

*Adapted from M. Islam, in *Handbook of Case Histories of Failure Analysis*, Vol 2, K.A. Esaklul, Ed., ASM International, 1993, p 235–239

**Adapted from L.J. Hodas, in *Handbook of Case Histories of Failure Analysis*, Vol 2, K.A. Esaklul, Ed., ASM International, 1993, p 307–310

steam system will be required to identify the source of the caustic. The operation of the steam drum/steam separators and the effectiveness of the plant water treatment program should be addressed. Additionally, the procedure of using saturated steam during start-up should be assessed with regard to the potential for inducting feedwater in to the stream turbine.

Fatigue Failure at Fillet-Welded Nozzle Joints in a Type 316L Stainless Steel Tank*

Upon arrival at the erection site, an AISI type 316L stainless steel tank intended for storage of fast breeder test reactor coolant (liquid sodium) exhibited cracks on its shell at two of four shell/nozzle fillet-welded joint regions. The tank had been transported from the manufacturer to the erection site by road, a distance of about 800 km (500 mi.). During transport, the nozzles were kept at an angle of 45° to the vertical because of low clearance heights in road tunnels. The two damaged joints were unsupported at their ends inside the vessel, unlike the two uncracked nozzles. Surface examination showed ratchet marks at the edges of the fracture surface, indicating that loading was of the rotating bending type. Electron fractography using the two-stage replica method revealed striation marks characteristic of fatigue fracture. The striations indicated that the cracks had advanced on many "mini-fronts," also indicative of nonuniform loading such as rotating bending.

Conclusion and Recommendations. Severe grinding marks were visible, and the cracks often followed these marks. Utmost care must be taken not to introduce grinding marks or other stress raisers near weld joints.

Fatigue failure was consistent with the type of loading present. The nozzles, positioned at angles of 45° to the vertical during transport of the tank, acted as cantilevers. Cyclic loading resulted from the usual jerks and vibrations associated with road journeys. Such loading could have been avoided had the two nozzles been supported at their ends inside the vessel, similar to two other nozzles that did not show failure. In addition to avoiding transport problems, this design modification would reduce fatigue loading that occurs in service due to vibration of the nozzles during filling and draining of the tank.

Corrosion Fatigue and Subsequent Rupture of a Yankee Dryer Roll on a Modified Paper Machine**

A steam-pressurized Yankee dryer shell ruptured during normal operation. The dryer was used for final drying of soft paper-like sheet material. Cracking had occurred around much of the circumference at the drive end of the shell, which measured 3.7 m

*Adapted from D.K. Bhattacharya, in *Handbook of Case Histories of Failure Analysis*, Vol 1, K.A. Esaklul, Ed., ASM International, 1993, p 256–258

**Adapted from M.H. Gilkey, in *Handbook of Case Histories of Failure Analysis*, Vol 1, K.A. Esaklul, Ed., ASM International, 1993, p 132–135

(12 feet) in diameter by 3.4 m (11 feet) long and had a head bolted to each end. The crack initiated at a 90° corner in contact with the edge of the head. The material was a hardened gray cast iron containing 2.8% Ni and 1.2% Mo. Visual, nondestructive, metallographic, and chemical analyses were conducted.

Conclusion and Recommendations. The Yankee dryer ruptured when a circumferential corrosion fatigue crack reached a critical size. Cracking occurred because of the combination of the corrosive environment inherent in the manufacturing process and the cyclic stresses of normal operation, such as dryer rotation and periodic startups. Significant static stress also was applied to the crack initiation site by the pressure of oxides forming at the surfaces of the already filled head-shell gap. The iron to hydrated iron oxide transformation involves a significant increase in volume, which resulted in a buildup of static stress in the 90° corner of the shell.

Ultrasonic examination of the shell, especially at the head-shell joint area, could have detected the crack before it reached a critical size. It is likely that the crack propagated for at least several months prior to the rupture incident. Annual or semiannual inspection may have detected the cracking.

The visibly corroded condition of the external surfaces adjacent to the head-shell joint also should have alerted the operators and/or inspectors of the dryer to the need for further inspection, including ultrasonic examination. It is known in the pulp and paper industry that such corrosion can indicate corrosion product buildup in the joint, which can lead to shell and head cracking.

In summary, it appears that the use of the derived methodology outlined earlier in this chapter would also have prevented the hydrocarbon processing service failures described here and in Chapter 1. Although different interpretations on the chosen prevention method exist, the fact remains that, by using this methodology, all of the preceding failures could have been prevented.

REFERENCES

1. T. Swift, Damage Tolerance in Pressurized Fuselages, *New Materials and Fatigue Resistant Aircraft Design,* 14th Symp. of the Int. Committee on Aeronautical Fatigue, Ottawa, Douglas Paper 7768, 1987, p 4–11
2. D. Broek, Concepts of Fracture Control and Damage Tolerance Analysis, *Fatigue and Fracture,* Vol 19, *ASM Handbook,* ASM International, 1996, p 410
3. M. Stone, Fatigue and Fail-Safe Design of a New Jet Transport Airplane, Fatigue Design Procedures, *Proc. of the Fourth Symp. of the Int. Committee on Aeronautical Fatigue,* E. Gassner and W. Schutz, Ed., Pergamon Press, London, 1969, p 1–65
4. *Damage Tolerance Assessment Handbook,* Report DOT-VNTSC-FAA-93.13.1, U.S. Department of Transportation, 1993, p 1–11
5. "Airplane Damage Tolerance Requirements," MIL-A-83444, U.S. Air Force

6. A.P. Berens, J.C. Burns, and J.L. Rudd, Risk Analysis for Aging Aircraft Fleets, *Structural Integrity of Aging Airplanes,* S.N. Alturri et al., Ed., Springer-Berlin, 1991, p 37–51
7. J.P. Gallagher, F.J. Giessler, A.P. Berens, and R.M. Engle, *USAF Damage Tolerant Design Handbook: Guidelines for the Analysis and Design of Damage Tolerant Aircraft Structures,* Report AFWAL-TR-82-3073, Wright Patterson Air Force Base, Dayton, OH, 1984
8. DOT-TSC-FAA 890-88-26, Proc. of the Int. Conf. on Aging Airplanes, Department of Transportation, June 1988
9. U.A. Hinders, "F-111 Design Experience—Use of High Strength Steel," paper presented at AIAA Second Aircraft Design and Operations Meeting, AIAA, 1988
10. M.P. Kaplan, "TACT-38 Wing Life Assessment, Program Management Review," Wright Patterson Air Force Base, Dayton, OH, 1973
11. United Airlines Inc. Aircraft Accident Report, N1819U, NTSB-AAR-9006, Nov 1990
12. M. Kaplan and T. Wolf, *Fatigue and Fracture,* Vol 19, ASM Handbook, ASM International, 1996, p 557
13. CFR Title 14, Part 25.609. Protection of Structure
14. "Corrosion Control for Aircraft," Federal Aviation Administration, Advisory Circular AC 43-4A, July 1991
15. M.L. Bauccia, Corrosion Control in the Aircraft Industry, *Corrosion,* Vol 13, *Metals Handbook,* 9th ed., ASM International, 1987
16. U.G. Goranson, "Damage Tolerance: Fact and Fiction," 14th Platerna Memorial Lecture, paper, presented at ICAF (Stockholm, Sweden), ICAF, June 1993
17. A. Hartman, "On the Effect of Water Vapor and Oxygen on the Propagation of Fatigue Cracks in an Aluminum Alloy," *Int. J. Frac. Mech,* Vol 1, 1965, p 167–183
18. M.R. Achter, "Effect of Environment on Fatigue Cracks," *Fatigue Crack Propagation,* STP 415, American Society for Testing and Materials, 1967, p 181–204
19. J. Schijve and D. Broek, "The Effect of Frequency on the Propagation of Fatigue Cracks," NLR TR-M-2094, National Aerospace Institute, Amsterdam, The Netherlands, 1961
20. J. Schijve and P. DeRijk, "The Effect of Temperature and Frequency on Fatigue Crack Propagation in 2024-T3," NLR TM-M-2138, National Aerospace Institute, Amsterdam, The Netherlands, 1963
21. R.N. Miller and F.H. Meyer, Computerized Corrosion Forecasting Model for C-5 Aircraft, AFWAL-TR-88-4128, *Proc. of the 1987 ASIP/ENSIP Conf.,* U.S. Dept. of Transportation, 1987
22. American Airlines, Inc. Aircraft Accident Report, DC-10-10, N110AA, NTSB-AAR-79-17, May 1979
23. W.R. Hendricks, The Aloha Accident—A New Era for Aging Aircraft, *Structural Integrity of Aging Aircraft,* S.N. Alturri et al., Ed., Springer-Verlag, Berlin, 1991, p 153
24. "Supplemental Structural Inspection Program for Large Transport Category Airplanes" Federal Aviation Administration Advisory Circular AC 91-56, May 1981

25. *The Continued Airworthiness of Older Airplanes,* Federal Aviation Administration Advisory Circular AC 91–60, June 1983
26. B.K. Young, Jr., Cold Expansion Techniques for Fatigue Enhancement, *Durability of Metal Aircraft Structures,* S.N. Alturri et al., Ed., Atlanta Technology Publications, 1992, p 405
27. P.R. Abelkis et al., "Use of Durability and Damage Tolerance Concepts in the Development of Transport Aircraft Continuing Structural Integrity Programs," paper presented at ICAF (Pisa, Italy), ICAF May 1985
28. *Nondestructive Evaluation and Quality Control,* Vol 17, *Metals Handbook,* 9th ed., ASM International, 1989

CHAPTER 3

Solutions to Equipment Failures Based on Fracture Mechanics

The basis that fracture mechanics provides in solutions to equipment failures is shown in the failure boundary of Fig. 3-1. Structures using tough materials and having only small cracks will be in the strength of materials regime. If a material is brittle (low toughness) and strong (high yield stress), the presence of even a small crack is likely to cause fracture (crack propagation).

However, fracture is not the only way a structure can fail. Buckling failure is controlled by the elastic modulus, and failure by plastic deformation is controlled by the yield stress. In the presence of a crack-like flaw, however, fracture is likely to be of most concern.

There are three ways in which a flaw/crack/crack-like defect can appear in a structure:

- Inherent defects that occur in the material such as inclusions
- Defects introduced during the fabrication of a component such as arc strikes or lack of fusion in a weld
- Damage incurred during the service life of the component, such as dents and cuts, fatigue, and environmentally assisted cracking

Within the confines of linear elastic fracture mechanics (LEFM), it makes no difference how a flaw is introduced. A crack-like flaw of a given size and position in a structure is assumed to obey the same fracture rules regardless of its origins. The basic capability that fracture mechanics provides can then be employed in either of two general ways. First, the maximum safe operating loads that an engineering structure can sustain for the sizes and locations of existing flaws can be determined. Such cracks might be those actually found during an inspection, whereupon the continued safe operation of the structure is in question. Second, for given loads, the largest crack size that can exist without fracture can be determined. This will provide specifications to be set in advance of an inspection.

Of most importance in regard to LEFM are applications where weight is a primary concern. When the resistance to yielding is high (e.g., in a high-strength steel), the fracture toughness tends to be low. Combined with the necessity to use highly stressed components, fracture mechanics analyses are essential to achieve a proper balance between performance and reduced failure rates. In fact, the use of fracture mechanics for structural integrity assessments is largely due to the importance of considering small flaws in aircraft, offshore, and other applications where high-strength materials are used to save weight.

To optimize resistance to failure, it is necessary to develop both high yield stress and high toughness. The ideal balance is the point where failures by gross yielding and failures by fracture are equally unlikely to occur.

Linear Elastic Fracture Mechanics

Stress Intensity at a Crack

For a through-thickness crack of length 2*a* in a plate, a panel, a vessel wall, or a line-pipe subjected to a remote tensile stress, the stress intensity at the tips of the crack, *K*, is given by (Ref 1):

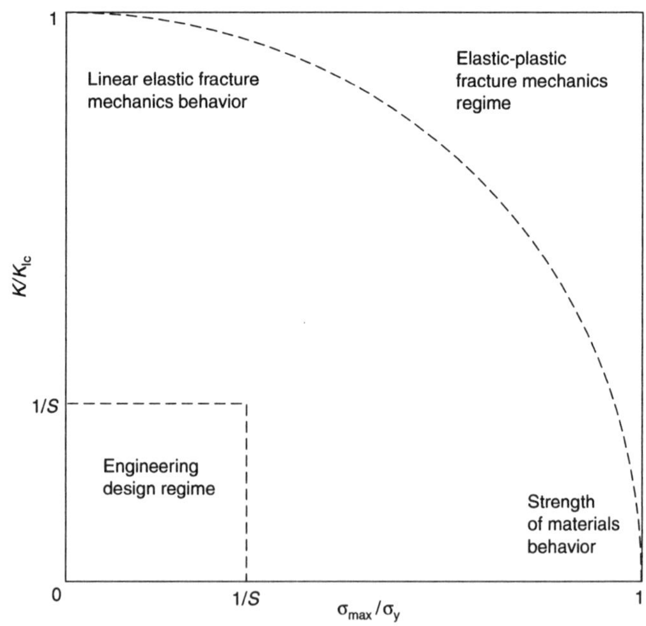

Fig. 3-1 Generalization of engineering structural integrity assessments provided by fracture mechanics

$$K = \beta\sigma\sqrt{\pi a}$$

where σ is the remote stress (such as the stress in a vessel wall due to the operating pressure of the vessel) and β is a dimensionless factor that describes the crack geometry and can include complex terms which may be difficult to derive. The stress intensity, K, is the similitude that exists between laboratory data and the service conditions. If the remote stress, taken now to be the operating stress, is increased to a level where it becomes the fracture stress, σ_c:

$$K_{Ic} = \beta\sigma_c\sqrt{\pi a}$$

where K_{Ic} is the elastic fracture toughness and the critical stress intensity at the onset of fracture. This relationship is the physical basis for modern LEFM. The subscript I means that the type of stress applied is tensile, and the subscript c means that the stress intensity is critical. The operating stress, therefore, can be the maximum permissible stress for a maximum permissible crack length, a_c.

β Values and Complex Geometries

For a through-thickness crack in a plate, $\beta = 1$. For an edge crack in the same plate, $\beta = 1.12$ (Ref 2). For an embedded elliptical crack in the plane shown in Fig. 3-2(a), where $2a$ is the minor diameter and $2c$ is the major diameter (Fig. 3-2c), K_{Ic} is given by (Ref 3):

$$K_{Ic} = \sigma_c \frac{\sqrt{\pi a}}{\phi}\left[\sin^2\alpha + \frac{a}{c}\cos^2\alpha\right]^{1/4}$$

where a is the angle shown in Fig. 3-2(g), and ϕ is the double elliptical integral, which depends on the crack aspect ratio, a/c. Values of ϕ are shown in Fig. 3-2(h), where ϕ varies from 1.0 to 1.51 for a/c ranging from zero (slender ellipse) to unity (circle). The stress intensity varies along the elliptical crack tip with the maximum value at the minor axis and the minimum value at the major axis. In fatigue, this crack would grow into a circle with a uniform stress intensity at all points. For the circular embedded crack (Fig. 3-2b), K_{Ic} is:

$$K_{Ic} = \sigma_c\sqrt{\pi a(2/\pi)}$$

Surface elliptical cracks tend to grow to other elliptical shapes due to the free surface effect. A general expression for the semielliptical surface crack (Fig. 3-2d) is:

$$K_{Ic} = \sigma_c\sqrt{\frac{\pi a}{\phi}}\, M_f \cdot M_b\,[\sin^2 a + (a/c)^2\cos^2\alpha]^{1/4}$$

where M_f is a front face correction factor and M_b is a back face correction factor. M_f and M_b are functions of α. For a semielliptical crack in a thick plate, K_{Ic} at the deepest point is approximated by:

$$K_{Ic} = \frac{1.12\sigma_c\sqrt{\pi a}}{\phi}$$

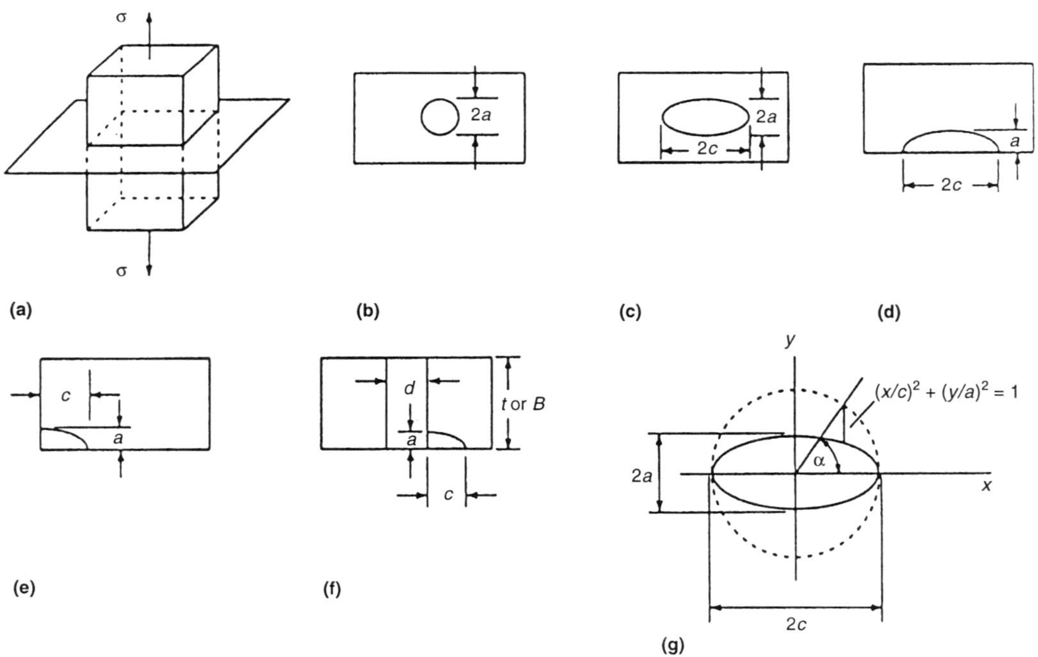

a/c	0	0.1	0.2	0.3	0.4	0.5	0.6	0.7	0.8	0.9	1.0
ϕ	1.0	1.016	1.051	1.097	1.151	1.211	1.277	1.345	1.418	1.493	1.571

(h)

Fig. 3-2 Elliptical and circular cracks. (a) Tensile loading and crack plane. (b) Embedded circular crack. (c) Embedded elliptical crack. (d) Surface half-elliptical crack. (e) Quarter-elliptical corner crack. (f) Quarter-elliptical corner crack emanating from a hole. (g) Elliptical crack parameters. (h) Values of ϕ

where $M_f = 1.12$ is analogous to the free edge correction of the single edge crack.

For the quarter-circular corner crack ($a/c = 1$, Fig. 3-2e) with two free edges, K_{Ic} is approximated by:

$$K_{Ic} = \frac{(1.12)^2 \sigma_c \sqrt{\pi a}}{\phi}$$

These approximations agree well with more complex expressions. For the crack at a hole shown in Fig. 3-2(g), no rigorous solution currently exists, and controversy exists over the variation of K_{Ic} along the crack front (Fig. 3-3), such that consensus dictates the K_{Ic} value chosen (Ref 4).

β values may be compounded (Ref 5) to give a total geometry factor. For example, a surface crack close to a weldment may have β compounded or a K_t value built into it. (Compounding β values means multiplying them together. Building in a K_t value means adding in an extra factor to increase the β value.)

The K_t for fatigue notch is given by (Ref 6):

$$K_t = 1 + \frac{K - 1}{1 + (a/r)}$$

where K = elastic stress concentration factor, r = root radius, and a = material parameter, which is a function of ultimate tensile strength. Other K_t values for different geometries are available in the literature (Ref 6, 7). K_{Ic} values may be added or subtracted, depending on geometry. For example, K_{Ic} for part of a geometry may be known, and K_{Ic} for the remaining part of the desired geometry may be known. The total value of K_{Ic} needed is:

$$K_{Ic}\ (\text{total}) = K_{Ic}\ (\text{part 1}) + K_{Ic}\ (\text{part 2})$$

The easy way to remember this is to add the K_{Ic} and multiply the β.

Crack Tip Plasticity

Even when the analysis is based in linear elastic fracture mechanics, there is always a region of plasticity in the form of a roll at the leading edge of the crack tip (Fig. 3-4). The radius of this roll of plastic deformation or plastic zone depends on the condition of constraint at the crack front. The most common situation is that of plane strain, and the plastic zone size or radius of the plastic roll (determined experimentally) is given by (Ref 8):

$$r_p = \frac{1}{6\pi} \times \left(\frac{K_{Ic}}{\sigma_y}\right)^2 \times \frac{1}{6\pi}$$

in plane stress:

$$r_p = \frac{1}{2\pi} \times \left(\frac{K_c}{\sigma_y}\right)^2 \times \frac{1}{2\pi}$$

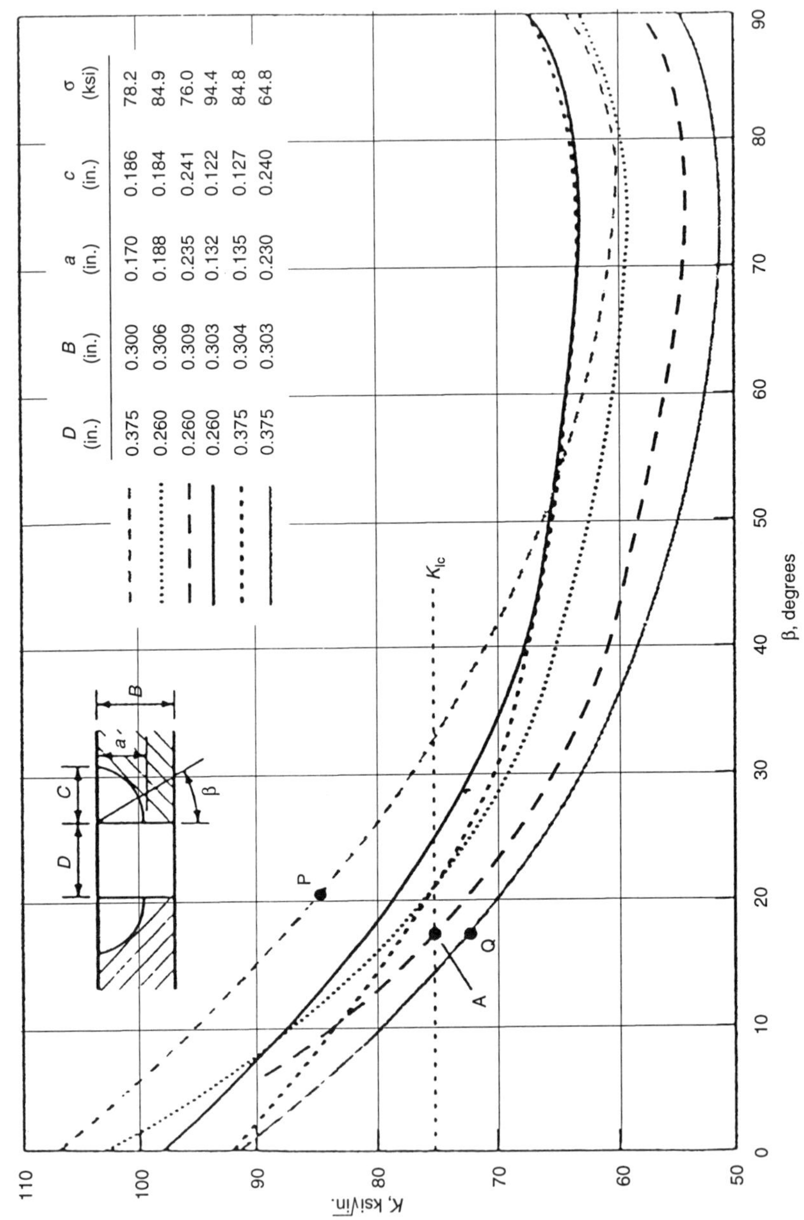

Fig. 3-3 Test data for corner cracks at holes

where K_c is plane-stress fracture toughness, and σ_y is the tensile yield stress. The degree of plasticity and accompanying plastic zone size increase as crack tip conditions change from elastic-plastic to plastic behavior.

Plane Strain and Plane Stress

Constraint at the crack tip is dependent on the plate thickness, B. In general, for plane strain (Ref 8):

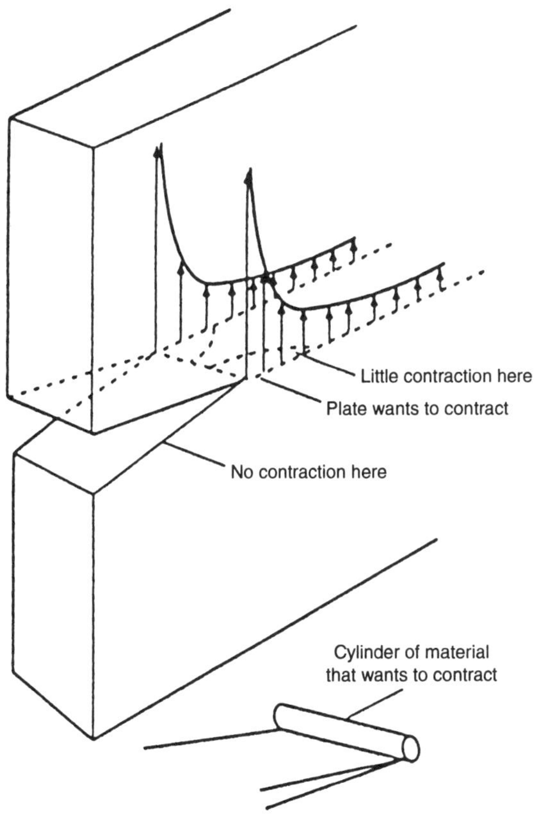

Fig. 3-4 Contraction and constraint from crack-tip plasticity. High stress and strain at the crack tip in the *x*- and *y*-directions cause a contraction in the *z*-direction. However, because the crack itself is stress free, no contraction occurs behind the crack front. Further away from the crack, the stresses are low; therefore, the contraction is small. Thus, there is a thin cylindrical-like zone of material at the crack tip wanting to contract a great deal, while the surrounding material does not need to contract. If this zone is long and thin, its contraction will be prevented because it is attached to the surrounding material. The surrounding material will constrain the contraction by exerting a tension stress upon the cylindrical zone in the *z*-direction so as to keep $\varepsilon = 0$.

$$B \geq \left(\frac{K_{Ic}}{\sigma_y} \right)^2 \times 2.5$$

and for plane stress:

$$B \leq \left(\frac{K_c}{\sigma_y} \right)^2 \times 0.3$$

However, there is no clear-cut transition point between these conditions. Cracks behave under conditions of either predominant plane strain or predominant plane stress.

Plasticity, Plane Strain, and Plane Stress

For an applied tensile stress remote from a crack tip, multiaxial stress fields are developed at the crack tip (Ref 8). For thick sections in plane strain, a triaxial stress field is developed at the center of the crack front:

$$\sigma_{11} > \sigma_{33} > \sigma_{22}$$

where σ_{11} is the tensile stress normal to the crack front, σ_{33} is the tensile stress parallel to the crack front, and σ_{22} is the tensile stress ahead of the crack front.

In thin sections or at the surfaces of a thick section, a biaxial stress field is developed:

$$\sigma_{11} > \sigma_{22} > \sigma_{33}\ (= 0)$$

Also, the Tresca yield criterion gives in plane strain:

$$\sigma_y = \sigma_{11} - \sigma_{22}$$

where σ_{11} is at a maximum at the plastic zone boundary, thereby yield spreads at a higher local stress in plane stress than in plane strain (i.e., at σ_{11} rather than $\sigma_{11} - \sigma_{22}$).

The plastic zone size is larger in plane stress than in plane strain. Less constraint and higher toughness are achieved in plane stress. If cracking occurs in plane strain, it is promoted more because of the high value of σ_{11} at the plastic zone boundary. Thus, the more conservative measure, K_{Ic}, is used in operating stress maps and calculations because plane-strain conditions promote conditions that, in general, are more likely to give rise to service failure.

Stress Maps

If operating equipment contains a crack of initial length a_1, and, during inspection intervals, the crack grows by some mechanism to length a_z (less than the critical length), the growth rate may be low enough to allow operation until the next shutdown before the critical crack length is achieved, thus allowing time for repairs to be made during the shutdown. Alternatively, the operating stress may be reduced, giving a larger critical crack length and, therefore, more time. The operating stress is given by:

$$\sigma_c = \frac{K_{Ic}}{\beta\sqrt{\pi a}}$$

where σ_c is the safe limit of operating stress for a given value of crack length.

If K_{Ic} and the operating stress on the equipment in service are known, then the critical crack length, a_c, at the operating stress, σ_c, can be determined from:

$$a_c = \frac{1}{\pi}\left(\frac{K_{Ic}}{\beta\sigma_c}\right)^2$$

Thus, by maintaining operating stress and crack length at "critical" levels, the maximum allowable (critical) stress for a maximum allowable (critical) crack length are known for a given critical stress intensity, K_{Ic}. This interchange of operating stress and crack length (or depth, depending on geometry) is the basis for deriving a stress map.

For a through-thickness crack in a plate geometry, an LEFM stress map is shown in Fig. 3-5. Because the curve would be asymptotic to the abscissa, it is reasonable to insert the tensile yield stress and, ⅔ of the tensile yield stress on the operating stress axis and draw in the appropriate tangent to the curve (Ref 9). Stress maps based on plastic fracture mechanics (PFM) and elastic-plastic fracture mechanics (EPFM) are described in this chapter.

In Fig. 3-6, the detected crack length, a_d, is drawn. If the growth mechanism is known, it may be possible to estimate the time available before the critical crack length, a_c, is reached. Alternatively, if the operating stress can be reduced to such a level that the critical crack length could never be achieved without loss, then the failure rate will be decreased. The stress maps can be viewed in the same way as a residual strength diagram in that a fixed value of applied load will tolerate a growing crack until the crack reaches criticality and the (operating) stress exceeds the maximum expected service stress. The requirements in defect tolerance analysis are such that a stress map or residual strength diagram is required to obtain the permissible crack size. A crack growth curve is required to provide a basis of calculating the inspection interval. The determination of the inspection interval is described in Chapter 2.

Crack Growth Curves

If cracks are presumed present or detected in service, then the growth law governing these cracks becomes critical. Some common mechanisms governed by growth laws will now be considered, such as fatigue, stress corrosion cracking, corrosion fatigue, and delayed hydride cracking.

Fatigue

Crack propagation depends on the alternating stress intensity factor, ΔK, defined as:

$$\Delta K = K_{max} - K_{min}$$

For intensities below ΔK_{th}, propagation does not occur. For intensities where K_{max} approaches K_{Ic}, fast fracture generally results. For intermediate ΔK values, growth can be predicted.

If growth is defined by the rate of change of crack length with the number of cycles, then

$$da/dN = c(\Delta K)^m$$

where c and m are constants (Fig. 3-7).

Alternatively,

$$da/dN = \frac{A(\Delta K)^n}{(1-R)K_{Ic} - \Delta K}$$

where A and n are empirical constants and $R = K_{min}/K_{max}$. This Forman equation (Ref 10) is general because R is represented. It has a vertical asymptote as $\Delta K = (1-R)K_{Ic}$ and

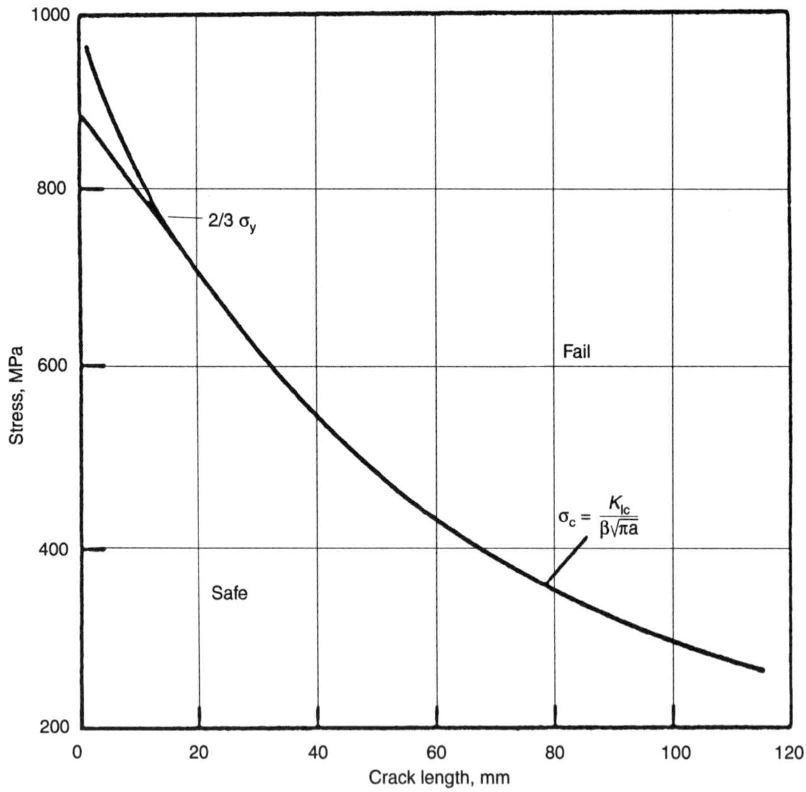

Fig. 3-5 Definition of quantites in constructing an operating stress map for a through-thickness crack ($\beta = 1$) in a plate of width W. When applied stress is at the critical level (σ_c), the crack of length $2a$ propagates across the plate of width W.

an asymptote at ΔK_{th}. Neither this nor the preceding Paris (Ref 2) equation, nor any other of the 32 rate equations, can be given any theoretical support. Any expression that fits the data is acceptable.

More sophisticated relationships that consider crack growth retardation and a variety of geometries for a wide range of materials are currently available. Such a system is the NASA/NASGRO computer program developed by Royce Forman at NASA, which provides a comprehensive package (Ref 11).

Barsom (Ref 12) evaluated the Paris equation for steels varying in yield stress from 250 to 2070 MPa and suggested the following conservative equations for steels of various microstructures.

For ferritic/pearlitic steels:

$$\frac{da}{dN}(\text{m/cycle}) = 6.9 \times 10^{-12}\,(\Delta K)^{3.0}(\text{MPa}\sqrt{\text{m}})$$

$$\frac{da}{dN}(\text{in./cycle}) = 3.6 \times 10^{-10}(\Delta K)^{3.0}(\text{ksi}\sqrt{\text{in.}})$$

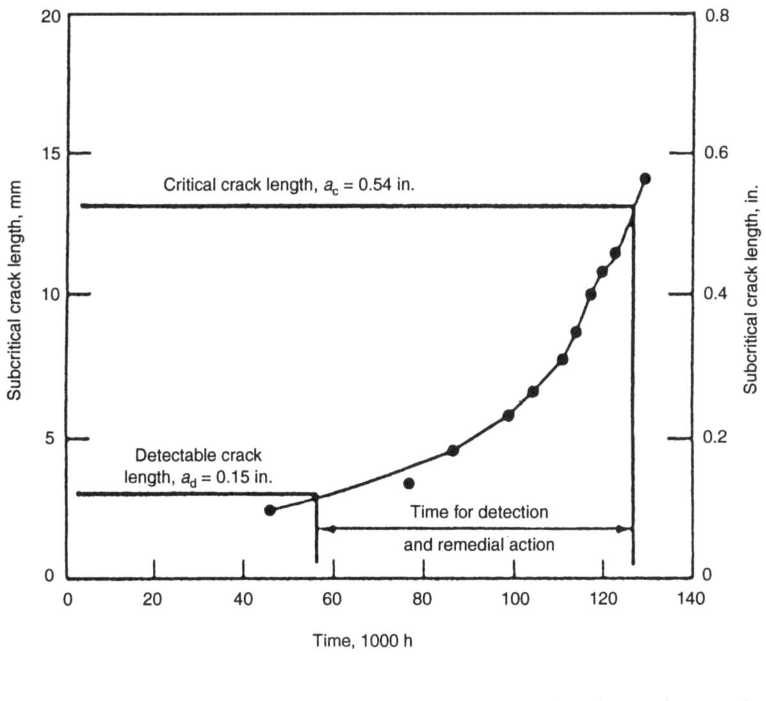

Fig. 3-6 Example of subcritical crack growth plot for fatigue crack growth of a surface crack in quenched-and-tempered steel

For martensitic steels:

$$\frac{da}{dN}(\text{m/cycle}) = 1.35 \times 10^{-10}(\Delta K)^{-2.25}(\text{MPa}\sqrt{\text{m}})$$

$$\frac{da}{dN}(\text{in./cycle}) = 6.6 \times 10^{-9}(\Delta K)^{-2.25}(\text{ksi}\sqrt{\text{in.}})$$

For austenitic steels:

$$\frac{da}{dN}(\text{m/cycle}) = 5.6 \times 10^{-12}(\Delta K)^{-3.25}(\text{MPa}\sqrt{\text{m}})$$

$$\frac{da}{dN}(\text{in./cycle}) = 3.0 \times 10^{-10}(\Delta K)^{-3.25}(\text{ksi}\sqrt{\text{in.}})$$

Crack Growth Life Integration Example

Assume a wide SAE 1020 cold-rolled plate is subjected to constant-amplitude uniaxial cyclic loads that produce nominal stress varying from σ_{max} = 200 MPa to σ_{min}

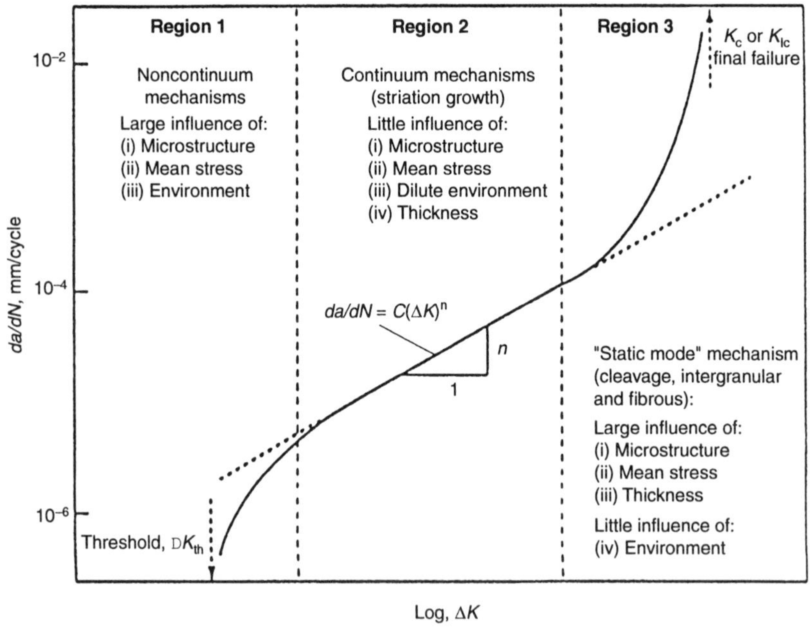

Fig. 3-7 Schematic plot of fatigue crack propagation rate, *da/dN*, versus stress-intensity range, ΔK, on a log-log scale

= –50 MPa. Also, consider the following: σ_y = 630 MPa, ultimate tensile strength (UTS) = 670 MPa, Young's modulus, E = 207 GPa, and K_{Ic} = 104 MPa $\sqrt{m}$.

To find the fatigue life that would be attained if an initial through-thickness edge crack were no greater than 0.5 mm in length:

$$K_{max} = (1.12)(200)\sqrt{\pi(0.0005)} = 9 \text{ MPa}\sqrt{m}$$

The final crack length, a_c, can be determined by setting K_{max} at fracture equal to K_{Ic}.

$$\Delta K = \Delta\sigma\beta\sqrt{\pi a}$$

$$\frac{da}{dN} = c(\Delta K)^m = c(\Delta\sigma\beta\sqrt{\pi a})^m = c(\Delta\sigma)^m(\pi a)^{m/2}\beta^m$$

$$a_c = \left(\frac{K_{Ic}}{\sigma_{max}\beta}\right)^2$$

Integrating the rate equation:

$$N_f = \int_0^{N_f} dN = \int_{a_i}^{a_c} \frac{da}{c(\Delta\sigma)^m\,(\pi a)^{m/2}\beta^m}$$

$$= \frac{1}{c(\Delta\sigma)^m(\pi)^{m/2}\beta^m}\int_{a_i}^{a_c}\frac{da}{a^{m/2}}$$

If $m \neq 2$, then:

$$\int_{a_i}^{a_c}\frac{da}{a^{m/2}} = \frac{a^{-(m/2)+1}}{-m/2+1}\Bigg|_{a_i}^{a_c}$$

$$= \frac{a_c^{(-m/2)+1} - a_i^{(-m/2)+1}}{-m/2+1}$$

Therefore:

$$N_f = \frac{a_c^{(-m/2)+1} - a_i^{(-m/2)+1}}{(-m/2+1)c(\Delta\sigma)^m(\pi)^{m/2}\beta^m}$$

which is the general integration of the Paris equation when β is independent of crack length, a, and when m is not equal to 2.

Because specific crack growth data were not given for the SAE 1020 steel, a reasonable first approximation could use the conservative empirical equation for ferritic-pearlitic steels (Ref 12):

$$\frac{da}{dN} = c(\Delta K)^m = 6.9 \times 10^{-12}(\Delta K)^{3.0}$$

Although this equation was developed for $R = 0$, the small compressive stress, 50 MPa, will not have much effect on crack growth and can be neglected. Thus,

$$\Delta\sigma = 200 - 0 = 200 \text{ MPa}$$

Also,

$$a_c = \frac{1}{\pi}\left(\frac{K_{Ic}}{\sigma_{max}\beta}\right)^2 = \frac{1}{\pi}\left(\frac{105}{200 \times 1.12}\right)^2$$

Substituting,

$$N_f = \frac{(0.068)^{-(3/2)+1} - (0.0005)^{-(3/2)+1}}{(-3/2+1)(6.9 \times 10^{-12})(200)^3(\pi)^{3/2}(1.12)^3}$$

$$= \frac{(0.068)^{-0.5} - (0.0005)^{-0.5}}{-2.1 \times 10^{-4}} = 189{,}000 \text{ cycles}$$

Alternatively, by using the code listed in Table 3-1, the problem can be solved readily in less than 60 seconds using a microcomputer. Results are shown in Table 3-2. Similarly, using the listing in Table 3-3 for through-thickness cracks and all the same data as in the preceding example, a solution can be reached quickly, as shown in Table 3-4.

Stress Corrosion Cracking

Stress corrosion cracking (SCC) includes both anodic dissolution processes, which preferentially remove stressed material by chemical means, and cathodic processes, such as the evolution of hydrogen that can diffuse into a material and cause failure ahead of a crack tip by a hydrogen embrittlement mechanism (Ref 13, 14).

SCC growth rates can depend on many factors including temperature, chemistry of the environment, microstructure, and the stress intensity at the crack tip. K_{ISCC} growth laws are similar to those described by fatigue crack extension (Fig. 3-8), but there is no single unified theory to explain SCC.

SCC degrades K_{Ic} to a level K_{ISCC}, below which crack growth does not occur. In between K_{ISCC} and K_{Ic}, a growth law will operate, but it will be peculiar to a given alloy and environment.

If it is assumed that circumstances have arisen under which SCC will be detected, it becomes prudent to generate operating stress maps for K_{ISCC} and K_{Ic}. Such maps are presented in Fig. 3-9 and 3-10, for 4340 steel from the data shown in Tables 3-5 and 3-6, which was reinterpreted from the results of Yokobori et al. (Ref 15). These maps are for surface crack geometries of various aspect ratios. The stress map in Fig. 3-9 indicates that 4340 with a K_{ISCC} of 40 MPa$\sqrt{m}$ can be safely operated at 25% of the yield strength with surface cracks of aspect ratio $a/c = 0.6$ present up to maximum depth of 10 mm. The environment used must be one that gives rise to the K_{ISCC} value used. Deeper cracks can be tolerated, but because the crack growth rate law is not known, the time-to-failure of such cracks cannot be predicted from the data available. A complete set of stress maps

Table 3-1 Computer code for determining change in crack length based on number of cycles

```
1030    REM FOR EDGE CRACKS ONLY; PARIS EQUATION ONLY;
1040    REM INITIALIZE
1200    DIM OP (3, 14)
1210    N = 0: PI = 3.1415
1300    REM INPUT
1310    CLS: PRINT “GIVE COEFFICIENT FOR PARIS EQUATION:”:;INPUT C
1320    PRINT:PRINT “GIVE EXPONENT FOR PARIS EQUATION:”;:INPUT M
1330    CLS:PRINT “GIVE PLATE WIDTH:”;:INPUT W
1340    PRINT:PRINT “GIVE DETECTED OR INITIAL CRACK SIZE:”;:INPUT AI
1350    PRINT:PRINT “GIVE CALCULATED CRITICAL CRACK LENGTH:”;:INPUT AF
1360    IF AF > W/2 THEN AF = .9*w/2’ AVOIDS DIVIDING BY ZERO IN 1530
1370    CLS:PRINT“GIVE STRESS RANGE DELTA SIGMA:”;:INPUT DS
1380    CLS:PRINT“ANY CHANGES? ANSWER YES OR NO;:
1390    INPUT A$:IF A$ = ”YES” GO TO 1310
1400    CLS: PRINT “READY PRESS ‘ENTER’ KEY TO GO”;:INPUT N$
1410    CLS: PRINT “O.K. I AM NUMBER CRUNCHING”
1500    REM MAIN PROGRAM
1510    IA = (AF-AI)/ 12::AC = AI: SA = .01*AC: DA = 0: OP (1,1) = AI: OP (2,1) = 0
1520    FOR I = 1 TO 12
1530    BE = 1.12
1540    DK = BE*DS*SQR (PI*AC)
1550    DD = C*DK^M
1560    DN = SA/ DD
1570    N = H + DN: AC = AC + SA: DA = DA + SA
1580    IF DA < IA THEN SA = .01*AC
1590    If DA < IA GO TO 1530
1600    GOSUB 2000
1610    DA = 0: SA = .01*AC
1620    NEXT I
1630    END
2000    REM SUB OUTPUT
2010    OP(1, I + 1) = AC:OP(2, I + 1) = N
2020    CLS: PRINT “RESULTS”: PRINT
2030    FOR J = 1 TO 78: PRINT “–”;:NEXT J: PRINT
2040    PRINT TAB (20) “CRACK LENGTH CYCLES”: PRINT
2050    FOR J = 1 TO 78: PRINT “–”;:NEXT J: PRINT
2060    FOR J = 1 TO I + 1
2070    PRINT TAB (20) USING “###.##”; OP (1, J);
2080    PRINT TAB (50) USING “#######”; OP (2, J)
2090    NEXT J
2100    FOR J = 1 TO 78: PRINT “–”;:NEXT J: PRINT
2110    IF 1 < 12 THEN PRINT “WORKING ON NEXT CRACK INCREMENT”
2120    IF 1 < 12 GO TO 2170
2130    PRINT “CONSTANTS IN PARIS EQUATION ARE: C = ”C”; M = ”M
2140    PRINT “GEOMETRY: AI = ”AI”; AF = ”AF
2150    PRINT “STRESS RANGE DELTA SIGMA = ”DS
2160    PRINT “JOB DONE”
```

may be generated for surface cracks of various aspect ratios, through-thickness cracks, and edge cracks.

Corrosion Fatigue

There is not yet a generally accepted growth law for corrosion fatigue (CF). The da/dN versus ΔK curves have a shape similar to those of corrosion fatigue and SCC curves (Fig. 3-11, 3-12, and 3-13). Testing in a similar environment under loading conditions similar to those expected in the application situation is necessary to characterize corrosion fatigue behavior. Some empirical relationships do exist, however, to aid in the estimation of the number of cycles to initiation in a given geometry and environment. Such a relationship is shown in Fig. 3-14, which indicates the number of cycles to initiation, N_i, from a notch of root radius, r, for a number of structural steels (Ref 16).

Unification of SCC and CF Growth Laws

Recent work has uncovered the possibility of life prediction from a unified approach to growth by SCC and CF, subcritical mechanisms. The approach is described in Chapter 4.

Delayed Hydride Cracking

The primary coolant containment in the current generation of CANDU (Canada Deuterium Uranium Atomic Energy of Canada Limited, Mississauga, Ontario, Canada) nuclear reactors is provided by cold-worked Zr-2.5Nb pressure tubes. Under certain conditions of stress, temperature, and hydrogen concentration, this alloy is susceptible to delayed hydride cracking (DHC). The mechanism of this cracking has been well described, both theoretically (Ref 17, 18) and experimentally (Ref 19, 20).

Table 3-2 Results from running code in Table 3-1

Crack length	Cycles
0.01	149,118
0.01	165,790
0.02	173,531
0.02	178,102
0.03	181,273
0.03	183,621
0.04	185,538
0.05	187,096
0.05	188,347
0.06	189,430
0.06	190,364
0.07	191,166
0.08	191,850

Constants for Paris equation: $c = 6.9E{-}12$: $M = 2$, Geometry: AI = 0.0005; AF = 6.8000001 $E{-}0.2$; $\Delta\sigma = 200$

Table 3-3 Computer code for determining through-thickness cracks based on number of cycles

```
1000    REM
1010    REM
1020    REM
1030    REM FOR CENTRE CRACKS ONLY; PARIS EQUATION ONLY;
1040    REM INITIALIZE
1200    DIM OP (3, 14)
1210    N, = 0: PI = 3.1415
1300    REM INPUT
1310    CLS: PRINT "GIVE COEFFICIENT FOR PARIS EQUATION:";:INPUT C
1320    PRINT: PRINT "GIVE EXPONENT FOR PARIS EQUATION:";: INPUT C
1330    CLS: PRINT "GIVE PLATE WIDTH:";: INPUT W
1340    PRINT: PRINT "GIVE DETECTED OR INITIAL CRACK SIZE:" ;:INPUT AI
1350    PRINT: PRINT "GIVE CALCULATED CRITICAL CRACK LENGTH:" ;:INPUT AF
1360    IF AF >W/2 THEN AF = .9*W/2' AVOIDS DIVIDING BY ZERO IN 1530
1370    CLS: PRINT "GIVE STRESS RANGE DELTA SIGMA:" ;: INPUT DS
1380    CLS: PRINT "ANY CHANGES? ANSWER YES OR NO:;
1390    INPUT A$: IF A$ = "YES" GO TO 1310
1400    CLS: PRINT "READY PRESS 'ENTER' KEY TO GO";: INPUT N$
1410    CLS: PRINT "O.K. I AM NUMBER CRUNCHING"
1500    REM MAIN PROGRAM
1510    IA = (AF-AI)/12::AC = AI: SA = .01*AC: DA = 0: OP (1, 1) = AI: OP (2, 1) = 0
1520    FOR I = 1 TO 12
1530    BE = SQR (1/COS (PI*AC/ W))
1540    DK = BE*DS*SQR (PI*AC)
1550    DD = C*DK^M
1560    DN = SA/ DD
1570    N = N + DN: AC = AC + SA: DA = DA + SA
1580    IF DA < IA THEN SA = .01*AC
1590    IF DA < IA GO TO 1530
1600    GOSUB 2000
1610    DA = 0: SA = .01*AC
1620    NEXT I
1630    END
2000    REM SUB OUTPUT
2010    OP (1, I + 1) = AC: OP(2, I + 1) = N
2020    CLS: PRINT "RESULTS": PRINT
2030    FOR J = 1 TO 78: PRINT "-";: NEXT J: PRINT
2040    PRINT TAB (20) "CRACK LENGTH CYCLES": PRINT
2050    FOR J = 1 TO 78: PRINT "-";: NEXT J: PRINT
2060    FOR J = 1 TO I + 1
2070    PRINT TAB (20) USING "###.##"; OP (1, J);
2080    PRINT TAB (50) USING "#######"; OP (2, J)
2090    NEXT J
2100    FOR J = 1 TO 78: PRINT "-";: NEXT J: PRINT
2110    IF 1 < 12 THEN PRINT "WORKING ON NEXT CRACK INCREMENT"
2120    IF I < 12 GO TO 2170
2130    PRINT "CONSTANTS IN PARIS EQUATION ARE: C = "C"; M = "M"
2140    PRINT "GEOMETRY: W = "W" AI = "AI"; AF = "AF
2150    PRINT "STRESS RANGE DELTA SIGMA = "DS
2160    PRINT "JOB DONE"
```

Zr-2.5Nb readily absorbs and retains hydrogen from gaseous and aqueous environments. The hydrogen is occluded exothermically, and when the concentration in solution exceeds the terminal solid solubility, the excess hydrogen precipitates as platelets of hydrides. The hydride phase is extremely brittle and can lead to a severe loss of ductility of the bulk metal. In addition, because of the high mobility of hydrogen in the zirconium lattice, redistribution of hydrides can occur, causing localized areas of embrittlement to form. For example, hydrogen will diffuse up the intense stress gradient ahead of a crack subjected to a tensile stress, resulting in the buildup of a high hydride concentration adjacent to the crack tip. If the presence of the hydride reduces the fracture toughness of the crack-tip region sufficiently, the crack will rapidly propagate through the hydride zone, and then will arrest in the ductile matrix. This process will repeat sequentially, resulting in stepwise crack growth in a stable fashion. The rate of hydride accumulation at the crack tip and, consequently, the crack velocity are controlled by the kinetics of hydrogen diffusion. When the crack has grown to some critical length characteristic of the bulk material, catastrophic failure will result.

Surface Cracks

Surface crack behavior and reliability can be analyzed successfully by calculations and stress mapping. Fatigue problems lend themselves to solutions by numerical integration and by the use of simple computer programs. Real industrial problems are presented.

In stress mapping, use is made of normalizing the operating stress by the yield stress to produce maps for various (cast) steels of various aspect ratios. For example, Fig. 3-15 and 3-16 are for cast 0.5%C-1%Cr and 1.5%Ni-Cr-Mo steels. Calculations based on complex and simple formulas are examined and compared to the result predicted by the stress map for Ni-Cr-Mo.

Table 3-4 Results from running code in Table 3-3

Crack length	Cycles
0.01	209,500
0.01	232,923
0.02	243,798
0.02	250,220
0.03	254,674
0.03	257,973
0.04	260,667
0.05	262,854,
0.05	264,612
0.06	266,133
0.06	267,445
0.07	268,572
0.08	269,532

Constants for Paris equation: $c = 6.9e - 12$; $M = 3$ Geometry: $W = 10$; AI = 0.0005; AF = 6.800001E – 02; $\Delta\sigma$=200

Surface Crack Geometries

During the water quenching of Christmas tree steel valve cast component, 30 mm section, the transformation stress generated was 120 MPa. The measured K_{Ic} was 25 MPa$\sqrt{m}$, and the 0.2% proof stress was 600 MPa. The maximum size of surface defect specified in production was 0.50 mm. Determine the tolerable crack size with the aspect ratio of the crack, $a/2c = 1/10$. What would happen if the transformation stress was close to the proof stress of the steel?

Case 1.

$$K_{Ic} = \frac{1.12\ \sigma_C\sqrt{\pi a}}{\Phi} \qquad \text{(Eq 3-1)}$$

for $a/2c = 0.1, \Phi = 1.05$:

$$a = \frac{1.05^2 \times 25^2}{1.12^2 \times 120^2\pi} = \frac{689}{56{,}548.6} = 12.2 \text{ mm}$$

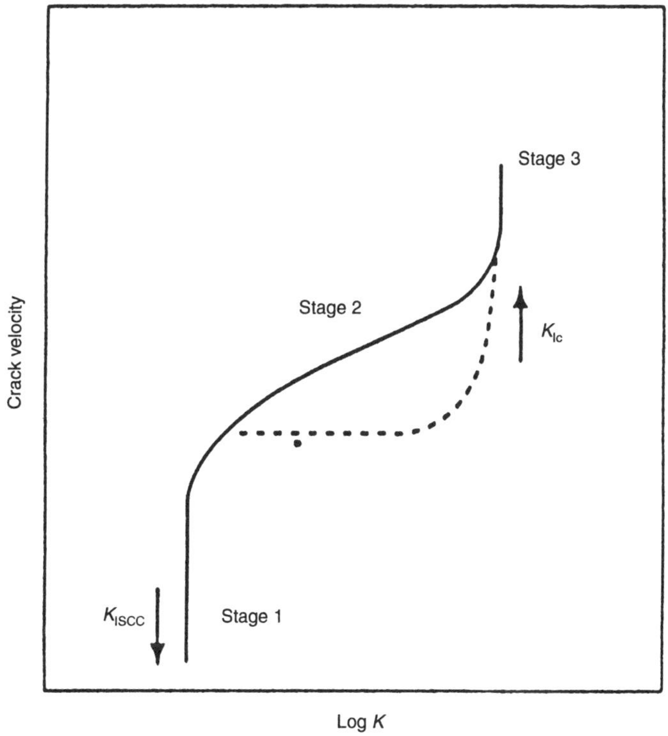

Fig. 3-8 Schematic plot of stress-corrosion cracking velocity against stress intensity factor

The size specified in production is safe because cracks would have to grow to a depth of 12.2 mm to be unsafe.

Case 2.

$$K_{Ic} = \frac{1.12\sigma_c\sqrt{\pi a}}{\Phi}$$

$$a = \frac{1.05^2 \times 25^2}{1.12^2 \times 600^2 \times \pi} = \frac{689}{1{,}413{,}716.7} = 0.48 \text{ mm}$$

The size specified must decrease to 0.25 mm, for example, because the critical depth in this case (0.48 mm) is too close to the specified size (0.50 mm) to be safe.

Case 3. The fracture of a pump casing during pressure testing showed that an internal crack had extended to 5 mm by 2 mm. The fracture stress was 1000 MPa. The steel had been subjected to heat treatment that led to a 0.2% proof stress of 1200 MPa and $K_{Ic} = 50$ MPa$\sqrt{m}$. Determine the operating stress necessary to cause fracture using the relationship for an embedded elliptical crack:

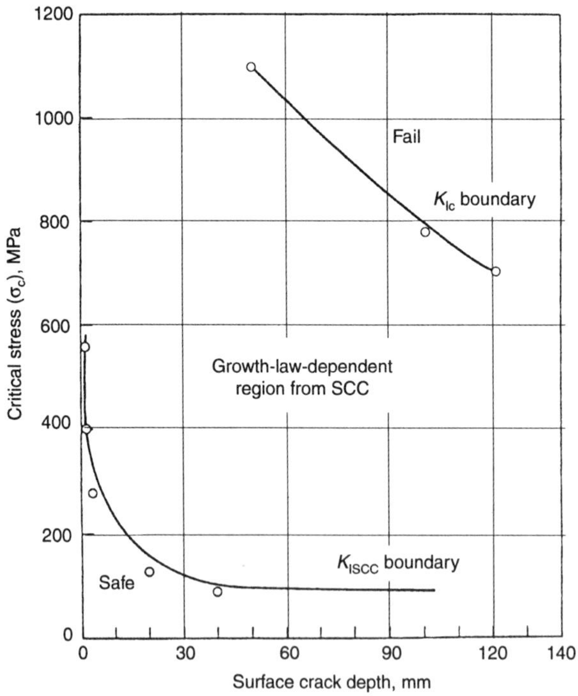

Fig. 3-9 Operating stress map for 4340 steel based on the data in Table 3-5. K_{Ic} = 127 MPa $\sqrt{m}$; σ_y = 1168 MPa; K_{ISCC} = 40 MPa$\sqrt{m}$

$$K_{Ic} = \frac{1.12\sigma_c\sqrt{\pi a}}{\Phi}$$

$$\sigma_c^2 = \frac{50^2 \times 1.05^2}{1.12^2 \times 0.0001 \times \pi} = \frac{2{,}756}{0.00393} \qquad \sigma_c = 838$$

The operating stress necessary to cause fracture of 838 MPa correlates to the observed fracture stress of 1000 MPa.

Weldments in castings can also cause problems because of easy propagation of surface cracks. Data presented in Table 3-7 indicate that recommended weld material is generally of lower toughness than the parent metal. In the example shown, as-cast

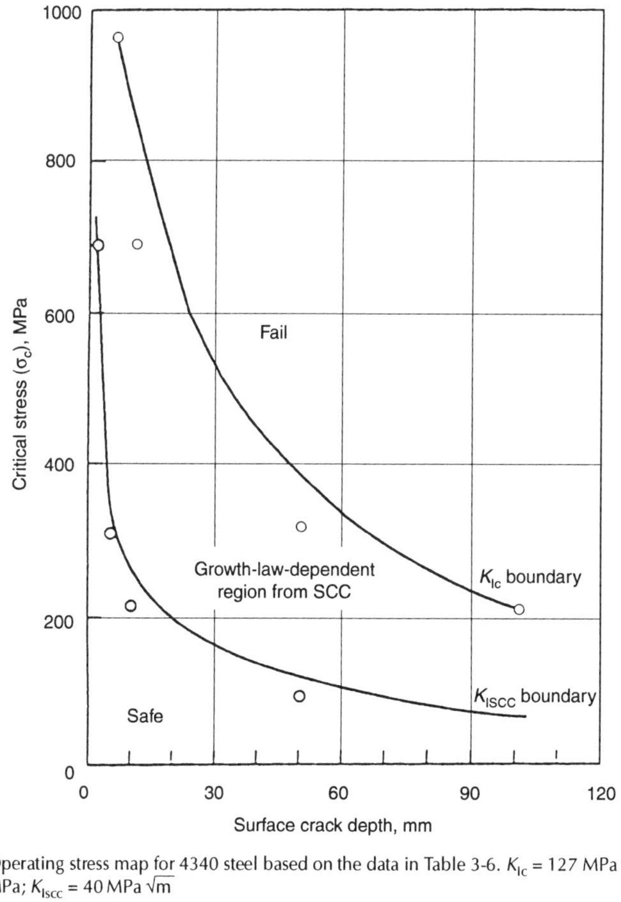

Fig. 3-10 Operating stress map for 4340 steel based on the data in Table 3-6. $K_{Ic} = 127$ MPa $\sqrt{m}$; $\sigma_y = 1168$ MPa; $K_{Iscc} = 40$ MPa $\sqrt{m}$

2.5%Ni-Cr-Mo can sustain surface cracks to a depth of 70 mm. 3%Ni-Cr-Mo B weld metal can sustain these cracks to a depth of 52 mm. Stress-relieved Jetweld LH110 weld metal (Lincoln Electric Company, Cleveland, OH) can sustain them to 31 mm. As-welded Jetweld LH110 in a high residual stress state, can only sustain surface cracks to a depth of 5 mm (Ref 21).

Stress maps for two cast steels are presented in Fig. 3-15 and 3-16 (Ref 21). The validity of the maps is evaluated by comparing the results of three types of calculation to solve surface crack problems. Case 4 demonstrates that both the stress map and the simplest calculation provide just as accurate a solution to the problem as the more complex calculations, in keeping with the intent of this work.

Table 3-5 Data for operating stress map in Fig. 3-9

Permissible crack, mm	Operating stress, MPa	Factor in K equation
For measured K_{Ic} = 127, a/c = 0.6, yield strength = 1168		59.963
10	2448	
50	1095	
100	774	
120	706	
200	548	
For measured K_{ISCC} = 40		785
2.5	560	
5	396	
10	280	
50	125	
100	89	

K_{Ic} = MPa$\sqrt{m}$ surface cracks

Table 3-6 Data for operating stress map in Fig. 3-10

Permissible crack, mm	Operating stress, MPa	Factor in K equation
For measured K_{Ic} = 127, yield strength = 1168		4704
5	970	
10	686	
50	397	
100	217	
For measured K_{ISCC} = 40		467
1	684	
5	306	
10	216	
50	97	
100	68	

K_{Ic} = MPa$\sqrt{m}$ surface crack or edge crack

Case 4. A casting made in 1.5%Ni-Cr-Mo steel is to be operated at half of its yield stress (740 MPa) and one application of load. It has a surface crack the major axis of which is 10 times the length of the minor axis. The fracture toughness (K_{Ic}) of the steel is measured at 86 MPa$\sqrt{m}$. The critical defect size may be calculated as:

$$a_{cr} = K_{Ic}^2 \left[\frac{\Phi - 0.212 \left(\frac{\sigma_c}{\sigma_y} \right)^2}{1.21 \pi \sigma_c^2} \right]$$

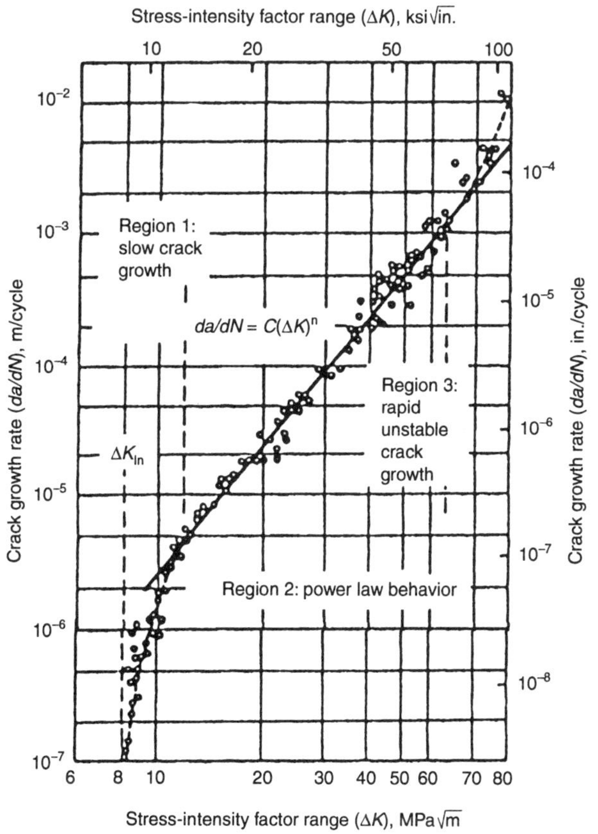

Fig. 3-11 Entire *da/dN* vs. Δ*K* plot for A533 steel showing asymptotic behavior at either end of the curve and a relatively linear portion in the center. Yield strength 470 MPa (70 ksi). Test conditions: $R = 0.10$; ambient room air, 24 °C (75 °F).

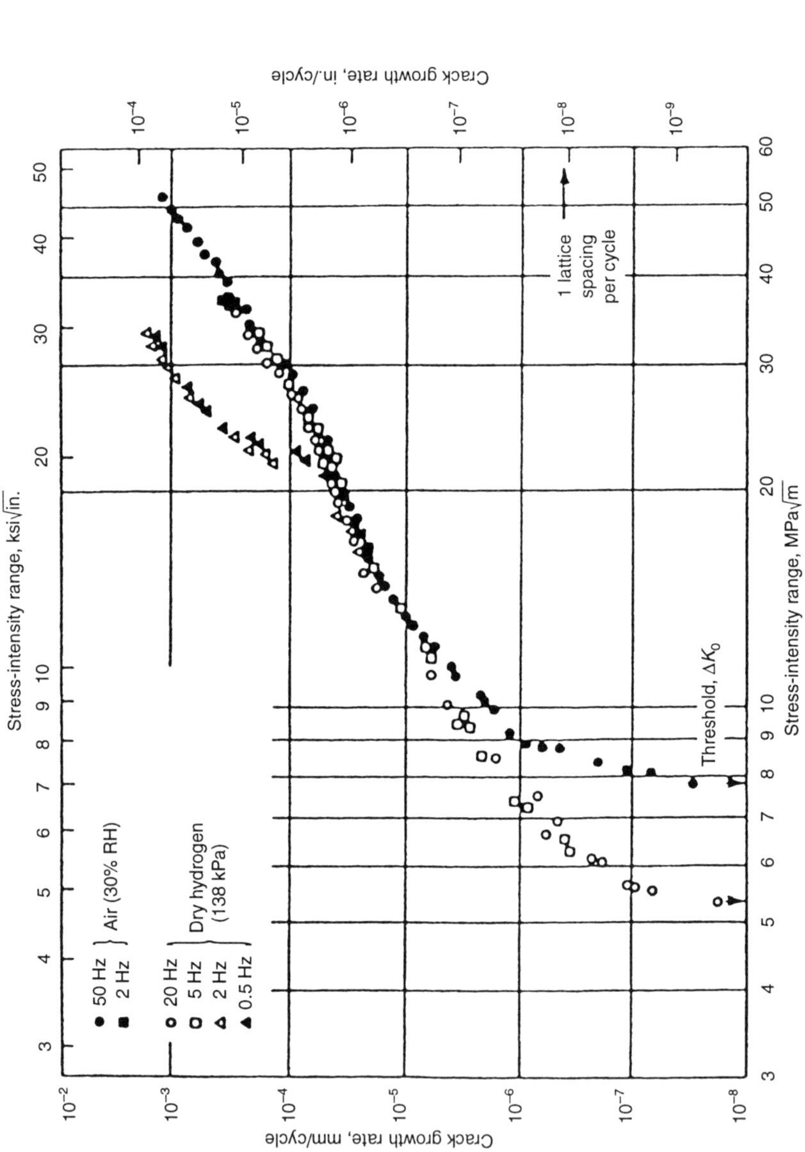

Fig. 3-12 Corrosion fatigue in 2.25Cr-1Mo pressure-vessel steel in hydrogen due to hydrogen embrittlement at high ΔK and reduced oxide-induced closure at low ΔK. Stress ratio = 0.05

where K_{Ic} is plane-strain fracture toughness, σ_c is gross working stress normal to the major axis of the crack, a_{cr} is critical depth of a surface flaw (i.e., half the width of an embedded crack), σ_y is 0.2% proof stress, and Φ is double elliptical integral. Let:

$$Q = \left[\Phi^2 - 0.212\left(\frac{\sigma_c}{\sigma_y}\right)^2\right]$$

For embedded cracks, the coefficient on the denominator is taken as unity.

To define the shape of the crack, $a/2c$ can be considered to represent the crack size aspect ratio, where a is the minor and $2c$ is the major axis of an ellipse (when $a/c = 1$, the ellipse becomes a circle). The relationship between Φ and $a/2c$ is given in Fig. 3-17 for easy reference.

In this example, $a/2c = 0.1$ and $\sigma_c = 740/2$.

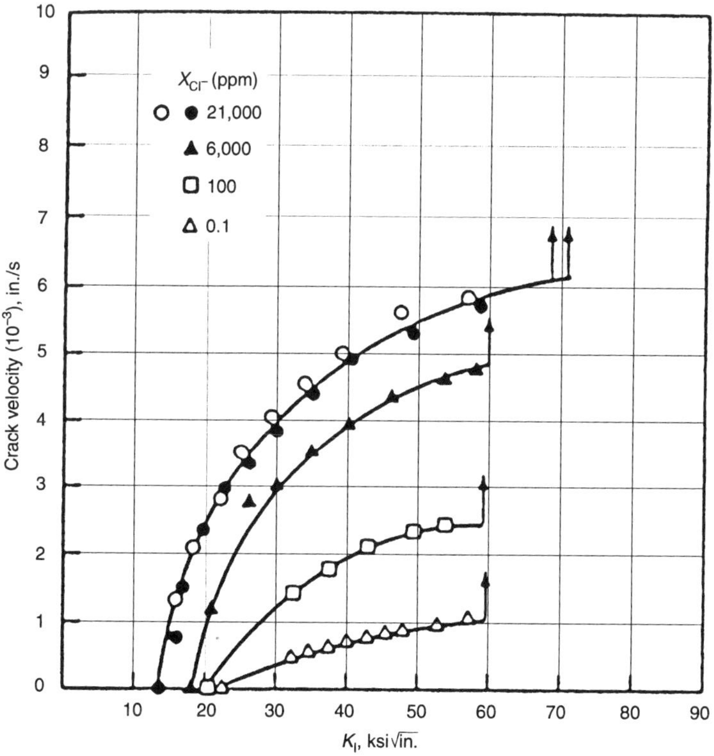

Fig. 3-13 Influence of chloride concentration on the SCC behavior of Ti-8Al-1Mo-1V in aqueous chloride solutions at 25 °C (77 °F)

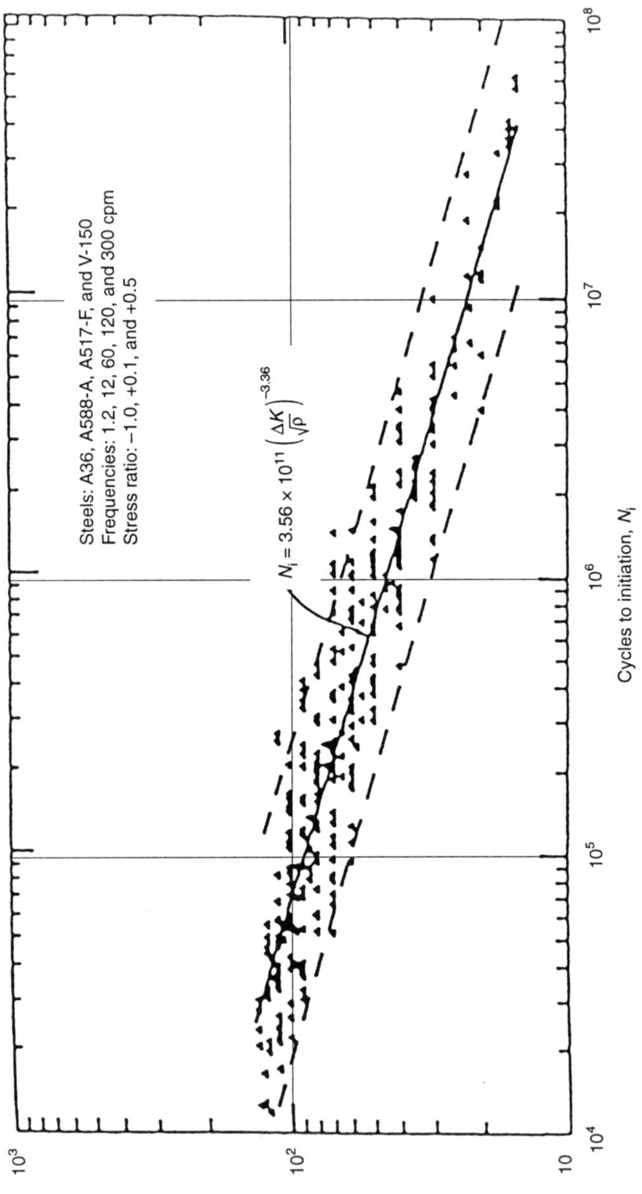

Fig. 3-14 Corrosion fatigue cracking initiation from notches

$$\left(\frac{a}{Q_{cr}}\right)=\frac{K_{Ic}^2}{\sigma_c^2}\times\frac{1}{1.21\pi}=\frac{86^2}{370^2}\times\frac{1}{3.80}=\frac{7{,}396}{136{,}900}\times 0.263=0.0540\times 0.263=0.0142$$

$$a_{cr}=Q\times 14.2\text{ mm} \qquad \text{(Eq 3-2)}$$

Now, for $a/2c = 0.1$ (from Fig. 3-17), $\Phi = 1.05$ and

$$Q=\Phi^2-0.212\left(\frac{\sigma_c}{\sigma_y}\right)^2=1.05^2-0.212(0.5)^2=0.971$$

Inserting this value for Q in Eq 3-2:

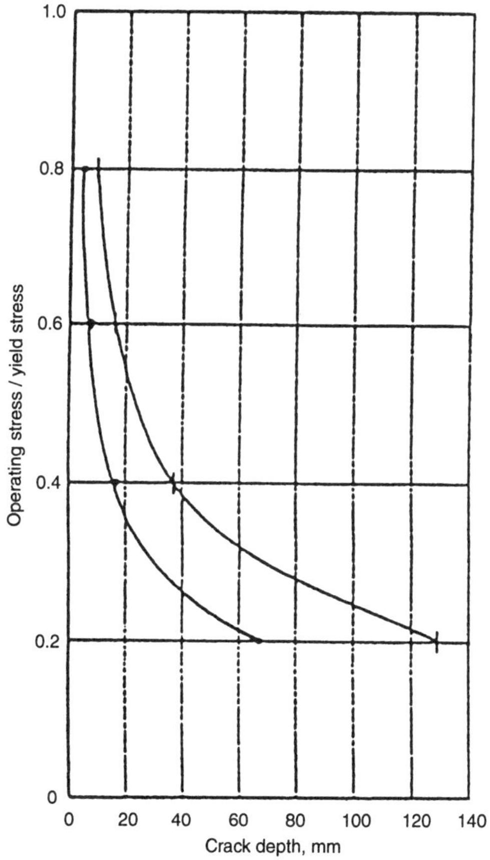

Fig. 3-15 Operating stress map for as cast 0.5%C-1%Cr steel with surface cracks. K_{Ic} = 46 MPa $\sqrt{m}$; σ_y = 480 MPa; $a/2c$ = 0.1; $a/2c$ = 0.5

$$a_{cr} = 0.971 \times 14.2 = 13.78 \text{ mm}$$

Consider this same calculation using the equation:

$$K_{Ic} = \frac{1.12\sigma_c\sqrt{\pi a}}{\Phi}$$

where $K_{Ic} = 86$ MPa$\sqrt{\text{m}}$; $\sigma_c = 370$ MPa; $\Phi = 1.05$ (from Fig. 3-17, $a/2c = 0.1$). Therefore:

$$a = \frac{86 \times 1.05}{1.12 \times 370 \times 3.142} = 15 \text{ mm}$$

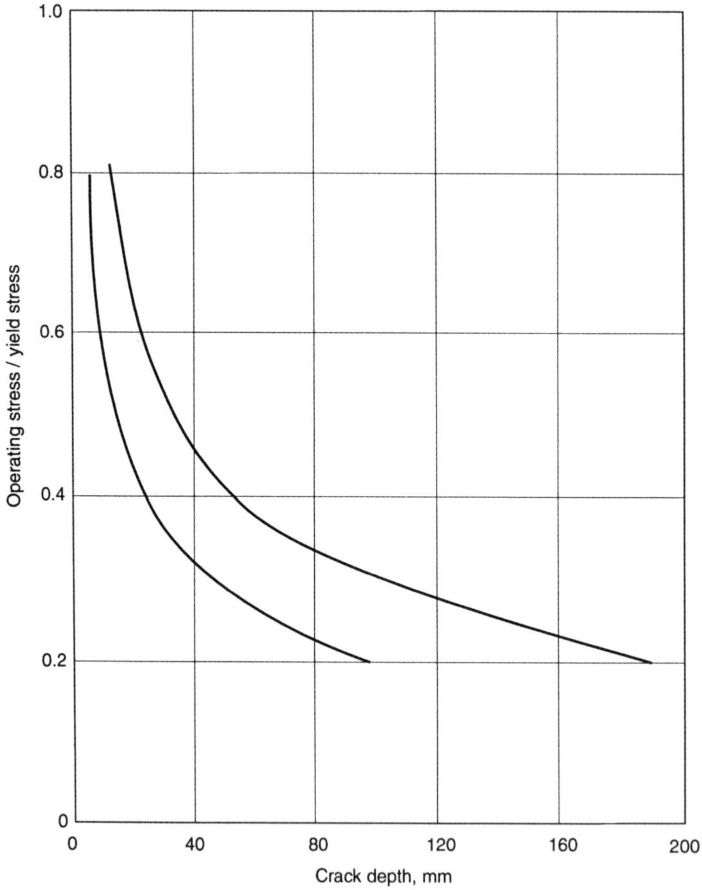

Fig. 3-16 Operating stress map for as-cast 1.5%Ni-Cr-Mo steel with surface cracks. $K_{Ic} = 86$ MPa $\sqrt{\text{m}}$;$\sigma_y = 740$ MPa; $a/2c = 0.1$; $a/2c = 0.5$

Compare this value to the 13.78 mm from the preceding calculation, and a difference of 1.22 mm or 8.1% results.

In Fig. 3-16, the stress map yields a value for a of about 14 mm for σ_c of 370 MPa. Also, if section thickness, B, is considered, and the back surface correction M_b is used, then for crack depths of up to a/B of 0.25 with $a/2c$ still at 0.1 (Fig. 3-18), the preceding equation holds true. Thus, equipment failure rate reduction from surface crack control can be achieved by using stress maps derived from a simplified equation rather than using more complex formulas.

Elastic-Plastic Fracture Mechanics (EPFM)

Elastic-plastic fracture mechanics (EPFM) considers larger crack tip plasticity than LEFM, together with some ductile, stable, slow crack growth taking place prior to the onset of instability. Note that the ASME Boiler and Pressure Vessel Code Section XI Flaw Evaluation Committees are developing EPFM. The theory for EPFM is based on nonlinear elastic behavior. In the measurement of J_{Ic}, a parameter for describing elastic-plastic toughness, the reality is that nonlinear elastic behavior does not exist. A compact specimen exhibiting elastic-plastic behavior shows a rising load curve similar to that shown in Fig. 3-19, with no obvious point indicating the onset of stable crack growth. The challenge then becomes to determine the conditions which adequately describe the onset of stable crack growth.

Compact specimens are machined from the material of construction. Note that these compacts are the same geometry as the specimens used for K_{Ic} determination. The slight difference is that compacts used in this testing may have lugs machined on them to accept constant current leads. The tensile specimen for each compact is always cut from the same material but at right angles to the machined notch in the compact. The reason for this is that the tensile behavior at right angles to the notch is assumed to be the same as the bending behavior about the notch. In other words, the compact specimens are not

Table 3-7 Allowable defect sizes in 2.5%Ni-Cr-Mo steel casting and weld metals

Material	0.2% proof stress, nodules/ mm^2	Design stress (a)/ proof stress	Fracture parameter: Applicable criterion	Fracture parameter: Value	Allowable defect in 100 mm thickness: Depth, mm	Allowable defect in 100 mm thickness: Length, mm
2.5%Ni-Cr-Mo casting	601	0.66	COD	0.72 mm	70	280
Metrode 3%Ni-Cr-Mo B welded metal	575	0.70	COD	0.44 mm	52	208
Jetweld LH110 weld metal						
Stress relieved	676	0.59	K_{Ic}	167 MNm$^{-3/2}$	31	124
As welded	676	0.59	K_{Ic}	189 MNm$^{-3/2}$	38	152
As welded	676	1.59 (b)	K_{Ic}	189 MNm$^{-3/2}$	5	20

COD = crack opening displacement. K_{Ic} = elastic fracture toughness. (a) Design stress = 400 nodules/mm^2. (b) Assuming yeild point residual stress

really tensile loaded at all. Loading is apparently tensile but, in fact, the notch is opened in bending. For this reason, ASTM altered the specimen designation from compact tensile specimens to simply "compacts."

Because toughness varies with notch acuity, a fatigue crack in the compacts at notch root should be grown. Frequencies of ~ 7 Hz are accurate, and the stress amplitude must be kept low enough not to sustain any structural damage. The levels are described in ASTM E 813 or E 1152 (the final *J* difference between these two methods is 7%). The fatigue cracks should be grown to give a/W of ~0.6.

Load the prefatigued compact into a suitable tensile load frame. Constant current of about 7 to 9 amps is fed through No. 14 gage copper wire screwed on the lugs by stainless steel sleeves that have been drilled and tapped. This current gives a power drop of about 7 mW. Potential drop leads (fine stainless steel wire insulated using ceramic

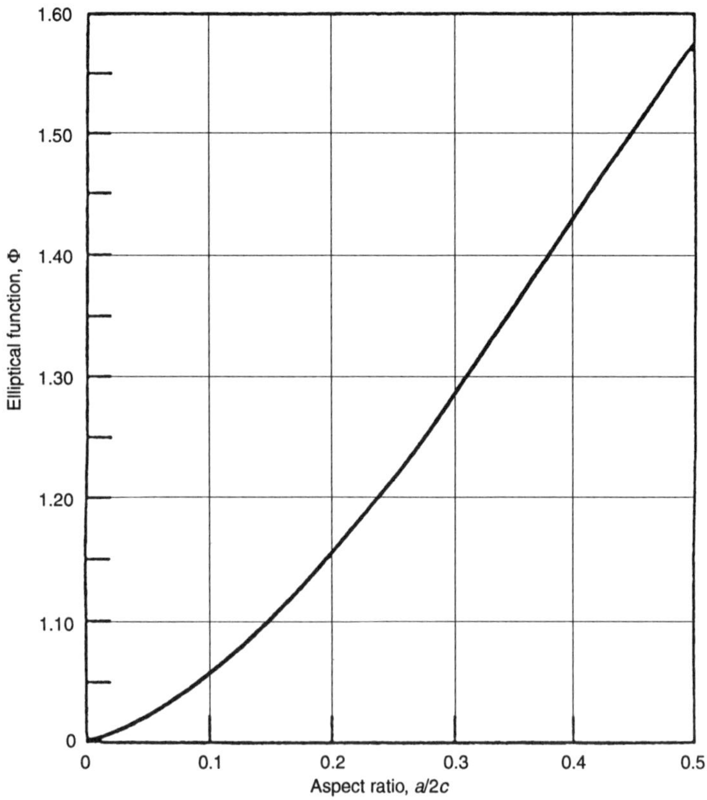

Fig. 3-17 Elliptical function vs. aspect ratio, $a/2c$

sleeves works well) are spot welded in situ to the appropriate points above and below the notch opening. A diagonal arrangement gives the best results.

Displacement of the loading holes is measured by linear voltage drop transducers placed at either side of the specimen. It is crucial that the loading pins are insulated from the specimen, and this is best done by oxidizing suitable tool steel pins. Set the cross head speed to a normal level.

Record the following data at the same time: load, load-pin displacement, potential drop (for 17 mm compacts, about 0.75 mV of crack growth will meet the ASTM requirements), and time. Use any data logging system, but make sure that load and load pin displacement units are in volts and potential drop units are in millivolts.

Now repeat the experiment on a specimen that does not contain a notch, and record the load and the load-pin displacement in volts. This experiment is critical in calibrating the load frame so that all the "slack" can be removed from the test data.

Four columns of data should result:

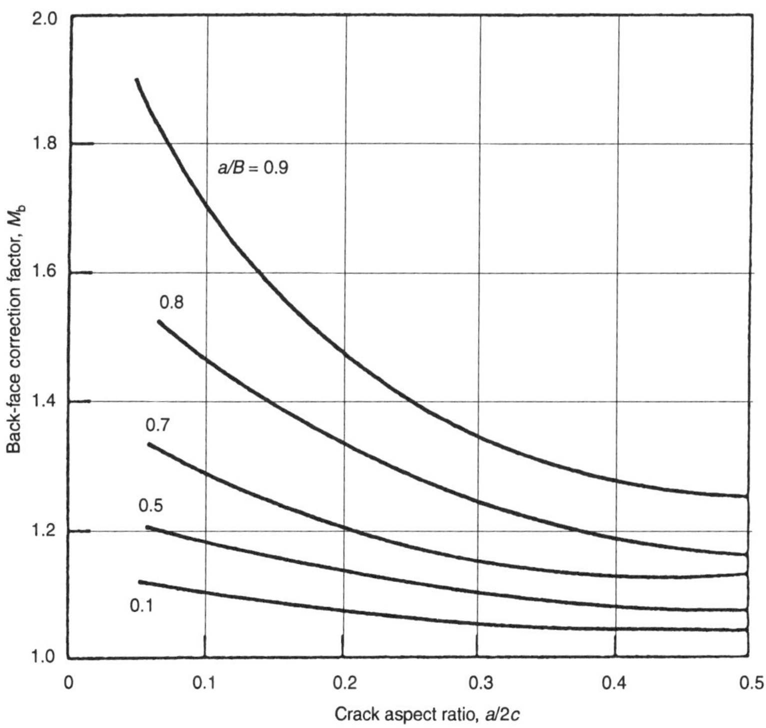

Fig. 3-18 Effect of back-face correction on aspect ratio

- data point number
- load in volts
- displacement in volts
- potential drop in millivolts

After the test is complete, carefully remove and post-fatigue the specimen. This marks on the surface of the specimen the extent of stable-crack growth that took place. After post-fatiguing, pull the specimen to separation, photograph the two halves, and measure the crack growth using the nine-point average method specified by ASTM E 813 or E 1152. Pull the appropriate tensile test specimen for the compact tested. From the tensile test, record the (offset) yield stress and the UTS.

Convert the raw data into usable numbers to derive a J_R curve (i.e., an energy input/crack extension curve). The methodology is described in ASTM E 813 or E 1152.

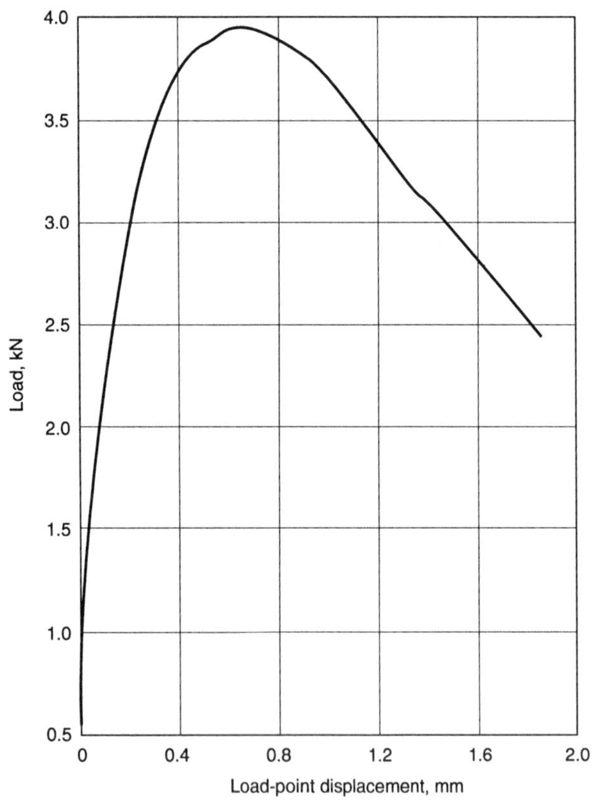

Fig. 3-19 A rising load curve similar to that shown by a compact specimen exhibiting elastic-plastic behavior

The "blunting line" at $J = 2\bar{\sigma}\Delta a$ is drawn together with the offset lines, and J_{Ic}, the crack initiation toughness, together with dJ/da, and the slope of the J_R curve or the crack growth toughness are derived (Fig. 3-20).

Crack Growth in Piping

J_R curves are frequently used for modeling axial crack growth in pipes. The implication is that a crack has grown through a pipe by some mechanism and that the crack will grow in a stable manner to some critical length in the axial direction of the pipe or vessel. The crack driving force curve is determined from the expression for the fracture toughness of a through-thickness axial crack (Ref 22):

$$K_Q = \left[\frac{8\, a_c\, \bar{\sigma}^2}{\pi} \ln\left(\sec \frac{\pi\, M\, \sigma_H}{2\bar{\sigma}} \right) \right]^{1/2}$$

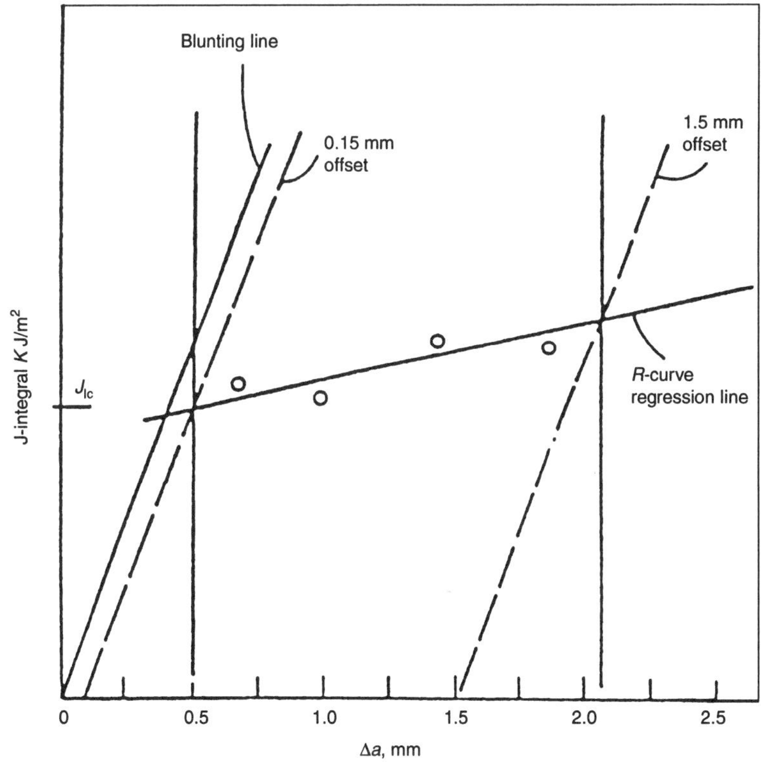

Fig. 3-20 Typical J_R curve; blunting line intercept yields J_{Ic}; *R*-curve regression line gives *dJ/da*

where $2a$ = critical crack length, $\bar{\sigma} = \frac{1}{2}$ (yield stress + UTS), and σ_H = applied hoop stress (i.e., the operating stress). Also M = Folias bulging factor (Ref 23) = $(1 + 1.61\ a^2/R_i t)^{1/2}$, where R_i = inner radius, t = wall thickness, and $2a$ = crack length. K_Q is converted to units of J through:

$$K_Q = \sqrt{EJ_I}$$

where E = Young's modulus, and J_I = first "pop-in" on rising J_R curve.

The superposition method described in the introduction to this section is then used (Fig. 3-21 and 3-22) to determine the critical crack length. Note that the specimens used model edge crack growth or growth of half the through-thickness crack length in the axial direction.

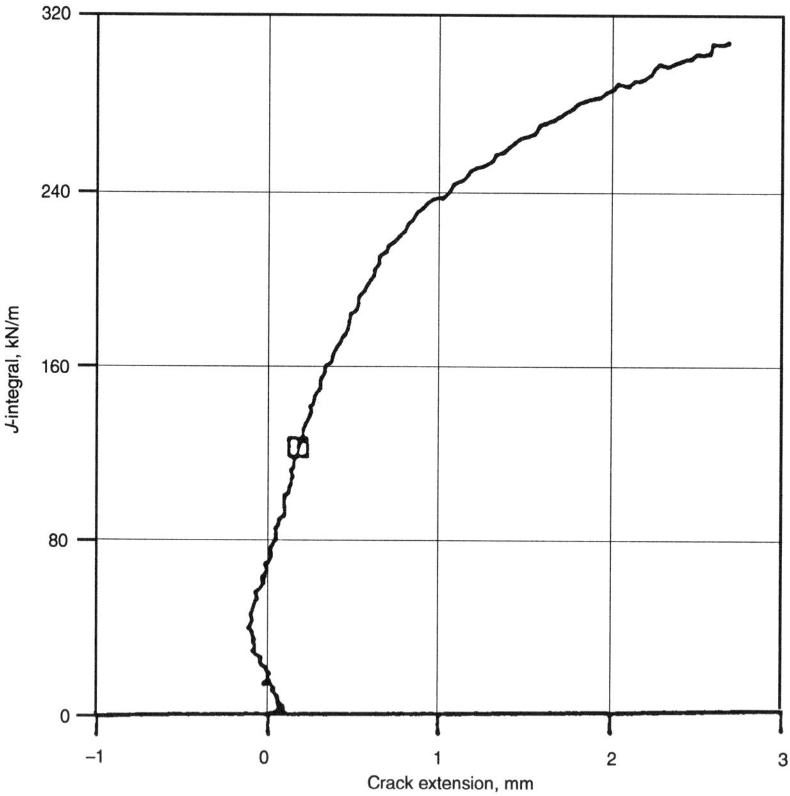

Fig. 3-21 J_R curve for determining critical crack length

The data in Table 3-8 were determined using the preceding approach. These data can be reinterpreted to give estimates of critical crack length if use is made of stress intensity equation:

$$K_{Ic} = \beta \sigma_c \sqrt{\pi a}$$

where $K_{Ic} = K_Q$ in Table 3-8; $\sigma_c = 90$ MPa (from Ref 24); and β for the edge crack specimen is multiplied by β for stable crack growth. Thus $\beta = (2 \times 1.12 \times (1/1.571))$ for a stable crack aspect ratio of 1.0, taken from ϕ values. The factor 2 takes into account the fact that there are two (edge crack and ellipse) geometries set edge-to-edge. Thus:

$$K_Q = 1.43 \times 90 \times \sqrt{\pi a}$$

For the K_Q values given in Table 3-8, an extra column is added for critical crack lengths determined by this solution, which employs the β compounding technique.

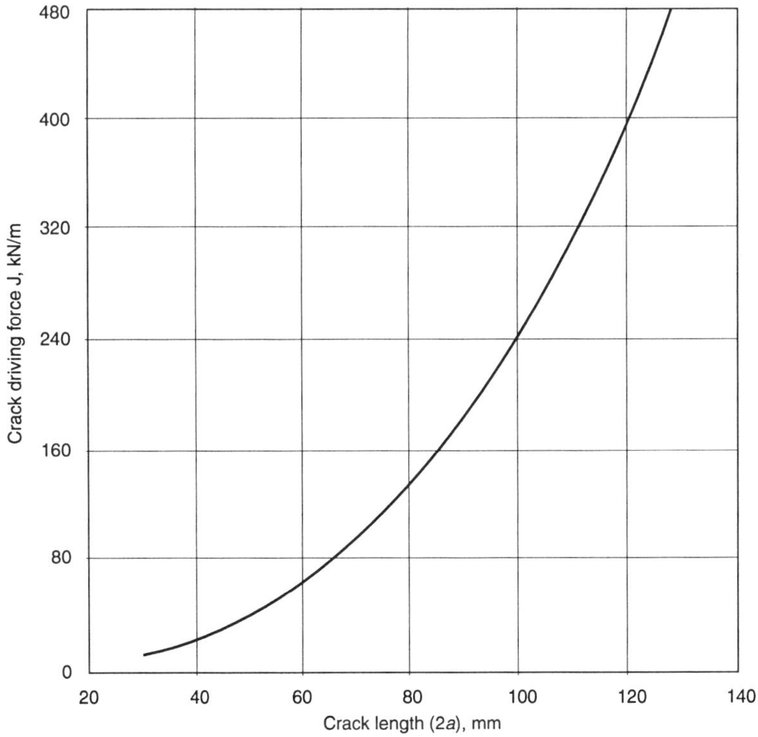

Fig. 3-22 Crack length vs. crack driving force

In the tough specimens in Table 3-8, the difference between the measured and the calculated critical crack lengths is a conservative 20%. In the brittle specimens, this difference is ~5%, reflecting the unchanged β value, which always considered stable crack growth. This modeling appears satisfactory, but it is intended as food for thought rather than a definitive approach.

It is possible to use EPFM in constructing a stress map in the same way that LEFM and PFM are used. The operating stress, σ, is derived from solving the following equation (Ref 25) for σ :

$$\pi\beta^2 \sigma^2 a/E + H\sigma^{N+1} a/F = J_c$$

where E is Young's modulus, F is the "nonlinear" modulus (Ref 25), and a is the crack size (Ref 25).

If β and H are simple functions or available in tabular form as Table 3-9, for example (Ref 25), then the preceding equation is solved for a range of a values by iteration to produce the required data for an EPFM stress map (Ref 25).

Plastic Fracture Mechanics

Net Section Yield

A tangent is drawn to the stress curve from an operating stress of zero and a crack length $2a$, equal to the plate width, W (Fig. 3-23 and Table 3-10). For β = 1, the tangent always hits the curve at $2a$ $W/3$. As the plate width becomes smaller, the tangent will move to the left until it coincides with the tangent from the yield stress to ⅔ the tensile

Table 3-8 Summary of fracture toughness results and specimen properties

Test temperature, °C	Tough or brittle	Fracture toughness, $K_Q \cdot \text{MPa}\sqrt{\text{m}}$	Yield stress σ_y, MPa	Critical crack length, mm $K_Q = 1.43\times 90\times\sqrt{\pi a}$	Measured critical crack length, mm	$2.5 K_Q^2/\sigma_y^2$, mm
280	T	54.0 (a)	662	56	71	...
280	T	63.6 (a)	662	78	102	...
280	T	69.7 (a)	662	93	108	...
280	T	56.9 (a)	662	62	88	...
280	T	63.6 (a)	662	78	95	...
280	T	67.3 (a)	662	87	103	...
280	T	66.0 (a)	662	84	99	...
280	T	55.4 (a)	662	59	75	...
230	B	48.7	719	46	49	11
230	B	53.6	719	55	53	14
230	B	50.3	719	49	50	7.7
280	B	46.0	662	41	47	13
230	B	40.0	719	31	41	19
230	B	48.5	719	45	59	27

Calculated from $K_Q = \sqrt{EJ_I}$ where J_I is read from the first pop-in on the J-resistance curve; E = 95.9 – 0.05 T (GPa), T = °C

yield stress (or 0.2% proof stress, whichever is most convenient). So all the operating stress data are on one line from the yield stress at $a = 0$ to the plate width at $W = 2a$, at zero operating stress. This line is given by:

$$\frac{\text{Plate width} - 2a}{\text{Plate width}} \times \text{Yield stress}$$

This is the condition for net section yield, so the plate would fail on reaching net section yield, regardless of crack length. The two tangents coincide at the point equal to an operating stress of ⅔ of the yield stress. Since the left tangent is always at a crack length of $2a = W/3$, the corresponding plate width is always three times the crack length at that point. At that point, the operating stress may be calculated from:

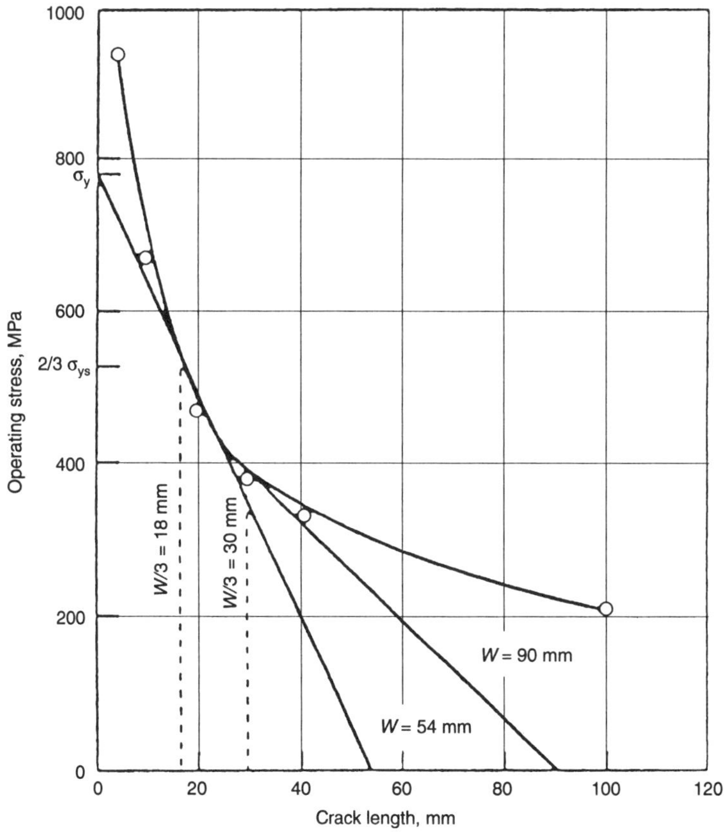

Fig. 3-23 Operating stress map for calculating net section yield. Data are shown in Table 3-10.

$$\text{Operating stress} = \frac{K_{Ic}}{\sqrt{\pi a}}$$

Because the operating stress is ⅔ the tensile yield stress and $a = W/6$, it follows that:

$$\frac{2\sigma_y}{3} = \frac{K_{Ic}}{\sqrt{\pi W/6}}$$

or

$$W = \frac{27}{2\pi}\left(\frac{K_{Ic}}{\sigma_y}\right)^2$$

These relationships (Ref 26) give the minimum plate width for which K_{Ic} applies. This value of plate width is also the maximum that will fail at net section yield, as will all smaller plates, irrespective of the crack length.

Leak-Before-Break

When a crack of surface length $2a$ grows through a vessel wall of thickness t, two outcomes are possible. Either crack grows no further, and the vessel leaks, or the crack grows catastrophically and the vessel breaks. Leak-before-break is the more desirable outcome.

The analysis proceeds as follows:

1. Perform all the relevant code calculations for the prevailing conditions of hoop stress, pressure, thickness, radius, and stress intensification factor (Iv).
2. Determine K_{Ic}.
3. Substitute appropriate values from 1 and 2.
4. Evaluate the results.

The following examples illustrate the procedure.

Case 1. A 304 stainless steel vessel cylindrical section is controlled by longitudinal seams. The vessel contains an innocuous liquid at room temperature. The vessel is 96 in. outside diameter (OD), 50 psig, with 0.25 in. wall thickness.

Table 3-9 Values for H for center-cracked panels in plane stress

	H (a) (b)					
a/W	$N = 1(\pi\beta^2)$ (c)	$N = 2$	$N = 3$	$N = 5$	$N = 7$	$N = 10$
0.0625	3.20 (3.20)	4.66	5.99	8.71	11.84	17.56
0.1250	3.38 (3.40)	5.28	7.44	13.48	23.30	50.79
0.1875	3.74 (3.77)	6.48	10.32	24.64	58.25	214.40
0.2500	4.42 (4.44)	8.80	16.48	57.92	208.64	1,464.32
0.3125	5.65 (5.63)	13.58	32.05	190.13	1,169.88	18,365.75
0.3750	8.28 (8.21)	27.36	93.44	1,239.04	17,694.72	909,115.39
0.4375	16.64 (16.10)	100.48	670.72	35,389.44	...	...

(a) For other values of a/w and N, use interpolation. (b) Where H is derived from Ref 25. (c) For $N = 1, H = \pi\beta^2$. Ref 25

$$SE = \frac{PR_0 - 0.4Pt}{t} = \frac{50 \times 48 - 0.4 \times 50 \times 0.25}{0.25} = 9580 \text{ psi}$$

where $E = 1$.

From the stress map (Fig. 3-24), the crack length at 9.580 ksi is about 140 in., giving rise to a wall thickness of 70 in. The actual value of wall thickness is 0.25 in. The situation for criticality is two orders of magnitude greater than the true vessel thickness. This vessel will always leak before break under the conditions specified.

Case 2. A 21 ft radius 304 stainless steel storage tank has pressure of 20.4 psig at the bottom. The lowest course of shell plates are $^5/_{16}$ in. thick. The contained fluid is innocuous at the temperature stored.

The stress on a vertical seam at the bottom course of the tank is:

$$SE = \frac{P(R + 0.6t)}{t} = \frac{20.4(21 \times 12 + 0.6 \times 5/16)}{5/16} = 16{,}463 \text{ psi}$$

where $E = 1$.

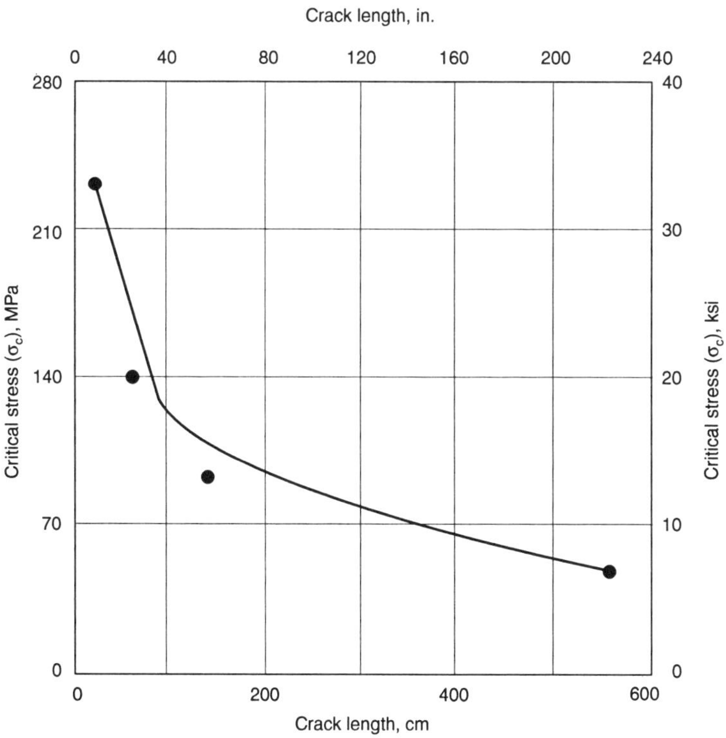

Fig. 3-24 Map of σ_c for cold-worked and annealed 304 stainless steel with a through-thickness crack ($\beta = 1$)

From inspection of the stress map (Fig. 3-24), the critical crack length at 16.463 ksi is ~50 in., which would give a wall thickness of 25 in. Because the true wall thickness is 5⁄16 in. (0.3125 in.), again two orders of magnitude greater than the true case, this storage tank would leak-before-break under the specified conditions.

Case 3. Consider a vessel in which there is a continuous discontinuity in the head in the head-to-shell weld, as shown in Fig. 3-25. The thickness in the weld region is 0.875 in. The discontinuity is 0.25 in. The minimum thickness is 0.625 – 0.25 = 0.4 in. from the diagram. In this case, the material is A285C with a K_{Ic} of 77 ksi and a yield stress of 30 ksi. The owner would operate the vessel safely with P = 50 psig, R = 48 in. OD, E = 0.7, t = 0.4 in. (minimum from Fig. 3-25).

$$t = \frac{PR}{2SE + 0.4P}$$

$$0.4 = \frac{50 \times 48}{2SE + 0.4 + 50}$$

Rearranging,

$$S = \frac{2392}{0.56} = 4271.4 \text{ psi}$$

Construct the stress map, and draw in the lines for net section yield with W = 0.875 (plate width for through-thickness crack of 0.25 in.).

Inspection of the stress map (Fig. 3-26) indicates that elastic fracture mechanics does not apply in this case because, at the plate width drawn in and the operating stress of 4.271 ksi, failure will always be by net section yielding. Thus, with a crack length of 2 × 0.25 = 0.5 in., at the operating stress of 4.271 ksi, the vessel is safe to operate. Consider now the state of affairs if one side of the weld is ground off. The wall thickness is now reduced to 0.45 – 0.25 = 0.2 in.

Table 3-10 Forged Ti-6Al-4V plate (data for operating stress map in Fig. 3-23)

Permissible crack, mm	Operating stress, MPa	Factor in K equation
For measured K_{Ic} = 83		2193
20	468	
30	382	
40	331	
80	234	
100	209	
10	662	
5	937	

K_{Ic} = MPa$\sqrt{m}$; σ_y = 790 MPa; through-thickness crack

$$t = \frac{PR}{2SE + 0.4\,P}$$

Rearranging and substituting,

$$S = \frac{2396}{0.28} = 8557.1 \text{ psi}$$

For the edge crack of 0.2 in., the operating stress of 8.557 ksi would be safe also. This example indicates the need to generate a stress map whenever possible and to always consider the net section yield situation where elastic fracture mechanics does not apply.

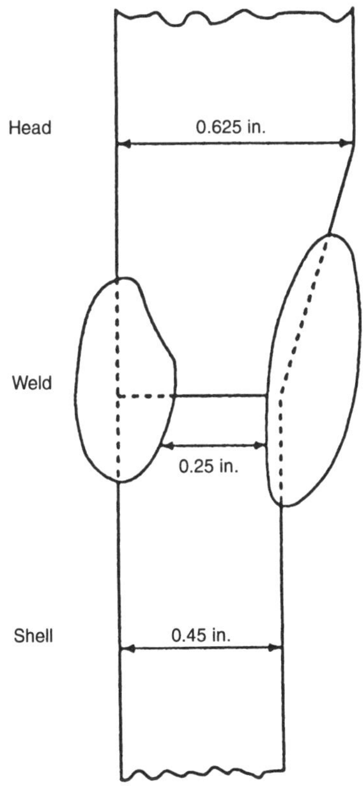

Fig. 3-25 Weld discontinuity in leak-before-break analysis

Prediction of Fracture Toughness

Prediction of Fracture Toughness from Charpy Data

The main reason for using the stress intensity approach in stress maps and defect tolerance is that the equations relate crack size to the stress, so the stress and dimension units are familiar. The less meaningful but inexpensive tests that have been used for years to describe toughness behavior are notched-bar impact tests, which provide measures of the energy absorbed on impact.

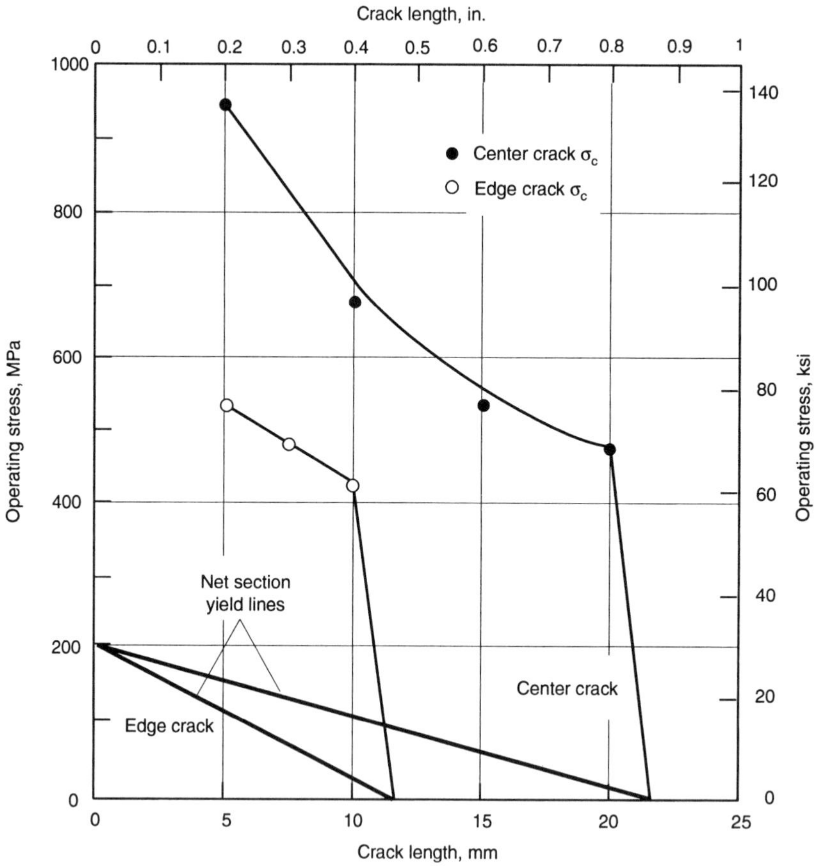

Fig. 3-26 Operating stress map for the vessel in Case 3

Empiricism and K_{Ic}

There are empirical relationships in the literature that relate impact data to the more meaningful but more expensive fracture toughness data. Unfortunately, there are about 36 different relationships available, and they must be used for specific conditions. However, two relationships are used quite commonly (Ref 26). For conditions approaching upper shelf toughness values:

$$\left(\frac{K_{Ic}}{\sigma_y}\right)^2 = 4\left[\left(\frac{CVN}{\sigma_y}\right) - 0.1\right] \text{ in.} \left(\frac{\text{ft–lb}}{\text{ksi}}\right)$$

or

$$\left(\frac{K_{Ic}}{\sigma_y}\right)^2 = 0.52\left[\left(\frac{CVN}{\sigma_y}\right) - 0.02\right] \text{ m} \left(\frac{\text{J}}{\text{MPa}}\right)$$

and for conditions toward the lower shelf (Ref 27):

$$K_{Ic} = 12\sqrt{CVN} \text{ (all units)}$$

where CVN = Charpy toughness.

The Charpy data presented in Fig. 3-27 may be reinterpreted to yield the elastic toughness values that are given in Table 3-11, by manipulation of the preceding equations which relate K_{Ic} to CVN.

There is a considerable amount of such Charpy data in the open literature that would lend itself to elastic fracture toughness prediction, to provide a cost-beneficial base for the derivation of stress maps for a wide range of alloys. Thus, for the measured K_{Ic} values examined (Ref 27), the empirical relationship for Charpy values toward the upper shelf predicts K_{Ic} values of ~70% of the measured values. For Charpy values toward the lower shelf, predicted values of K_{Ic} ~of 30% of the measured values. The predicted values are considered to be acceptably conservative in view of the scatter normally observed in Charpy data (~20%) and K_{Ic} data (~15%), particularly in cast steels, where properties tend to be anisotropic.

A stress map for cast 8630 steel is constructed from measured K_{Ic} and predicted K_{Ic} data from Table 3-12 in Fig. 3-28. Through-thickness geometry was chosen in the example shown, which is applicable to reworked hot tears in valve bodies, for example.

Prediction of Fracture Toughness from Microstructure

The prediction of fracture toughness from microstructure provides an extremely potent design and failure-rate controlling tool for most sectors of industry. In their review of 200 papers, Garrison and Moody (Ref 28) conclude that in the absence of nucleation strains, the initiation toughness will, for a fixed fine-scale microstructure, be a function of the interparticle spacing. The selection of the appropriate particle spacing can only be determined on the basis of experience and detailed experiments.

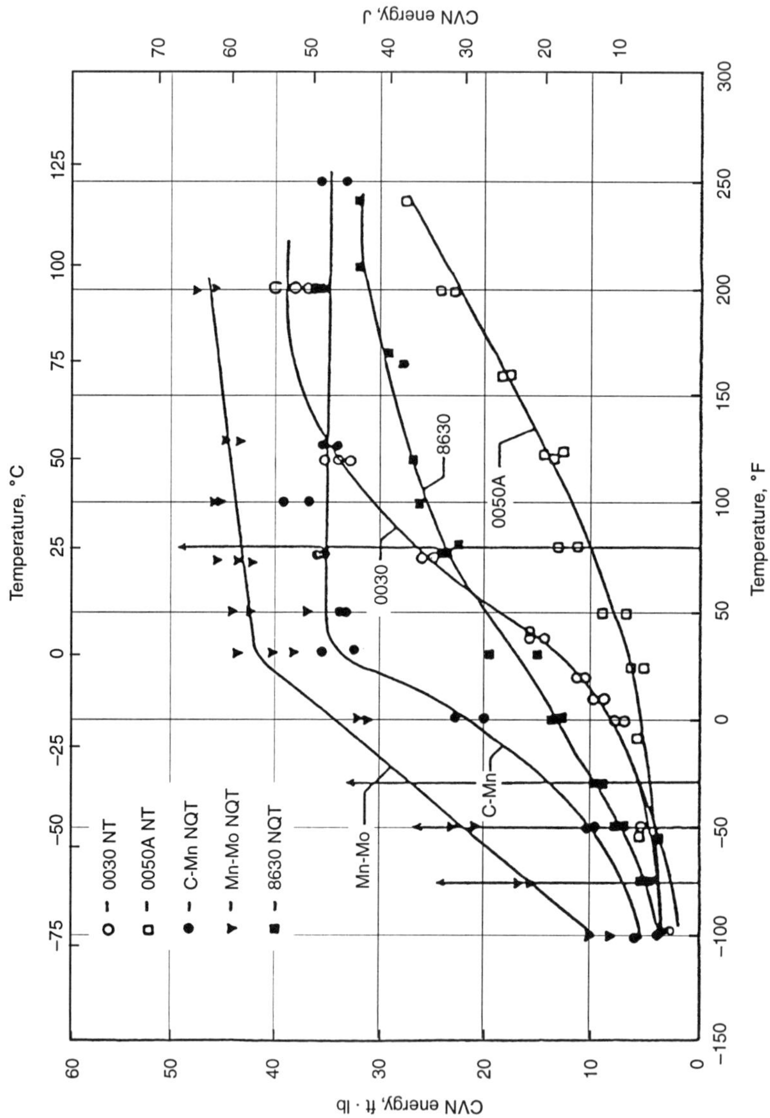

Fig. 3-27 Charpy data for determining values in Table 3-11

The conclusions of Garrison and Moody (Ref 28) are substantiated by a suggestion made by Firrao (Ref 29) insofar as the measurement of "dominant" microstructural features governs the macroscopic toughness parameter.

For example, the following describes specific microstructures in two different commercial alloys and demonstrates the potency of the "dominant" microstructural feature, in predicting toughness in sharp-cracked and blunt-notched tests. Details of the test procedures have already been reported (Ref 30, 31). Tests were performed on commercial grades of various compositions of pearlitic iron and commercial grade rail steel of near-eutectoid composition The compositions of pearlitic ductile irons are shown in Table 3-13 (Ref 30).

In an early investigation into the effect of microstructural variables on the elastic fracture toughness of pearlitic ductile cast irons, Holsworth and Jolley (Ref 30) reported

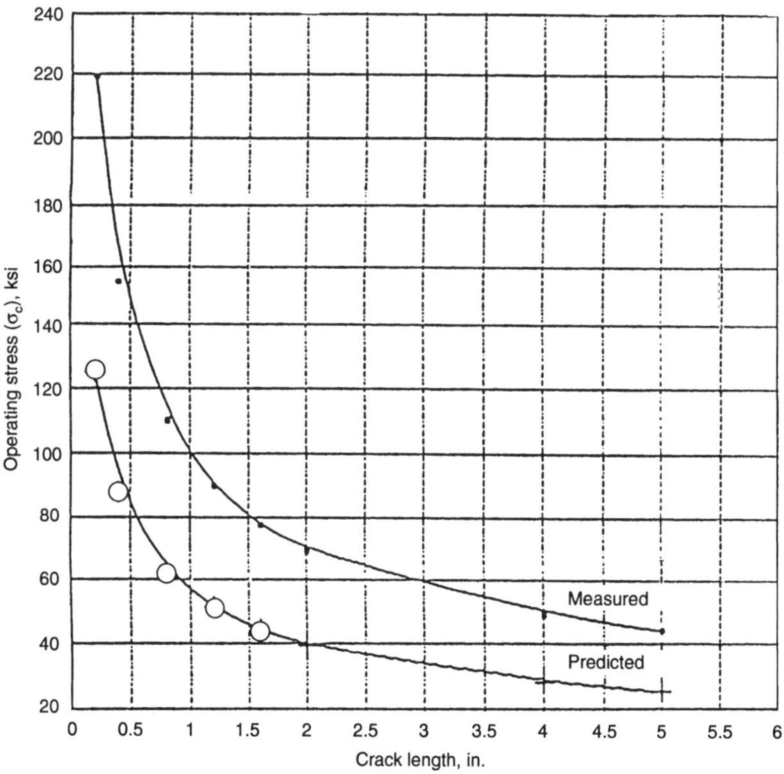

Fig. 3-28 An operating stress map for 8630 NQT cast steel with through-thickness cracks constructed from the data in Table 3-12. K_{Ic} = 123 (measured); K_{Ic} = 70 (predicted)

values of K_{Ic} for several such irons. Their K_{Ic} values together with their microstructural information on the spheroidal graphite nodules/mm^2 for the five irons investigated are shown in Table 3-14. The pearlitic irons were produced by air cooling after holding for six hours at 1172 K.

The toughness results were obtained on fatigue pre-cracked specimens 10 mm thick and 20 mm wide, loaded in three-point bending. The pearlitic ductile irons gave valid K_{Ic} data at 298 K as defined in ASTM E 399. The method of measuring the area fraction of graphite nodules was not described, but presumably a standard metallographic technique was employed.

The validity of their data presented in Fig. 3-29 is in the linear nature of the relationship between the number of nodules/mm^2 and the elastic fracture toughness at 298 K. The predominant feature in the microstructure which has the greatest influence on the toughness is the area fraction of graphite nodules, rather than the interlamella spacing of the pearlite.

The dynamic, lower shelf, Charpy impact toughness of rail steels of various manganese sulfur inclusion levels and various interlamella pearlite spacing was investigated at 233 K (Ref 31). The lower shelf level was found to be independent of nonmetallic inclusion content but dependent upon the interlamella spacing of the pearlite in steels of nominal eutectoid composition.

The available data show the dependence of microhardness on pearlite spacing in Fig. 3-30, which reflects the lower-shelf toughness dependence in rail steels (Ref 31); that is, the absolute level of the lower shelf increases with decreasing spacing of the pearlite.

The dominant microstructural feature which governs the plane strain elastic fracture toughness of pearlitic ductile iron is, without doubt, the area fraction of graphite nodules. This is true irrespective of the precise mechanisms taking place in some

Table 3-11 Cast steels from Fig. 3-27

Steel	σ_y	K_{Ic}	Temperature, °C	CVN	Predicted K_{Ic}
C-Mn	85	126	25	35	95
Mn-Mo	102	163	25	44	117
8630	166	123	25	24	70
C-Mn	89	120	−45	10	40
Mn-Mo	110	151	−45	22	56

Steel	Temperature, °C	K_{Ic} predicted / K_{Ic} measured (a), %	Criterion
C-Mn	25	75	$\left(\frac{K_{Ic}}{\sigma_y}\right)^2 = 4\left(\frac{CVN}{\sigma_y} - 0.1\right)$
Mn-Mo	25	72	As above
8630	25	60	As above
C-Mn	−45	33	$K_{Ic} = 12\sqrt{CVN}$
Mn-Mo	−45	37	As above

CVN = ft · lb; σ_y = ksi; K_{Ic} = ksi$\sqrt{in.}$ (a) Predicted K_{Ic} from Charpy data versus measured K_{Ic}

"process zone" ahead of the advancing crack tip and irrespective of some notion of "macroscopic" toughness (Ref 32). By the same token, the dominant microstructural feature which governs the Charpy blunt-notch lower-shelf toughness of rail steels is the interlamella spacing of the pearlite.

Both of these microstructures contained large volumes of different phases. The pearlitic ductile iron contained pearlite and nodules of graphite. The near-eutectoid composition rail steels contained massive manganese-sulfur stringers and pearlite. For these two materials, which failed in a brittle manner, the "dominant" microstructural features are clearly defined graphite in the iron and pearlite in the rail steel.

From a predictive capability standpoint, the toughness/dominant microstructural feature relationships have distinct advantages in the areas of design and failure control. In pearlitic ductile iron, consider the exploitation of the relationship in Fig. 3-29 in the defect tolerance analysis of a valve body. Let the number of graphite nodules/mm^2 be 20, the residual stress in the valve body 120 MPa, and the proof stress 550 MPa. From Fig. 3-29, the toughness, K_{Ic} is 30.5 MPa$\sqrt{m}$.

Determine the tolerable defect size with the aspect ratio of the crack,

$$\frac{a}{2c} = \frac{1}{10} \qquad \phi = 1.05$$

$$K_{Ic} = \frac{1.12\sigma_c\sqrt{\pi a}}{\phi}$$

Table 3-12 As-cast 8630 data for operating stress map in Fig. 3-28

Permissible crack, in.	Operating stress, ksi	Factor in *K* equation
For measured K_{Ic} = 123		4815
0.2	219	
0.4	155	
0.8	110	
1.2	90	
1.6	78	
2.0	69	
4.0	49	
5.0	44	
For predicted K_{Ic} = 70		1560
0.2	125	
0.4	88	
0.8	62	
1.2	51	
1.6	44	
2.0	39	
4.0	28	
5.0	25	

K_{Ic} = ksi$\sqrt{in.}$; σ_y = 166 ksi; through-thickness crack

Therefore,

$$a = \frac{1.05 \times 1.05 \times 30.5 \times 30.5}{120 \times 120 \times 1.12 \times 1.12 \times \pi} = \frac{1025.6}{56{,}755.07}$$

$$a = 18.07 \text{ mm}$$

The defect size must be more than 18.07 mm deep before fracture would occur. However, the customer has specified that the safe defect size must be greater than 20.0 mm. Thus, from Fig. 3-29, the microstructure giving rise to an increased elastic fracture toughness must be employed. A microstructure containing ~70 nodules/mm^2 would give a toughness level of 33 MPa$\sqrt{m}$.

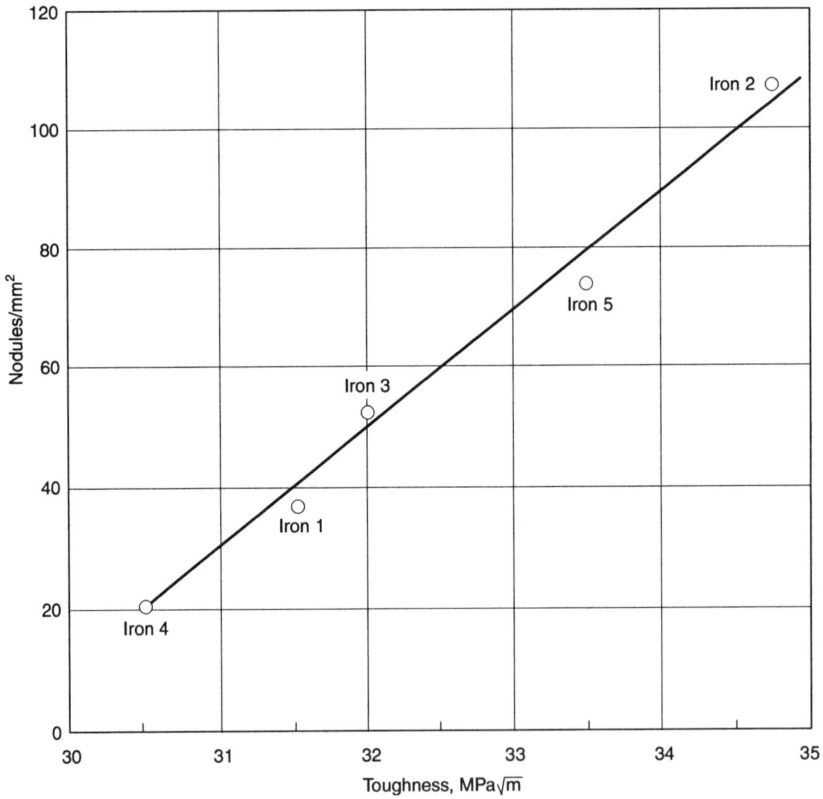

Fig. 3-29 Effect of area fraction of graphite nodules on toughness of ductile iron air cooled at 1650 °F

As before,

$$a = \frac{1.05 \times 1.05 \times 33.0 \times 33.0}{120 \times 120 \times 1.12 \times 1.12 \times \pi} = \frac{1200.6}{56,755.07}$$

$$a = 21.15 \text{ mm}$$

Thus the safe defect size is now 21.15 mm, in keeping with the customer specifications of 20.00 mm minimum.

In rail steels, consider what would be specified for the safest rail operating in, for example, a Canadian winter, where temperatures can fall to 233K (–40 °C). Clearly, the rail with the smallest interlamella spacing of pearlite, say 0.10 μm, from Fig. 3-30, should be selected.

Failure Assessment Diagrams

The British C.E.G.B. (Central Electricity Generating Board) R6 method uses a Failure Assessment Diagram (FAD). The coordinates are K_R, which considers elastic failure, and S_R, which considers plastic collapse. Failure will not occur, provided the assessment point for the flawed geometry lies within the area bounded by the axes and the assessment curve.

The physical basis for the FAD is the Dugdale strip-yield model (Ref 33). In this model, a through-thickness crack in a panel subjected to an applied stress σ, will open to an amount δ, the crack opening displacement, by:

$$\delta = \frac{8}{\pi}\frac{\sigma_y}{E} a \ln\left[\sec\left(\frac{\pi\sigma}{2\sigma_y}\right)\right]$$

Table 3-13 Composition of peralitic ductile irons

Irons	Composition, %							Inoculant
	C	Si	Mn	S	P	Ni	Mg	
1	3.64	2.31	3.35	0.016	0.024	0.88	0.072	¼%Fe/Si
2	3.62	2.41	0.34	0.016	0.024	0.83	0.64	½%Fe/Si+B
3	3.63	2.42	0.32	0.016	0.024	0.083	0.71	½%Fe/Si
4	2.58	2.16	0.45	0.020	0.025	0.081	0.054	½%Fe/Si
5	3.99	2.34	0.39	0.014	0.025	0.078	0.078	½%Fe/Si

Table 3-14 Toughness/area fracture spheroidal graphite for pearlitic ductile irons

K_{Ic}, MPa$\sqrt{m}$	Graphite, nodules/mm^2	Iron No. (from Table 13)
30.5	20.6	4
31.5	37.2	1
32	52.4	2
33.5	73.8	5
34.75	106.8	2

At applied stresses greater than 80% of the yield stress, σ_y, plastic deformation will spread from the crack tips to the edges of the panel; i.e., net section yielding occurs.

Harrison et al. (Ref 34) postulated that a modified version of the Dugdale expression could be used to interpolate between linear elastic fracture ($\sigma_f/\sigma_y << 1$) at one extreme and large scale plastic yielding or collapse at the other extreme ($\sigma_f/\sigma_y \rightarrow 1$). If the yield stress were replaced by the plastic collapse stress σ_c in order to accommodate other flawed configurations, the resulting failure assessment curve is:

$$\left(\frac{K_I}{K_c}\right)_f = \frac{\sigma_f}{\sigma_c}\left(\frac{8}{\pi^2}\right)\ln\left[\sec\frac{\pi\,\sigma_f}{2\sigma_c}\right]^{-1/2}$$

However, σ_f/σ_c is now replaced by S_R, defined as σ/σ_c, where σ is the remote stress, and K_I/K_c is replaced by K_R This approach has been incorporated into the British Standards Institute's Public Document 6493 (Ref 35).

The appealing aspect of the FAD is its simplicity. Only the LEFM stress intensity factor and the plastic limit or collapse stresses are required in the analysis. However, the question of what value should be used for the flow stress when determining S_R becomes increasingly difficult to answer for materials which exhibit appreciable strain hardening.

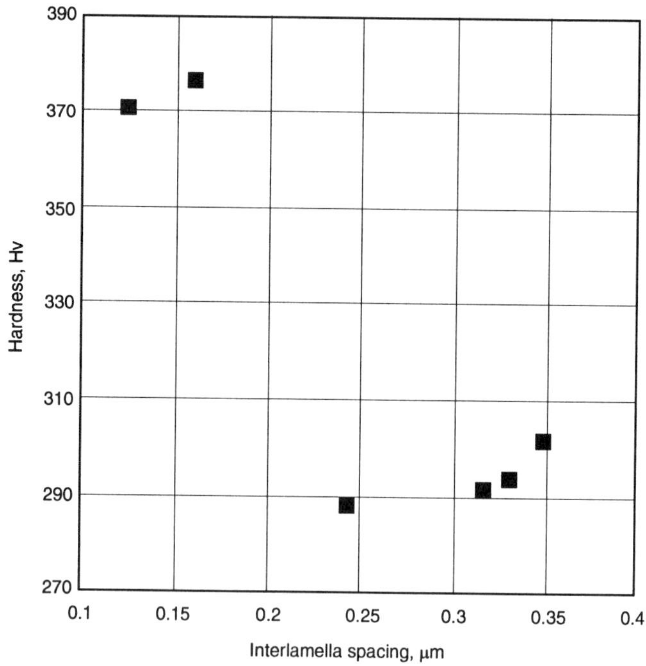

Fig. 3-30 Change in hardness with interlamella spacing of pearlite in rail steels

For the purposes of evaluation, consider an example, similar to that originally set by Spiekhout (Ref 36), in the C.E.B.G. R6 format.

Example. A cast steel plate has a surface flaw of depth $a = 7$ mm and length $2c = 35$ mm. The plate thickness, $B = 108$ mm, $\sigma_y = 430$ MPa, UTS = 589 MPa, $K_{Ic} = 32\ \text{MPa}\sqrt{\text{m}}$. The operating stress is 170 MPa. Define a safety margin against failure. From R6 - Rev. 2 (Ref 35),

$$\sigma_{Mc} = \frac{\sigma_y + \text{UTS}}{2} = 510\ \text{MPa}$$

Also,

$$S_R = \frac{\sigma}{\sigma_{Mc}} = \frac{170}{510} = 0.33$$

Further,

$$K_R = \frac{K_I^P}{K_{Ic}}$$

where K_R is derived from PD6493 (Ref 35), Fig. 3-31, as follows:

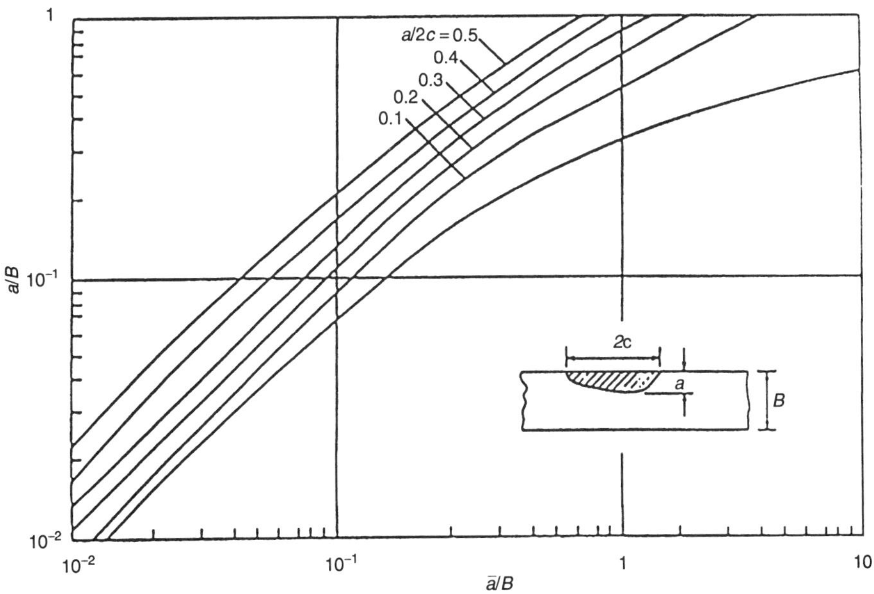

Fig. 3-31 Relationship between actual flaw dimensions and the parameter $\bar{a}$ for surface flaws

$$\frac{\text{crack depth}}{\text{plate thickness}}\ \frac{a}{B}=\frac{7}{108}=0.065$$

and

$$\frac{\text{crack depth}}{\text{crack length}}\ \frac{a}{2c}=\frac{7}{35}=0.2=\text{aspect ratio}$$

$$\frac{\bar{a}}{B}=0.06$$

$$a=6.5\ \text{mm}$$

Also,

$$K_I^P=\beta\sigma_c\sqrt{\pi a}$$

$$K_I^P=170\sqrt{\pi\times 6.5}$$

$$K_I^P=24.3\ \text{MPa}\sqrt{\text{m}}$$

$$K_R=\frac{24.3}{32}=0.76$$

From the FAD, Fig. 3-32, the point falls within the curve, so failure will not occur. The safety factor against failure

$$\frac{OX}{OY}=1.3$$

In Rev. 3 of R6 (Ref 37), S_R is changed to L_R and the limits imposed on L_R have been reduced. In Rev. 2, S_R was never > 1.

Consider this same problem using a stress map, shown in Fig. 3-33, from the following data.

Data. First reconsider Eq 3-1:

$$K_{Ic}=\frac{1.12\sigma_c\sqrt{\pi a}}{\Phi}$$

with aspect ratio:

$$\frac{7}{35}=0.2$$

From Ref 3-33, $\Phi = 1.016$ because $\frac{a}{2c}=0.2$. Then,

$$32^2=\frac{1.12^2\times 170^2\pi\times a}{1.016^2}$$

$$a_c = \frac{32^2 \times 1.046^2}{1.12^2 \times 170^2 \times \pi} = \frac{1057.03}{113{,}904.29} = 0.00928 \text{ m}$$

$$a_c = 9.3 \text{ mm}$$

$$a \cdot \sigma_c^2 = \frac{1057.03}{3.941} = 268$$

where a and σ_c are varied to produce the stress map.

Increasing the value of K_{Ic} to 101.25 MPa$\sqrt{m}$, for the FAD

$$K_R = \frac{24.3}{101.25} = 0.24$$

Thus, with $S_R = 0.33$ and $K_R = 0.24$, the point on the FAD, Fig. 3-32, is such that failure will not occur and the safety factor against failure is

$$\frac{OB}{OA} = 2.64$$

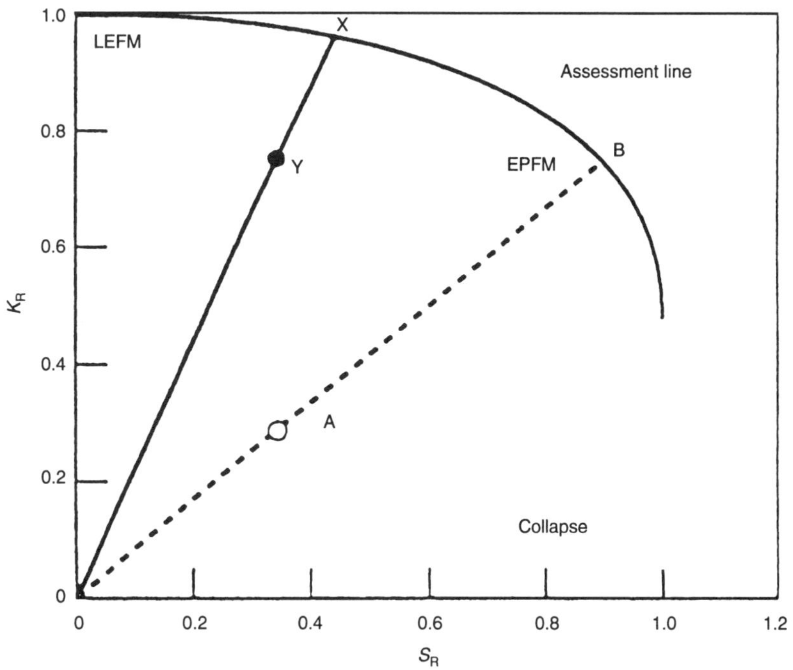

Fig. 3-32 Failure-assessment diagram

With a stress map,

$$a_c = \frac{101.25^2 \times 1.016^2}{1.12^2 \times 170^2 \times \pi} = \frac{10{,}582 \times 24}{113{,}904.29} = 0.09290\,\text{m}$$

$$a_c = 92.9\ \text{mm}$$

$$a \cdot \sigma_c^2 = \frac{10{,}582.24}{3.941} = 2685.15$$

The resulting curve is also shown on Fig. 3-33.

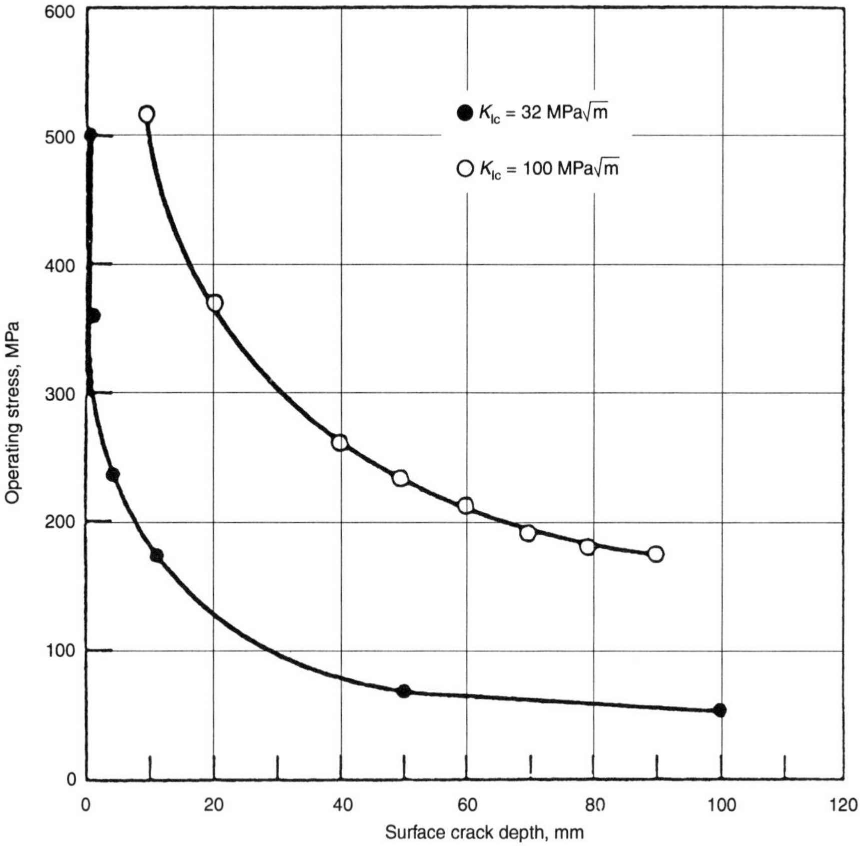

Fig. 3-33 Operating stress map. For a given K_{IC}, points below the curve represent values of operating stress and crack depth that will not cause failure.

The stress map shows the safety envelope in which to operate in terms of actual operating stress and actual crack size. The FAD gives a proximity of fracture number as a "safety margin against failure."

Further reference is made to the British Standards Institute Public Document 6493 in Chapter 5, where "engineering critical assessment" is reviewed in the context of solving equipment failures in weldments. PD 6493 may be issued as a British Standard after some revisions in the near future. The fundamental limitations that the Dugdale strip-yield model presents, however, must be reconciled.

REFERENCES

1. G.R. Irwin, Fracture, *Encyclopedia of Physics,* Vol 6, Springer-Verlag, Berlin, 1958
2. P.C. Paris and G.C. Sih, STP 381, ASTM, 1965, p 30
3. J.M. Svoboda, Steel Founders' Society of America Research Report 94A, Oct 1982
4. L.R. Hall and W.L. Engstrom, "Fracture and Fatigue-Crack-Growth Behavior of Surface Flaws and Flaws Originating at Fastener Holes," AFFDL-TR-7447, 1974
5. D. Broek, *Elementary Engineering Fracture Mechanics,* Noordhoff, 1978
6. R.E. Peterson, in *Metal Fatigue,* Sines and Waisman, Ed., McGraw-Hill, 1959
7. R.E. Peterson, "Analytical Approach to Stress Concentration Effect in Fatigue of Aircraft Structures," WADS Symposium, Wright Air Development Center, Dayton, OH, Aug 1959
8. J.F. Knott, *Fundamentals of Fracture Mechanics,* Butterworths, 1973
9. C.E. Feddersen, "Evaluation and Prediction of Residual Strength of Center Cracked Tension Panels," STP 486 ASTM, 1971, p 50–78
10. R.G. Forman, V.E. Kearney and R.M. Engle, Numerical Analysis of Crack Propagation in Cyclic-Loaded Structures, (Trans. ASME), *J. Basic Eng.,* Vol 89 (No. 3), 1967, p 459
11. R.G. Forman et al., Development of the NASA/FLAGRO Computer Program, STP 945, ASTM, 1988, p 781
12. J.M. Barsom, Fatigue-Crack Propagation in Steels of Various Yield Strengths, *J. Eng. Ind. Trans.,* Series B, No. 4, 1971, p 1190
13. P.F. Timmins, *Solutions to Hydrogen Attack in Steels,* ASM International, 1997
14. P.F. Timmins, Fracture Mechanics for Engineers, *Oil and Gas Journal,* 26 April 1982
15. T. Yokobori, et al., "Evaluation of K_{ISCC} Testing Procedure by Round Robin Tests on Steels," STP 945, ASTM, 1988, p 843
16. J. Barson and S. Rolfe; Fracture Mechanics in Failure Analysis, STP 945, ASTM, 1988, p 443
17. L.A. Simpson and M.P. Puls, *Metall. Trans. A,* Vol 10, 1979, p 1093
18. M.P. Puls, L.A. Simpson, and R. Dutton, in *Fracture Problems and Solutions in the Energy Industry,* L.A. Simpson Ed., Pergamon Press, Oxford, 1982, p 13
19. L.A. Simpson and K. Nuttall, in "Zirconium in the Nuclear Industry," STP 633, ASTM, 1977, p 608

20. L.A. Simpson, in *Mechanical Behaviour of Materials III*, K.J. Miller and R.F. Smith, Ed., Pergamon Press, Oxford, 1980, p 445
21. W.J. Jackson, "Fracture Toughness in Relation to Steel Castings Design and Application," Steel Founders' Society of America, Rocky River, OH, Aug 1978
22. J.F. Kiefner, W.A. Maxey, R.J. Eiber, and A.R. Duffy, Progress in Flaw Growth and Fracture Toughness Testing, STP 536, ASTM, 1973, p 461
23. E.S. Folias, A Finite Line Crack in a Pressurized Cylindrical Shell, *Eng. Fract. Mech.*, Vol 2, 1970, p 151
24. C.K. Chow and L. Simpson "Analysis of the Unstable Fracture of a Reactor Pressure Tube," STP 918, ASTM, 1986, p 78
25. D. Broek, *ASM Handbook*, Vol 19, *Fatigue and Fracture*, ASM International 1996, p 430
26. C.E. Feddersen, STP 486, ASTM, 1971, p 50
27. J.M. Svoboda, Steel Founders' Society of America Research Report, No. 94A, Oct 1982
28. W.M. Garrison and N.R. Moody, *Phys. Chem. Solids*, Vol 48, 1987, p 1035–1074
29. D. Firrao, I.C.F. 7, Houston, personal communication, March 1989
30. S.R. Holdsworth and G. Jolley, *Source Book on Ductile Iron*, ASM Engineering Bookshelf Series, American Society for Metals, 1977, p 305
31. P.F. Timmins, "Executive Summary of the Clean Steel Program," CN Rail Technical Research Centre, St. Laurent, Montreal, Canada, June 1987
32. P. Bowen, S.G. Druce, and J.F. Knott, *Acta Metall.*, Vol 35, 1987, p 1735
33. O.S. Dugdale, Yielding of Steel Sheets Containing Slits, *J. Mech. Phys. Sol.*, Vol 8, 1960, p 100–108
34. R.P. Harrison, K. Loosemore, I. Milne, and A.R. Dowling, "Assessment of the Integrity of Structures Containing Defects," Central Electricity Generating Board, R/H/R6-Rev. 2
35. "Guidance on Methods for Assessing the Acceptability of Flaws in Fusion Welded Structures," British Standards Institute Public Document 6493, 1991
36. Spiekhout, Fitness-for-Purpose Assessment of Weld Flaws: Application of Various Fracture Mechanics Codes, *Welding J.*, Sept 1988, p 55
37. I. Milne, R.A. Ainsworth, A.R. Dowling, and A.T. Stewart, "Assessment of the Integrity of Structures Containing Defects," Central Electricity Generating Board, R/H/R6-Rev. 3

CHAPTER 4

Solutions to Equipment Failure Based on Predicting Corrosion Behavior

Corrosion is understood mechanistically, and logical mitigation actions and design decisions can be formulated using the techniques of hazard identification outlined in Chapter 1. The intent of this chapter is to provide scenarios in design that can act as guidelines to facilitate these problem-solving techniques. Further, there is an additional requirement in sectors of industry that the lifetime of various components be predicted and that the technical (and economic) benefits of life extension actions be quantitatively defined. Such requirements for life extension are seen in aviation (outlined in Chapter 2), nuclear power, gas processing, and refinery sectors of industry.

To meet these particular design criteria, it is necessary to derive accurate and verifiable life prediction methods. One such approach described in this chapter attempts to unify life prediction of stress corrosion cracking (SCC), strain-induced cracking (SIC), and corrosion fatigue (CF).

Designing against Corrosion

The basic factors that most influence design for corrosion resistance are summarized in Table 4-1 (Ref 1). Each factor plays a unique, yet not always unrelated, role with other factors, and the following illustrates some aspects of design detail that can accelerate corrosion and may be considered in hazard identification (Ref 2).

Exposure to winds and airborne particulates can lead to deterioration of structures. Designs that leave structures exposed to the elements should be carefully reviewed, because atmospheric corrosion is significantly affected by temperature, relative humidity, rainfall, and pollutants. Also important are the season and location of onsite fabrication, assembly, and painting. Codes of practice must be adapted to the location and the season. This aspect of corrosion is discussed in Chapter 2, where, in the aviation sector, ground locations are carefully documented to include this facet of corrosion in the overall performance of a given aircraft.

Table 4-1 Corrosion factors that can influence design considerations

Environment	Natural
	Chemical
	Storage/ transit
Stress	Residual stress from fabrication
	Operating stress—static, variable, alternating
Shape	Joints, flanges
	Crevices, deposits
	Liquid containment and entrapment
Compatibility	Metals with metals
	Metals with other materials
	Quality control
Movement	Flowing fluids
	Parts moving in fluids
	Two-phase fluids
Temperature	Oxidation, scales
	Heat-transfer effects
	Molten deposits
	Condensation and dewpoint
Control	Surface cleaning and preparation
	Coatings
	Cathodic protection
	Inhibitors
	Inspection
	Planned maintenance

The shape or geometrical form of equipment is basic to design. The objective is to minimize or avoid situations that worsen corrosion. These situations can range from stagnation (e.g., retained fluids and/or solids, or contaminated water used for hydrotesting) to sustained fluid flow (e.g., erosion/cavitation in components moving in or contacted by fluids, as well as splashing or droplet impingement).

Common examples of stagnation include nondraining structures, dead-ends, badly located components, and poor assembly or maintenance practices (Fig. 4-1). General problems include localized corrosion associated with differential aeration (oxygen concentration cells), crevice corrosion, and deposit corrosion.

Fluid movement need not be excessive to damage a material. Much depends on the nature of the fluid and the hardness of the material. A geometric shape may create a sustained delivery of fluid or can locally disturb a laminar stream and lead to turbulence. Replaceable baffle plates or deflectors are beneficial where circumstances permit their use; they eliminate the problem of impingement damage to the structurally significant component. Careful fabrication and inspection should eliminate or reduce poor profiles (e.g., welds, rivets, bolts), rubbing surfaces (e.g., wear, fretting), and galvanic effects due to the assembly of incompatible components. Figure 4-2 shows typical situations in which geometric details influence flow.

Compatibility in plant environments is often necessary when using different materials in close proximity. Sometimes components that were designed in isolation can end up in direct contact in the plant (Fig. 4-3). In such instances, the ideal of a total design

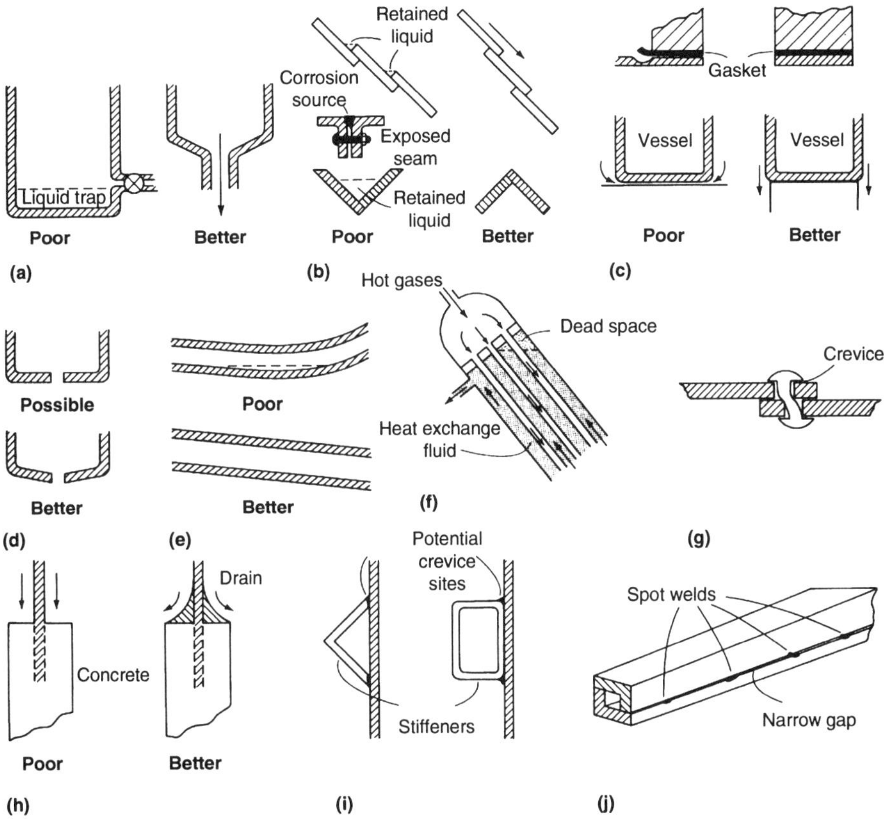

Fig. 4-1 Examples of how design and assembly can affect localized corrosion by creating crevices and traps where corrosive liquids can accumulate. (a) Storage containers or vessels should allow complete drainage; otherwise, corrosive species can concentrate in vessel bottom, and debris may accumulate if the vessel is open to the atmosphere. (b) Structural members should be designed to avoid retention of liquids; L-shaped sections should be used with open side down, and exposed seams should be avoided. (c) Incorrect trimming or poor design of seals and gaskets can create crevice sites. (d) Drain valves should be designed with sloping bottoms to avoid pitting of the base of the valve. (e) Nonhorizontal lubing can leave pools of liquid at shutdown. (f) to (j) Examples of poor assembly that can lead to premature corrosion problems. (f) Nonvertical heat exchanger assembly permits dead space that is prone to overheating if very hot gases are involved. (g) Nonaligned assembly distorts the fastener, creating a crevice that can result in a loose fit and contribute to vibration, fretting and wear. (h) Structural supports should allow good drainage, use of a slope at the bottom of the member allows liquid to run off, rather than impinge directly on the concrete support. (i) Continuous weld for horizontal stiffeners prevents traps and crevices from forming. (j) Square sections formed from two L-shape members need to be continuously welded to seal out the external environment.

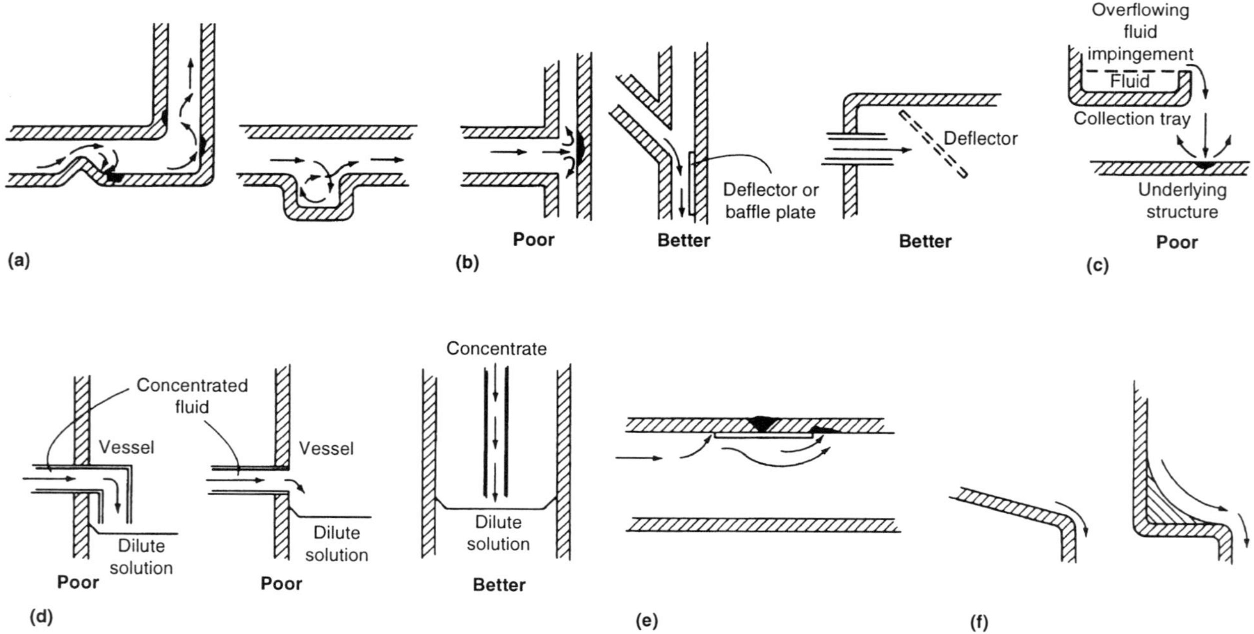

Fig. 4-2 Effect of design features on flow. (a) Disturbances to flow can create turbulence and cause impingement damage. (b) Direct impingement should be avoided; deflectors or baffle plates can be beneficial. (c) Impingement from fluid overflowing from a collection tray can be avoided by relocating the structure, increasing the depth of the tray, or using a deflector. (d) Splashing of concentrated fluid on container walls should be avoided. (e) Weld backing plates or rings can create local turbulence and crevices. (f) Slope or modified profiles should be provided to permit flow and minimize fluid retention.

concept becomes clear but usually in hindsight. Direct contact of dissimilar metals introduces the possibility of galvanic corrosion, and small anodic (corroding) areas should be avoided wherever this contact is taking place. In aircraft design Chapter 2, this facet is very carefully documented in terms of corrosion protection to extend life.

Weld decay and sensitization can generally be avoided by material selection or suitable fabrication techniques. Less obvious instances of localized attack occur because of end-grain attack and stray-current effects, which can render designs ineffective.

End-grain attack, or preferential attack of grains exposed by cross-cutting through a metal plate or rod Fig. 4-3, occurs in many corrosive fluids. Stray-current effects are common on underground steel pipelines that are located close to electrical supply lines, or where stray currents can cause active corrosion at preferred sites. For example, stray-current attack caused a titanium flange spacer to become anodic during service due to inappropriate grounding of weld equipment on an adjacent part of the plant structure.

The most common design details relating to galvanic corrosion include jointed assemblies (Fig. 4-1 and 4-3). Where dissimilar metals will be used, some consideration should be given to incompatible materials known to have similar potential. Care should be exercised because the galvanic series is limited and refers to specific environments, usually seawater. Where noncompatible materials are to be joined, it is advisable to use a more noble metal in a joint (Fig. 4-3). Effective insulation can be useful if it does not lead to crevice corrosion.

The relative surface areas of anodic and cathodic surfaces should not be underestimated, because corrosion at a small anodic zone can be several hundred times greater than that for the same bimetallic components of similar area.

Galvanic corrosion can occur when ion transfer results in the deposition of active and noncompatible deposits on a metal surface. For example, an aluminum stirrer plate used in water was extensively pitted because the water bath was heated by a copper heater coil (Fig. 4-3). The pits resulted from deposition of copper ions from the heater element. Anodic components can on occasion be over-designed (made thicker) to allow for the anticipated corrosion loss. In other instances, easy replacement is a cost-effective option (Table 4-2).

Where metallic coatings are used, there may be a risk of galvanic corrosion, especially along the cut edges. Rounded profiles and effective sealants or coatings are beneficial. Transition joints can be introduced when different metals are to be in close proximity. These and other situations are illustrated in Fig. 4-3. Another possibility is coating of the cathodic material for corrosion control. Ineffective painting of an anode in an assembly can significantly reduce the desired service lifetime, because local defects (anodes) effectively multiply the risk of localized corrosion.

Insulation represents another area for potential corrosion attack, although most problems arise because of poor installation. Insulation types and properties vary considerably, and expert advice from suppliers is recommended. The most common corrosion problems include crevice corrosion (where insulation and/or adhesives are tightly held against a metal surface, for example when straps or ties are too tight) and pitting

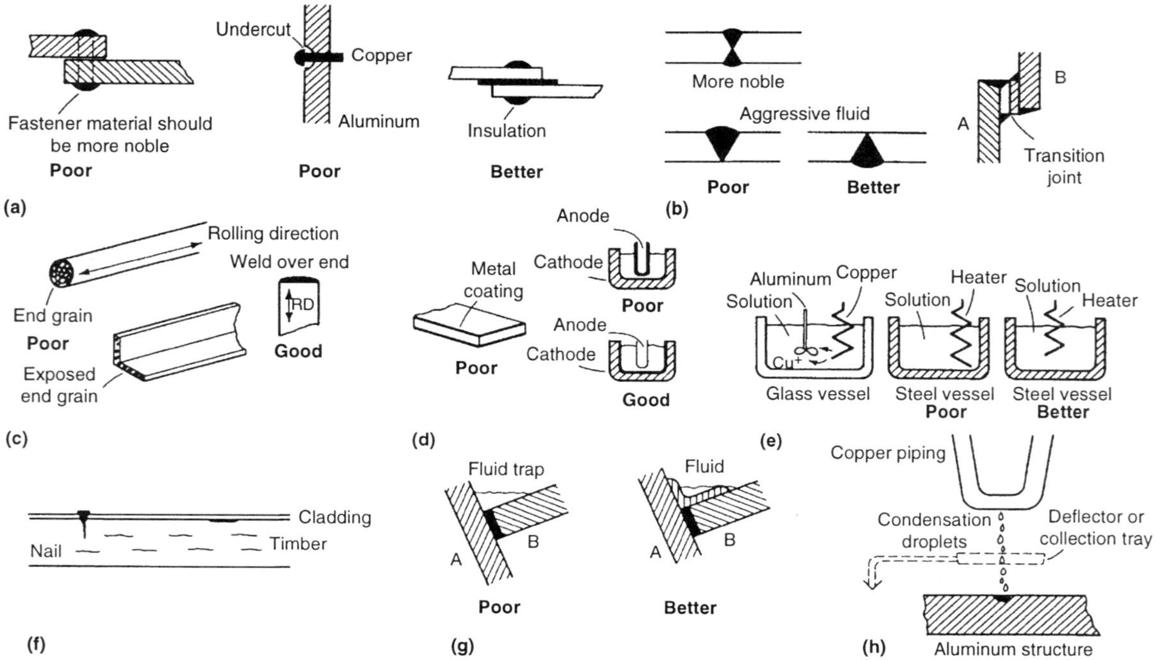

Fig. 4-3 Design details that can affect galvanic corrosion. (a) Fasteners should be more noble than the components being fastened; undercuts should be avoided, and insulating washers should be used. (b) Weld filler metals should be more noble than base metals. Transition joints can be used when a glavanic couple is anticipated at the design stage, and weld beads should be properly oriented to minimize galvanic effects. (c) Local damage can result from cuts across heavily worked areas. End grains should not be left exposed. (d) Glavanic corrosion is possible if a coated component is cut. When necessary, the cathodic component of a couple should be coated. (e) Ion transfer through a fluid can result in galvanic attack of less noble metals. In the example shown at left, copper ions from the copper heater coil could deposit on the aluminum stirrer. A nonmetallic stirrer would be better. At right, the distance from a metal container to a heater coil should be increased to minimize ion transfer. (f) Wood treated with copper preservatives can be corrosive to certain nails, especially those with nobility different from that of copper. Aluminum cladding can also be at risk. (g) Contact of two metals through a fluid trap can be avoided by using a collection tray or a deflector.

corrosion (where moisture condenses on the metal, usually because the insulation barrier was too thin or improperly installed). Moisture-absorbing tendencies vary from one insulation to another (Ref 3).

Wet-dry cycling has been known to lead to concentration effects (e.g., chloride ions from calcium-silicate insulation). Instances have been reported of chloride SCC in certain stainless steel pipes and vessels, or pitting of these and other materials, such as aluminum, when contacted by insulation. The early instances of SCC failure were mainly attributable to high chloride levels (500 to 1500 ppm) associated with asbestos-type materials. Chloride has been significantly reduced in recent years to a level that is not expected to cause SCC. Standards and tests to evaluate insulation materials are now available (Ref 4). A parallel German standard calls for zero nitrite content and <0.2% ammonia levels in elastomeric insulation that is used for copper and copper-alloy piping to reduce the risk of SCC (Ref 5).

Figure 4-4 illustrates situations in which design and installation procedures could have been improved. Additional problems occur when insulation is torn or joints are misaligned or incorrectly sealed with duct tape or similar bandaging, none of which is recommended by insulation suppliers.

Table 4-2 Design aspects for reliability

Approach	Factors to be taken into account
"Avoid failure" approach	
Control product duty	Fail safe. Visible instrumentation.
Keep product simple	Fewer functions. Fewer joints.
Predict reliability	Maintain in fixed limits. Overdesign (thicker), etc.
Study variable interfaces	Contact faces, bearings. Lubrication (wear, friction). Seals. Consult experts and test work.
Instill care for familiar items or designs	Complacency. Negligence.
Study Innovation	Consult experts and test work. Records of operation. Records of failure mode/frequency.
Redundant items	Stand-by units in good repair. Storage. Handling. Vibration.
Recognize human limits	Misreading of instructions. Sequence of controls (hypnosis?) Adequate housekeeping. User attitudes.
Test product design	Proving the design. Variables. Safety limits. Pilot plant and real-life testing.
Control malfunctions	Human. Automatic. Mixture. Important for software/hardware.
"Keep it working" approach	
Maintainability	Access.
Short-term items	Replace routinely before failure. Regular inspection. Monitoring.
Spares	Identity correctly coded. Location known. Easy access replacement. Avoid identical mating systems.
Built-in adjustments	Corrects for progressive deterioration. Manual/automatic.
"Let it fail then repair" approach	
Defects	Subtle or catastrophic.
Spares	Need quantified reliability (difficult for corrosion).
Minimize off period	Keep repair time low. Speedy fault diagnosis. Speedy removal of failed part(s). Speedy replacement. Speedy check of assembly. Experts, inspectors.

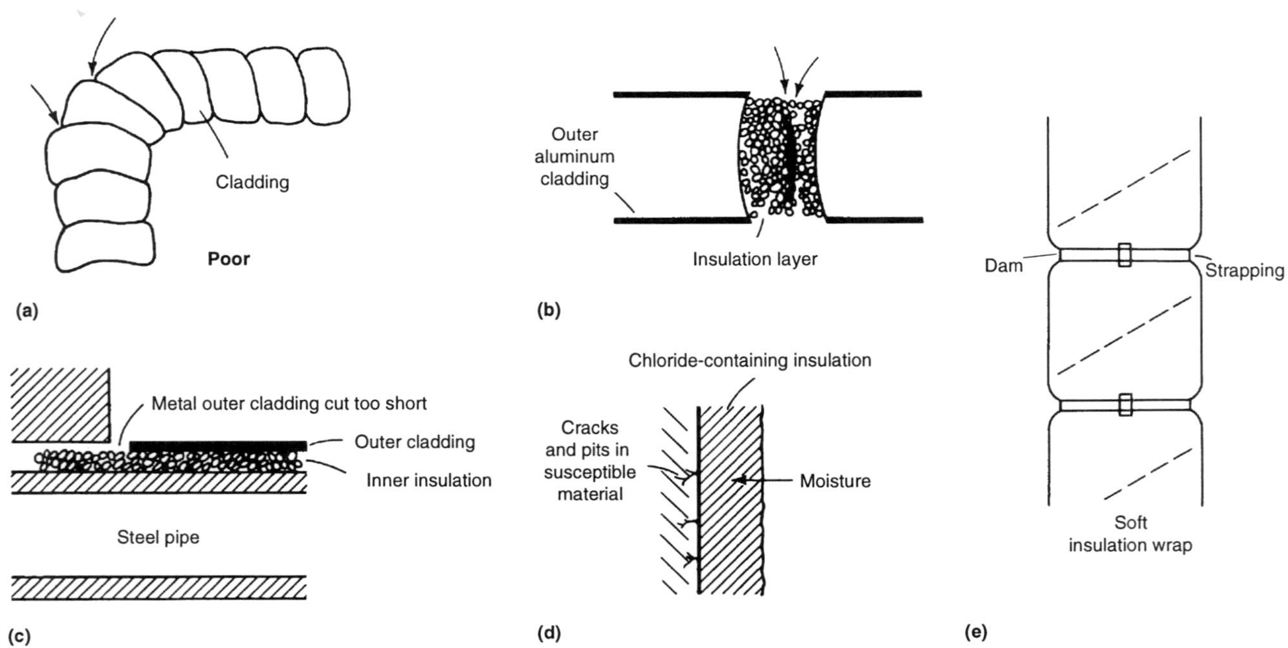

Fig. 4-4 Corrosion problems associated with improper use of insulation and lagging. (a) Incorrect overlap in lobster-back cladding does not allow fluid runoff. (b) Poor installation left a gap in the insulation that allows easy access to the elements. (c) Outer metal cladding was cut too short, leaving a gap with the inner insulation exposed. (d) Insufficient insulation can allow water to enter; chloride in some insulation can result in pitting or stress-corrosion cracking of susceptible materials. (e) Overtightened strapping can damage the insulation layer and cause fluid "dams" on vertical runs.

If stresses are involved, environments that promote metal dissolution can be considered more damaging from a general design philosophy. In such circumstances, materials can fail catastrophically and unexpectedly. Safety and health may also be significantly affected.

Figure 4-5 shows examples of using design detail to minimize stress. Perfection is rarely attained in general practice, and some compromise on materials limitation, both chemical and mechanical, is necessary. Mechanical loads can contribute to corrosion, and corrosion (as a corrosive environment) can initiate or trigger mechanical failure. Designs that introduce local stress concentrations directly or as a consequence of fabrication should be carefully considered.

Of particular importance in design are stress levels for the selected material: the influence of tensile, compressive, or shear stressing; alternating stresses; vibration or shock loading; service temperatures (thermal stressing); fatigue; and wear (fretting, friction). Profiles and shapes contribute to stress-related corrosion, especially if material selection dictates the use of materials that are susceptible to failure by SCC or corrosion fatigue (Ref 6). Materials selection is especially important wherever critical components are used. Also important is the need for correct procedures at all stages of operation, including fabrication, transport, storage, startup, shutdown, and normal operation.

Corrosion is a surface phenomenon, and the effects of poorly prepared surfaces, rough textures, and complex shapes and profiles can be expected to be deleterious (Ref 7). Figure 4-6 shows some examples in which design details could have considerably reduced the onset of corrosive damage resulting from ineffective cleaning or painting. Design should provide for the following: surfaces that are free from deposits; access to remove retained soluble salts before painting; free-draining assemblies; proper handling of components to minimize distortion, scratches, and dents; and properly located components relative to adjacent equipment (to avoid carryover and spillages). Further recommended procedures for coating constructional materials are shown in Fig. 4-7 (Ref 8).

Neglect and poor (or no) maintenance caused localized pitting on the underside of a type 304 stainless steel vessel lid that was exposed to high humidity, steam, and chloride vapors. Access in this example was possible but not used. Common engineering structural steelwork requires regular preventive maintenance, and restricted access makes this impossible. Figure 4-6 shows situations in which surface cleaning and/or painting is difficult or impossible. Condensation in critical areas can also contribute to corrosion. Typical structures susceptible to this phenomenon include automobile exhaust systems and chimneys or exhausts from high-temperature plants, such as boilers, kilns, furnaces, or incinerators.

Painting and surface-coating techniques have advanced in recent years and sophisticated products require careful mixing and application. Maintenance procedures frequently require field application where some control (use of trained inspectors) is essential, as in offshore oil and gas rigs. Inspection codes and procedures are available, and total design should incorporate these wherever possible. In critical areas, design for online monitoring and inspection will also be important. The human factor in mainte-

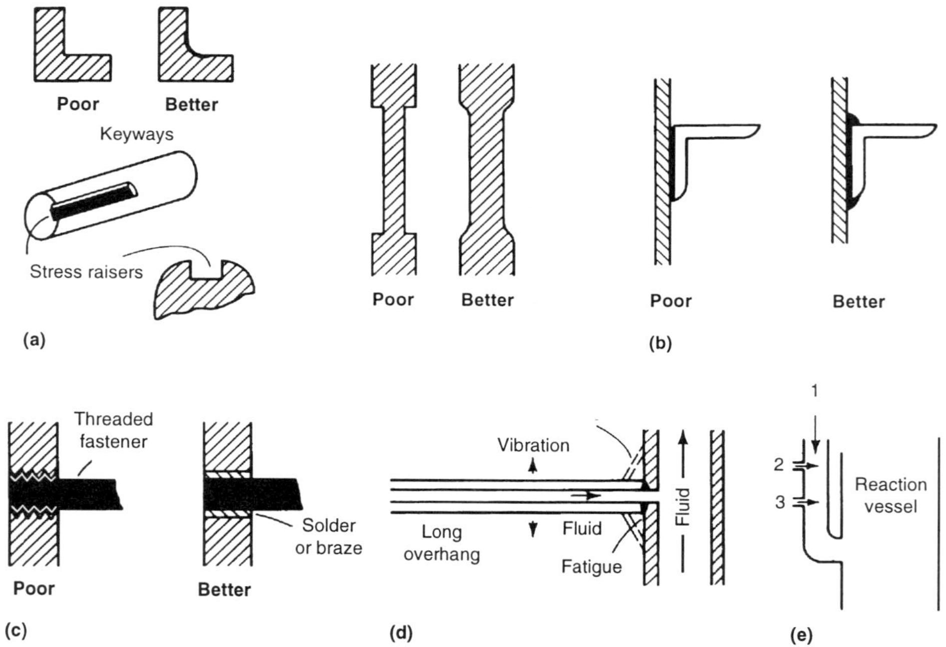

Fig. 4-5 Design details that can minimize local stress concentrations. (a) Corners should be given a generous radius. (b) Welds should be continuous to minimize sharp contours. (c) Sharp profiles can be avoided by using alternative fastening systems. (d) Too long an overhang without a support can lead to fatigue at the junction. Flexible hose may help alleviate this situation. (e) Side-supply pipework may be too rigid to sustain thermal shock from a recurring sequence that involves (1) air under pressure, (2) steam, and (3) cold water.

nance procedures is often questionable. Adequate training and motivation are of primary importance to ensure that design details are appreciated and implemented.

Prediction and Extension of Corrosion Life

It has become increasingly important to attempt to derive accurate and verifiable life prediction methods. An example follows of such a derivation for environmental degradation in light-water nuclear reactors (LWRs) (Ref 2).

Environmentally assisted cracking is one of the severest forms of life-limiting phenomena, but it also represents an area where "fundamental" approaches to resolving the design/prediction problems are most advanced. There have been well-documented instances of environmentally assisted cracking (e.g., stress corrosion and strain-assisted cracking and corrosion fatigue) in high-temperature water of austenitic stainless steels, nickel-base alloys, low-alloy and carbon steels, and their weld metals in various subcomponents of LWRs. These events have had a considerable economic impact (Ref 9) and are occurring at a time when competitive and deregulation pressures are forcing nuclear utilities (in the United States) to reduce costs (Ref 10). Thus, there is an incentive to develop validated life prediction methods (a) to evaluate the extent of a problem (i.e., is an incident an isolated occurrence or a precursor to a more widespread generic problem?) and (b) to quantitatively define the benefit of a specific mitigation action and evaluate its cost-effectiveness.

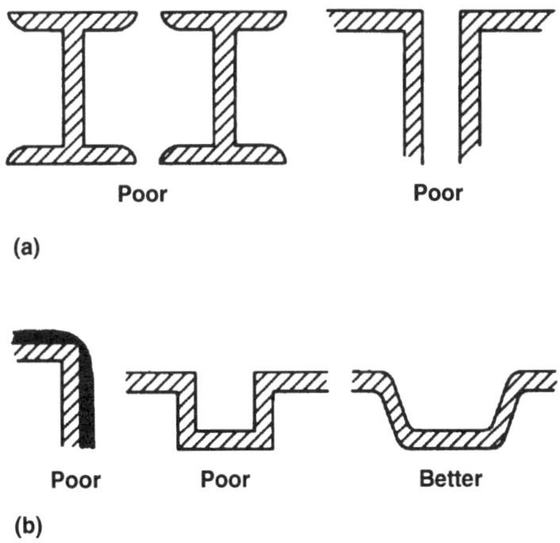

Fig. 4-6 Effects of design on effectiveness of cleaning or painting. (a) Poor access in some structures makes surface preparation and painting difficult; access to the types of areas shown should be maintained at a minimum of 45 mm (1¾ in.), or one-third of the height of the structure. (b) Sharp corners and profiles should be avoided if the structure is to be painted or coated.

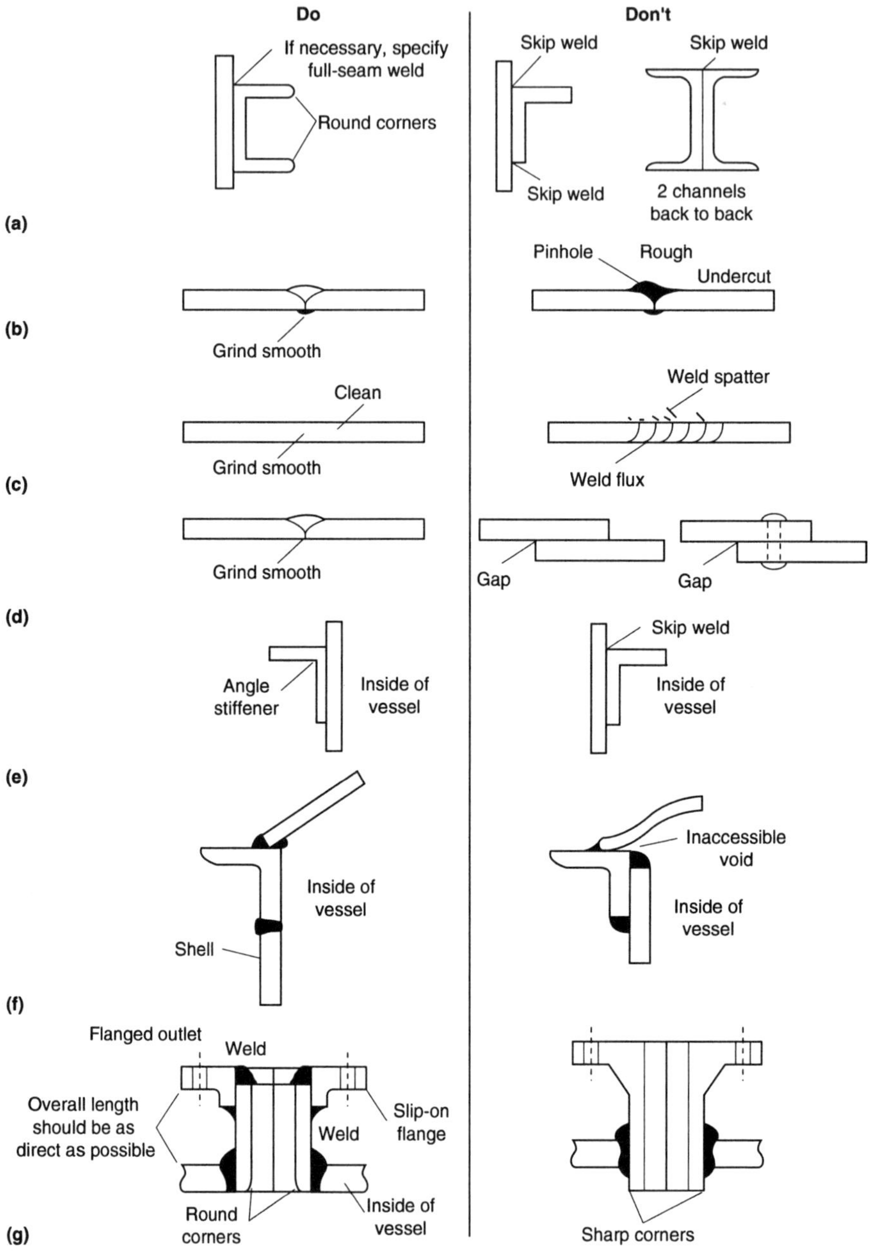

Fig. 4-7 Suggestions for steel construction to be coated. (a) Avoid pockets or crevices that do not drain or cannot be cleaned or coated properly. (b) Joints should be continuous and solidly welded. (c) Remove weld spatter. (d) Use butt welds rather than lap welds or rivet joints. (e) Keep stiffeners to outside of tank or vessel. (f) Eliminate crevice (void) at roof-to-shell interface in nonpressure vessel. (g) Outlets should be flanged or pad type, not threaded. Where pressure limits allow, slip-on flanges are preferred, because the inside surface of the attaching weld is readily available for radiusing and grinding. Source: Ref 8

The scatter in data from which such design or life prediction codes can be formulated is extreme. This scatter applies to data obtained from both field experience (Ref 11) and laboratory experiments. This latter observation is illustrated in Fig. 4-8 by crack-propagation rate/stress-intensity relationships for stainless steels (Ref 12). Similar data scatter has been noted for nickel-base alloys (Ref 13) and low-alloy steels (Ref 13) in 288 °C high purity water. In other instances, a large database does not exist, so it is impossible to define a usable life prediction algorithm. An example of this is irradiation-assisted SCC of stainless steels in reactor cores.

Consequently, the current life prediction codes for environmentally assisted cracking of structural materials in 274 to 345 °C water (representative of the coolant used in boiling-water reactors and pressurized-water reactors) usually represent upper bounds of the database available at a given time. A single crack-propagation-rate/stress-intensity relationship (e.g., ASME Section XI for life prediction/evaluation) or cyclic-stress/cycles-to-crack-initiation relationship (e.g., ASME Section III for fatigue design) is usually quoted with no account taken of the different cracking susceptibilities that are associated with the wide range of material/environment conditions in LWR systems. This conservative "upper bound" scenario can put unreasonable constraints on the continued operation of specific plants that have been operating under system conditions far better than those on which the design or life evaluation code was originally based.

Various statistical (extreme-value, pattern-recognition) or neural-network approaches can be applied to the available cracking database to develop an algorithm for component life or crack depth as a function of all the reactor operating parameters. This "classical" approach is limited, however, because there are comparatively few datasets for crack propagation. Data are rare that have been obtained under adequate control of system conditions (e.g., water purity, loading) and use adequately sensitive techniques that continuously monitor propagation rates relevant to design lives from 40 to 60 years for LWRs.

This situation is offset by the increase in mechanistic understanding of environmentally assisted cracking, an understanding which has accelerated in the last 20 years due to the availability of experimental and analytical procedures that allow quantification and validation of various cracking hypotheses. Because the precise shape of the crack depth/operational time relationship must be a function of the specific material, environmental, and stressing conditions, it follows that if there is a range in the actual system operating conditions, then there will be a predictable range in the observed cracking susceptibility, with the distribution of the range in cracking mirroring the distribution of system conditions that affect the propagation process. This predictable range in observed cracking susceptibility provides a bridge between deterministic and statistical/probabilistic life prediction methods (Ref 12, 14).

Historically, environmentally assisted cracking has been divided into the "initiation" and "propagation" periods. To a large extent, this division is arbitrary, because in most investigations, "initiation" is defined as the time at which a crack is detected, as in damage tolerance, or when the load has relaxed a specific amount (in a strain-controlled

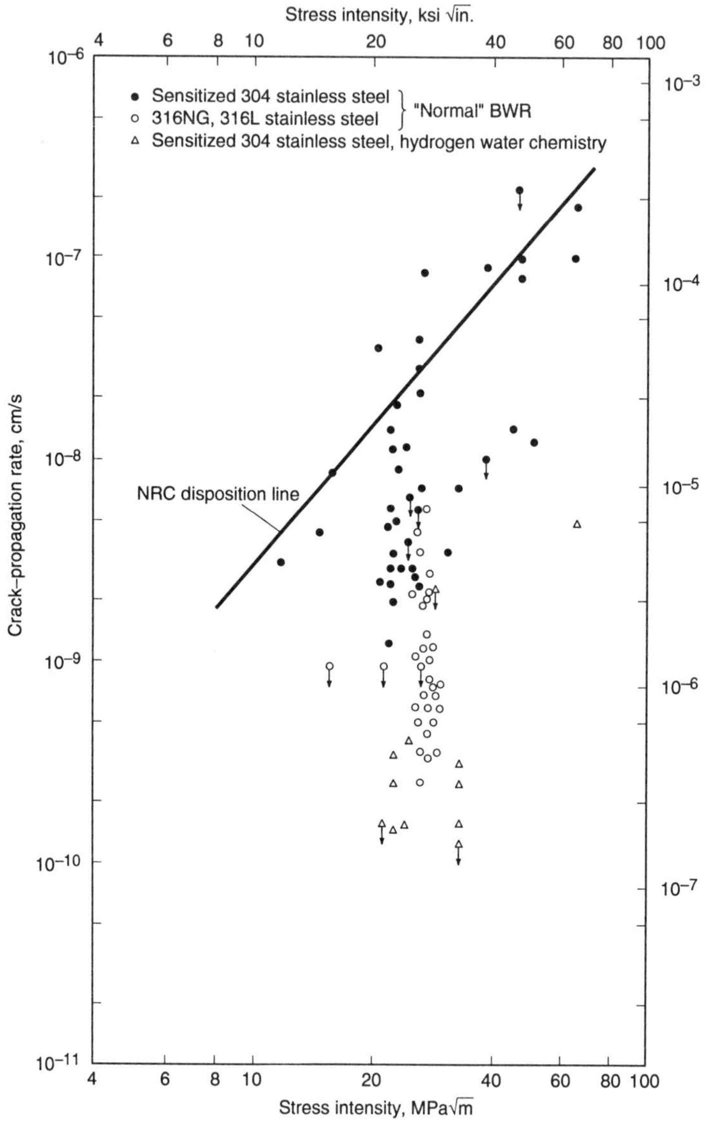

Fig. 4-8 Crack-propagation rate vs. stress-intensity data for stainless steel in 288 °C water. Also shown is the Nuclear Regulatory Commission disposition line.

situation). Such a definition of initiation, however, can correspond to a crack depth of significant metallurgical dimensions (e.g., >2 mm). Thus, for the purpose of lifetime modeling, it is proposed in Fig. 4-9 that initiation is associated with microscopic crack formation at localized corrosion or mechanical defect sites and is generally related to pitting, intergranular attack, scratches, weld defects, or design notches. It is further proposed that the probability is high for such microscopic initiation sites to exist or develop relatively early in the life of the component. Then the problem of life prediction becomes one of understanding the growth of small cracks from these geometrically separated initiation sites, and the coalescence of these small cracks to form a major crack, which can then accelerate or arrest, depending on the specific material, environment, and stress conditions.

Attention has been focused primarily on the propagation rates for "deep" cracks (e.g., approximately one grain diameter and greater), recognizing that the resultant life prediction can be conservative because the microscopic crack initiation and crack coalescence periods are not being recorded. Numerous mechanisms have been advanced for crack propagation in ductile alloy/aqueous environment systems. The usual procedure in developing a quantitative predictive model for crack propagation is to advance a hypothesis for the cracking mechanism and then independently quantify the relevant parameters in that mechanism. The resultant prediction algorithms are then validated by comparison with observed crack-propagation data. Once validated, these algorithms are used for design-life extension decisions. (Ref 15, 16).

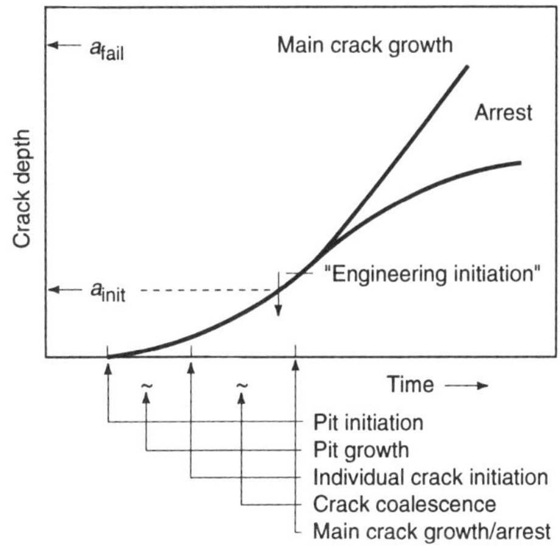

Fig. 4-9 Proposed sequence of crack initiation, coalescence, and growth for steels undergoing subcritical cracking in aqueous environments

Life Prediction Algorithms and Their Application

The overall comparison between the observed and theoretical crack-propagation rates in type 304/316 stainless steels in 288 °C water is shown in Fig. 4-10. The laboratory database upon which this comparison was made was obtained under a wide range of stressing (static, monotonically increasing, and cyclic load), material (solution annealed versus various degrees of sensitization), and water composition. There is a reasonable agreement between observation and prediction (Ref 12).

Changes in corrosion potential within the range expected in BWRs can have a significant effect on the cracking susceptibility of type 304/316 stainless steels, especially under constant-load conditions. This predicted and observed effect is illustrated in Fig. 4-11 for furnace-sensitized type 304 stainless steel under constant stress intensity (25 ksi$\sqrt{\text{in.}}$) in water with the conductivity in the range 0.1 to 0.3 μs/cm. Over the corrosion potential range – 550 mV_{SHE} to +250 mV_{SHE} (spanning "hydrogen-water" conditions to those under "normal" core conditions), the crack-propagation rate can change three orders of magnitude (Ref 2). From an operational design viewpoint, therefore, considerable benefit can be predicted by developing actions that lower the corrosion potential of the stainless steel structures, thereby highlighting remedial actions that lower the effective concentration of oxidants (oxygen, hydrogen peroxide) in the coolant. Solution conductivity is also predicted to have an effect on the cracking susceptibility, as indi-

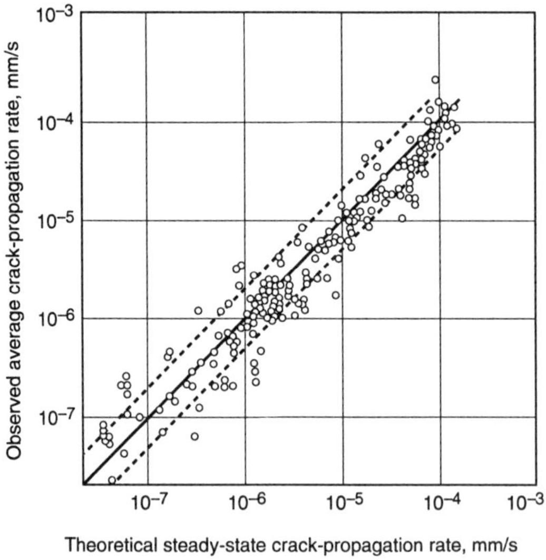

Fig. 4-10 Comparisons between observed and theoretical crack-propagation rates for type 304/316 stainless steels in 288 °C water. This database represents a wide combination of stressing material and environmental conditions.

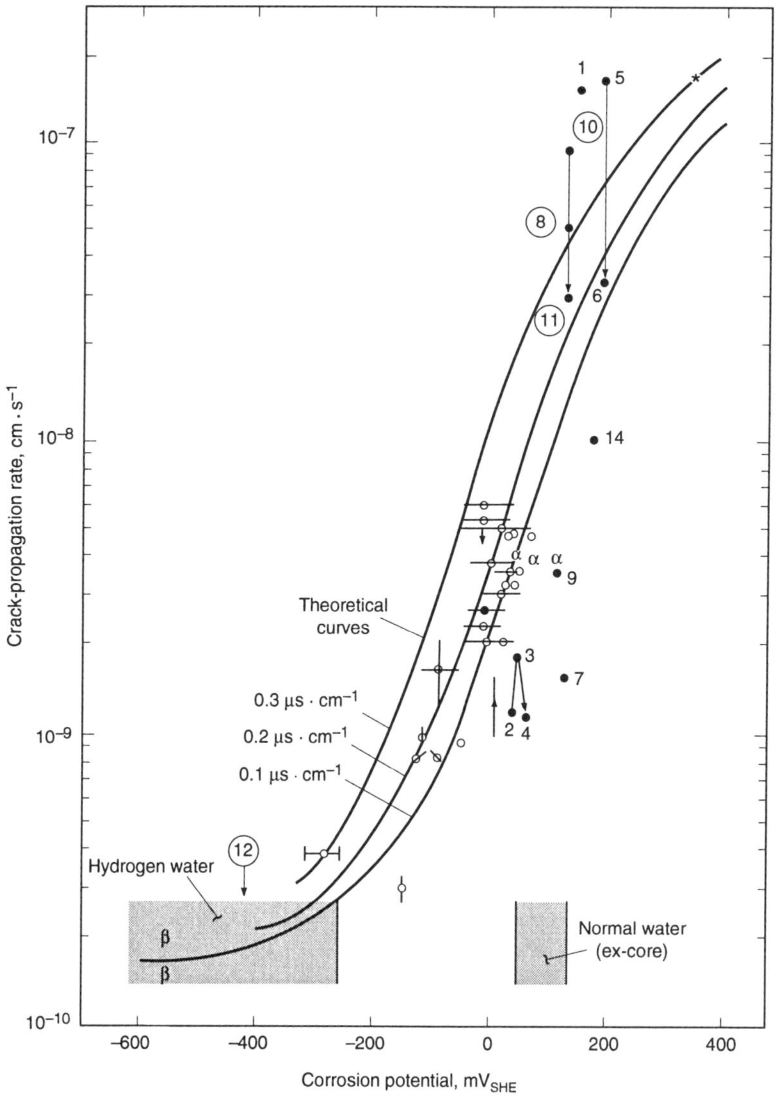

Fig. 4-11 Observed and predicted sensitivity of stress-corrosion-cracking sensitivity to corrosion potential for sensitized type 304 stainless steel in 288 °C water. The data points are measurements made in the laboratory or in reactors. The curves are the predicted relationships for the indicated conductivities. The numbered data points were obtained at the Harwell variable-energy cyclotron. The circled numbers were with the proton irradiation turned on, and the uncircled numbers with the irradiation off. Similarly the data point * was obtained under fast neutron irradiation in a boiling-water-reactor core.

cated by the three theoretical relationships shown in Fig. 4-11, thereby highlighting the quantitative value of maintaining water purity control.

Thus far, the comparison between observation and theory have centered on material/environment systems variables. The effect of stressing/straining conditions on the cracking susceptibility occurs primarily through the effect on the crack-tip strain rate. Because the crack tip does not recognize how the strain rate is maintained, the cracking susceptibility for a given material/environment condition should adhere to the same crack-propagation rate/crack-tip strain-rate relationship, regardless of the stressing/straining mode. This is shown in Fig. 4-12, which shows the theoretical and observed crack-propagation rate strain-rate relationship of a severely sensitized type 304 stainless steel in 8 ppm O_2, 0.5 μs/cm water. Movement along the strain-rate axis has been achieved by increasing stress intensity under constant-load conditions, increasing applied strain rate under monotonically increasing strain conditions, or cyclic loading under a variety of stress-intensity amplitude, mean stress, and loading frequency conditions. The single theoretical relationship line in Fig. 4-12 adequately predicts the cracking under this wide range of loading modes, indicating that the prediction method applies to SCC, SIC, and CF (Ref 2).

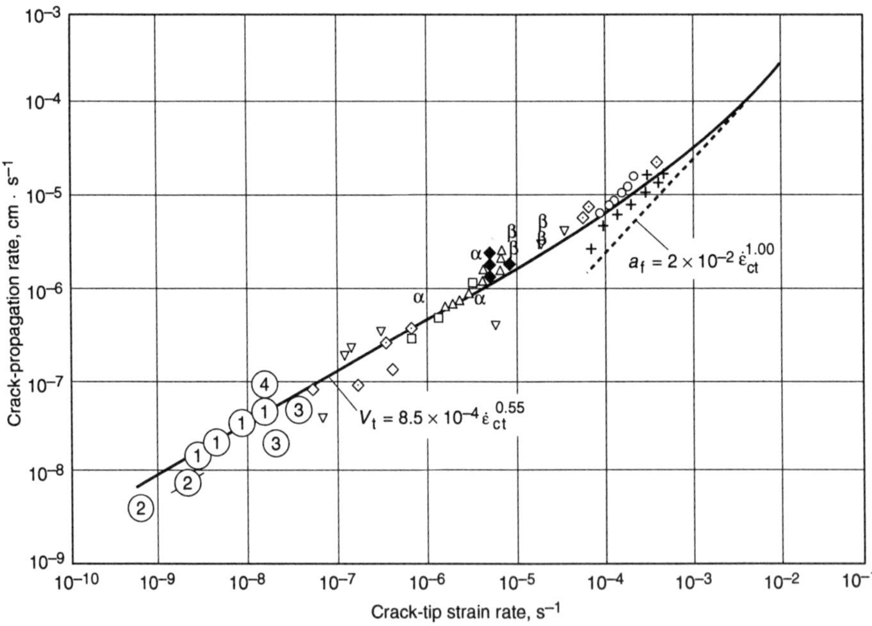

Fig. 4-12 Predicted and observed crack-propagation rate/crack-tip strain-rate relationships for sensitized type 304 stainless steel in 8 ppm oxygenated, 0.5 μs · cm^{-1} purity water at 288 °C

The belief that these types of cracking (SCC, SIC, CF) are separate phenomena with, by implication, different mitigation or design modification needs is probably incorrect. For instance, the sensitivity of the cracking susceptibility to the crack-tip strain rate will be a function of the material/environment conditions. Thus, the slope of the crack-propagation-rate/strain-rate relationship will be relatively shallow for severe environmental and material conditions (e.g., high dissolved oxygen, impure water, and high degrees of grain boundary sensitization), and the relationship will be steep for less severe material/environment conditions. This predicted and observed change (Fig. 4-13) in propagation-rate/strain-rate dependency on system conditions is significant when evaluating the validity of accelerated tests that are often used for development of design codes. For instance, increasing the crack-tip strain rate, and hence cracking susceptibility, by using the "slow-strain-rate test" is a valid test-acceleration procedure (because it is accelerating one of the rate-determining steps in the cracking mechanism), but the factor of improvement between a reference condition and a proposed mitigation condition will be less in this test than at the lower stress or strain-rate conditions expected in the operating plant. The relationship shown in Fig. 4-13 also gives an

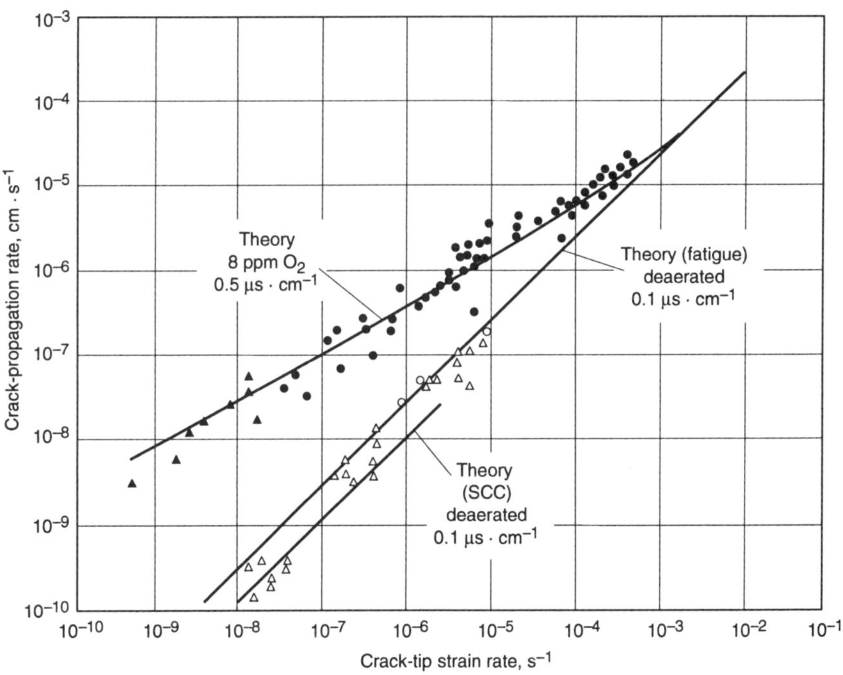

Fig. 4-13 Predicted and observed crack-propagation rate/crack-tip strain-rate relationships for stainless steels in a variety of material/environment systems

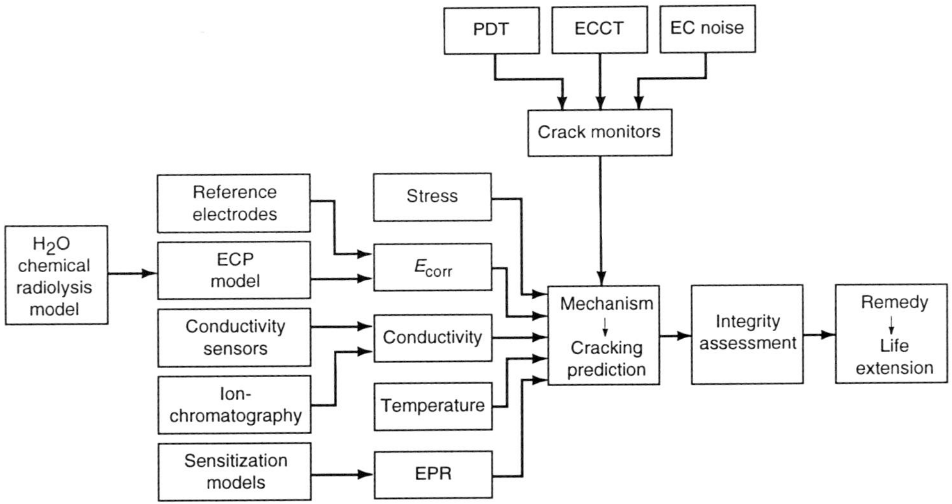

Fig. 4-14 The integration of system monitors, sensors, and environmental/material models as inputs to a crack-propagation-rate model. PDT, potential drip technique; ECCT, environmentally controlled cracking technique; EC, electro-chemical; EPR, electro-passivation rate

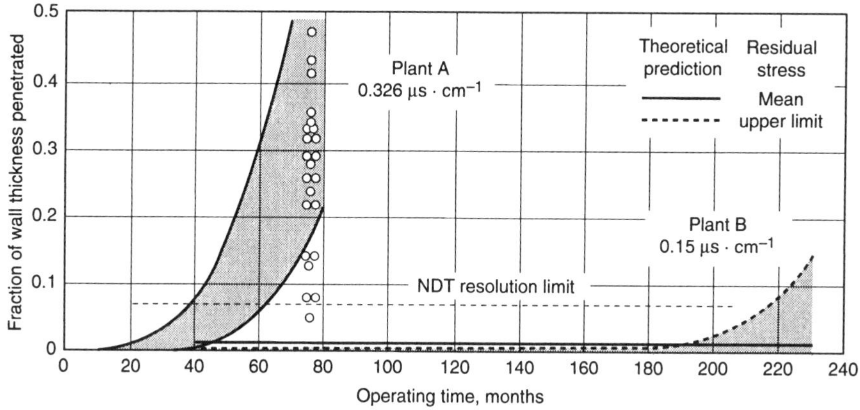

Fig. 4-15 Theoretical and observed intergranular stress corrosion crack depth vs. operational-time relationships for 28 in. diameter schedule 80 type 304 stainless steel piping for two boiling-water reactors operating at different mean coolant conductivities. Note the bracketing of the maximum crack depth in the lower-purity plant by the predicted curve, which is based on the maximum residual-stress profile and the predicted absence of observable cracking in the higher-purity plant (in 240 operating months).

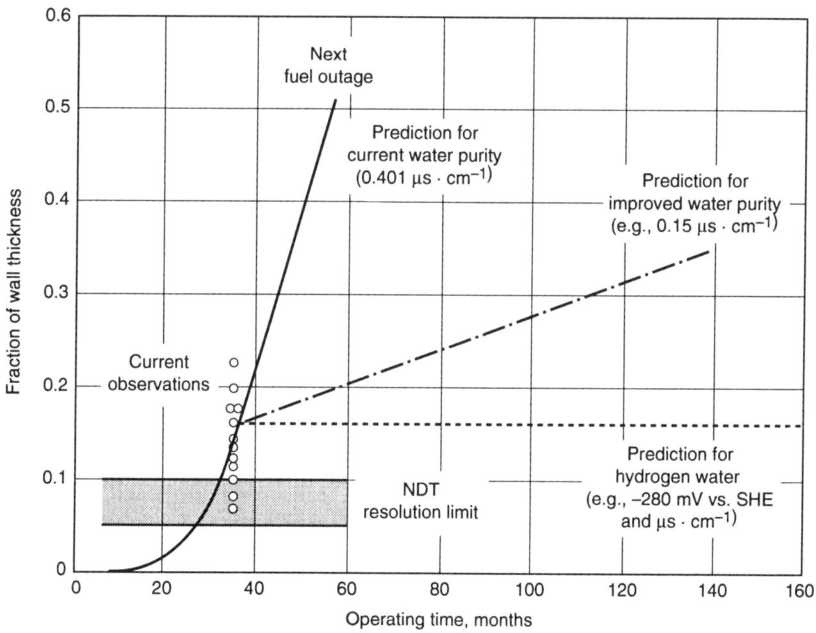

Fig. 4-16 Predicted crack depth vs. time response for defected 28 in. diameter schedule 80 recirculation piping in a given boiling-water reactor to defined changed in water purity. Also shown is the crack-depth limit that can be resolved by nondestructive testing (NDT).

explanation for the belief that the cracking susceptibility is more dependent on the specific environmental conditions under constant-load stress corrosion conditions than under CF conditions (Ref 2).

The crack-prediction algorithms are able to quantitatively explain the changes in crack-propagation rates for type 304/316 stainless steel in water at 288 °C for a wide combination of water composition (corrosion, potential, conductivity), material sensitization, and stressing (constant load/displacement, cyclic load) conditions. However, because the cracking response is so sensitive to changes in combinations of system conditions, it is necessary to combine the predictive method with system-defining sensors and models (Fig. 4-14). Provided this combining is done, it is then possible to make predictions of the extent of cracking in specific plant components (Ref 2) (Fig. 4-15) and the increase in life associated with specific system changes (Fig. 4-16). The implications of these findings are of paramount importance in fracture mechanics (Chapter 3), to the extent that the prediction of such growth can be integrated into those methods of assessing defect behavior. The potency of these techniques in establishing crack growth laws will be echoed in their application to other sectors of industry, such as gas processing facilities, refineries, petrochemical complexes, and hydrocarbon processing as a whole.

REFERENCES

1. *Engineering Design Guides,* Design Council, British Standards Institute, Council of Engineering Institutions, Oxford University Press, 1975–1979
2. F.P. Ford et al., Design for Corrosion Resistance, *ASM Handbook,* Vol 20, *Materials Selection and Design,* ASM International 1997, p 546
3. W. Pollock, *Corrosion under Wet Insulation,* NACE International, 1988
4. "Specification for Wicking-Type Thermal Insulation for Use Over Austenitic Stainless Steel," C 795, *Annual Book of ASTM Standards,* ASTM, 1997
5. "Codes of Practice for Drinking Water Installations (TRWI)," 628.1.033:696.11:620.193, DIN, Teil 7, 1988
6. H.H. Uhlig, *Corrosion and Corrosion Control,* 2nd ed., John Wiley & Sons, 1971, p 314
7. C.G. Munger, *Corrosion Prevention by Protective Coatings,* NACE International, 1984
8. R.S. Treseder, R. Baboian, and C.G. Munger, Ed., *NACE Corrosion Engineers Reference Book,* 2nd ed., NACE, 1991
9. R.L. Jones, "Corrosion Experience in U.S. Light Water Reactors—NACE 50th Anniversary Perspective," Paper 168, presented at Corrosion '93, NACE, 1993
10. R.L. Jones, "Critical Corrosion Issues and Mitigation Strategies Impacting the Operability of LWRs," Paper 103, presented at Corrosion '96, NACE, 1996
11. D.A. Hale, C.W. Jewett, and C.S. O'Toole, "BWR Coolant Impurities Program," First Annual Progress Report, Report NP2293, Electric Power Research Institute, Nov 1985

12. F.P. Ford, D.F. Taylor, P.L. Andresen, and R.G. Ballinger, "Corrosion Assisted Cracking of Stainless Steel and Low Alloy Steels in LWR Environments," Report NP5064S, Electric Power Research Institute, Feb 1987
13. P.L. Andresen, Environmentally-Assisted Cracking in Low Alloy Steels, *Corrosion '91*, NACE, 1991, p 917–938
14. F.P. Ford, P.L. Andresen, M.G. Benz, and D. Weinstein, On-Line BWR Materials Monitoring and Plant Component Lifetime Prediction, *Proc. Nuclear Power Plant Life Extension*, American Nuclear Society, Vol 1, June 1988, p 355–366
15. F.P. Ford, "Mechanisms of Environmental Cracking Peculiar to the Power Generation Industry," Report NP2589, Electric Power Research Institute, Sept 1982
16. F.P. Ford, Stress Corrosion Cracking, Corrosion Processes, R.N. Parkins, Ed., *Applied Science*, 1982

CHAPTER 5

Solutions to Equipment Failures at Welded Joints

As with the fastened structures of aircraft described in Chapter 2, fatigue cracking is by far the most common failure mechanism in welded structures. Unstable fracture (although rare) is perhaps the most dramatic, occurring without warning and often leading to serious consequences (see Fig. 15, Chapter 1).

Almost all welded joints contain stress concentrators other than welding flaws such that the incidence of these flaws may or may not be of consequence, depending on their severity. In practice, the majority of cases of fatigue cracking in service are associated with the inherent stress concentration features of the basic welded joint geometry and not with welding flaws.

A rational basis for the assessment of imperfections is engineering critical assessment (ECA) based on fitness for purpose, whereby a flaw can be accepted provided it can be shown that its presence will not reduce the fatigue strength of the welded joint below that required. The avoidance of unnecessary repair of welding flaws on the basis of ECA and consequent savings in fabrication costs and delays to projects has saved millions of dollars, in both the offshore and power generation industries. The ECA procedures used in the British Standards Institute Public Document 6493 (Ref 1) are reviewed here as an extension of Chapter 3 into the realm of providing solutions to equipment failures at welded joints. Methods of improving resistance to fatigue in welded joints are considered, together with a commentary on the nature of flaws and the ease by which weldments in steels can become embrittled by hydrogen.

ECA Procedures for Weld Imperfections

The scheme referred to applies to imperfections in fusion welded joints in material at least 10 mm thick and is basically applicable to structural ferritic steels, austenitic stainless steels, and aluminum alloys. Welding flaws are classified as either planar (cracks, incomplete penetration, lack of fusion) or nonplanar (slag inclusions, porosity). A third category is shape imperfections, which covers weld toe undercut and misalignment of a joint. The basis of the recommendations is different for the various types of imperfection. However, the fatigue strength of joints containing any of these types of imperfection can be expressed in the same way, using a grid of *S-N* curves relating to a range of quality categories.

Quality Categories

The quality category *S-N* curves are shown in Fig. 5-1. The diagram has two stress range scales, one for steels and the other for aluminum alloys. The quality category required in a particular instance is identified by locating the point on the diagram which corresponds to the stress range and life already determined in the design of the joint or joints to be incorporated. If this point lies between two curves, the required fatigue strength is then defined by the *S-N* curve above the point. Thus, if a steel joint is required to withstand 4×10^6 cycles of a stress range of 60 N/mm^2, the required quality category is Q2 and the *S-N* curve between Q1 and Q2 is relevant.

In practice, the fatigue strength required for the flawed joint being considered may be limited by adjacent details subjected to the same stress conditions or other stress concentrations in the same joint. This situation is illustrated in Fig. 5-2, which shows a steel girder detail assumed to be subjected to uniform bending moment so that the welds in the flange can be taken to be in the same stress field. The fatigue strength of flaws in the splice butt weld need be no higher than that of the nearby fillet welds, which here fall into Class F. In the absence of the fillet welds, the limiting case may be the butt weld

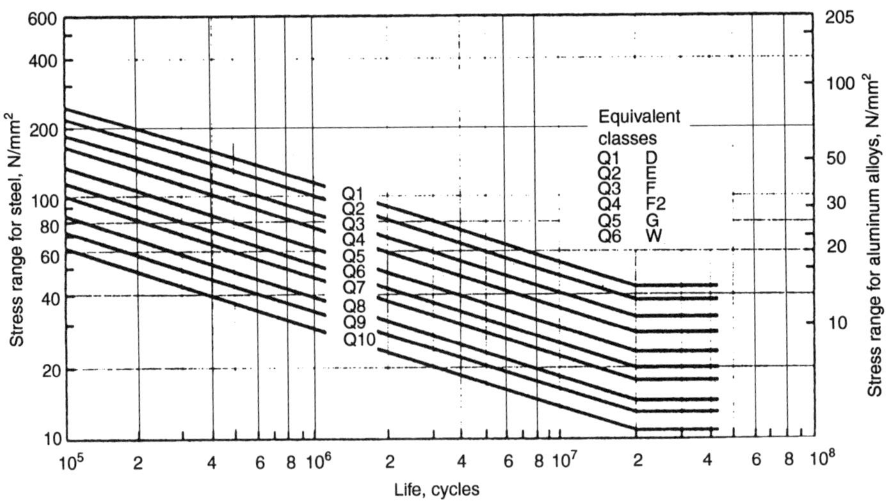

Fig. 5-1 Weld quality categories

toes (Class E or D). To facilitate comparisons between the quality categories and the design *S-N* curves, the *S-N* curves for quality categories Q1 to Q6 coincide with the Class D to W design curves respectively. Thus, a fatigue strength requirement corresponding to Class F, for example, is equivalent to Q3. Categories below Q6 are included because in some cases the required fatigue strength can be below that corresponding to the lowest design class.

Shape Imperfections

Misalignment. The welded joints most susceptible to misalignment are butt and cruciform joints (Fig. 5-3). Misalignment is a source of stress concentration in such joints, and the fatigue test data obtained from aligned and misaligned welded joints are found to correlate in terms of the sum of the applied axial stress and the induced secondary bending stress. Thus, the total stress can be used directly in conjunction with the appropriate design *S-N* curve for the aligned joint to assess the effect of misalignment. Table 5-1 includes formulas for calculating the stress magnification factor K_m in a variety of types of misaligned joints. Alternatively, in an ECA of misalignment, allowable values of K_m resulting from all types of misalignment present in the joint, to achieve fatigue strengths corresponding to the various quality categories, are given in Table 5-2. No

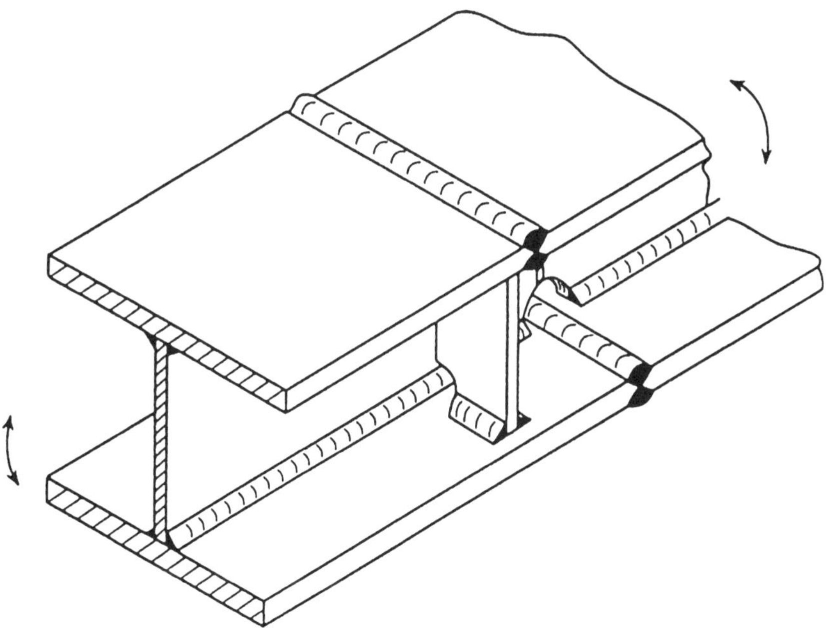

Fig. 5-2 Steel girder with splice butt weld

misalignment (i.e., $K_m = 1$) is allowed for the quality category which corresponds to the normal design class for the joint, and the allowances for lower quality categories are simply the factors by which the stress ranges corresponding to those categories must be multiplied to give the design class quality category. For example, in a Class E butt weld, $K_m = 1$ for Q2, for which $S = 218$ N/mm^2, then $K_m = 218/185 = 1.18$ for Q3, $218/163 = 1.34$ for Q4, etc.

In the case of a cruciform fillet welded joint, the effect of misalignment should be assessed with respect to possible toe failure based on K_m at the weld toe in the plate, and possible weld throat failure, based on K_m at the weld root related to the stress range on the weld throat see Table 5-1. As misalignment only gives rise to secondary bending stresses in joints subjected to axial loading, joints subjected to bending are not sensitive to the imperfection. Similarly, in joints subjected to a combination of axial and bending

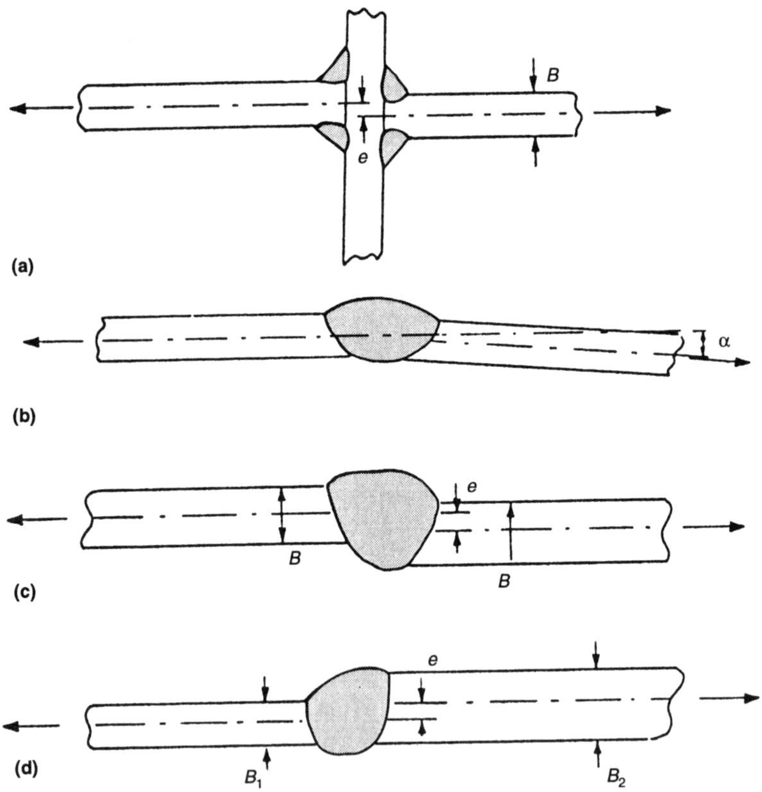

Fig. 5-3 Examples of misalignment. (a) Axial, in cruciform joint. (b) Angular, in butt weld. (c) Axial, in butt weld. (d) Axial, stemming from thickness change at butt weld

stresses, K_m applies only to the axial stress component. Misalignment may be present in any joint which is being assessed for other types of flaw. In such cases, the stress range used in the assessment should include any secondary bending stress due to misalignment, assuming that this bending stress varies linearly through the section thickness and is zero at the neutral axis. Thus, it will be important to consider misalignment in assessments of surface flaws.

Undercut. Fatigue test data for fillet and butt welds with known levels of undercut have been used to deduce the acceptance levels corresponding to each quality category given in Table 5-3. Few test data are available, therefore, the recommendations are deliberately conservative, with the requirement that undercut >1 mm must be assessed as a planar flaw.

Longitudinally Loaded Joints. Because misalignment and undercut are only important from a fitness-for-purpose point of view in joints loaded in the transverse direction,

Table 5-1 Formulas for calculating the secondary bending stress S_b (at the weld toe unless stated otherwise) due to misalignment in joints under transverse axial stress S_a

Note that, if required, $K_m = 1 + S_b/S_a$

Type	Figure	Bending stress, S_b
Axial misalignment in butt or cruciform joints between flat plates of:		
Equal thickness	5-3 (a) and (c)	$\frac{3e}{B} \cdot S_a$
Unequal thickness	5-3 (d)	$\frac{6e}{B_1}\left[\frac{B_1^{1.5}}{B_1^{1.5}+B_2^{1.5}}\right]S_a$ where $B_1 < B_2$
Axial misalignment in fillet welded cruciform joints (a)	5-3 (a)	$\frac{e}{B+H} - S_w$ where S_w = stress on weld throat, H = weld leg length
Axial misalignment in butt welded seam of pressurized vessel or pipe (b)	5-3 (c) and (d)	$\frac{6e}{B_1(1-\nu^2)}\left[\frac{B_1^n}{B_1^n+B_2^n}\right]S_a$ where ν = Poisson's ratio, $B_1 \leq B_2$
Angular misalignment in butt or cruciform joints between flat plates. Boundary conditions equivalent to:	5-3 (b)	
Fixed ends		$\frac{3}{4} \propto \frac{L}{B} \cdot S_a$
Pinned ends		$\frac{3}{2} \propto \frac{L}{B} \cdot S_a$
Angular misalignment in seam of pressurized vessel or pipe (c)		$\frac{6d}{B} \cdot S_a$

(a) Relevant to fatigue failure in weld throat from root. (b) $n = 1.5$ for circumferential seams and seams in spheres. $n = 0.6$ for longitudinal seams. (c) d = deviation from true circular shape. Assumes pinned end boundary conditions.

no limits to the extent of these imperfections are given for continuous welds stressed in the longitudinal direction.

Nonplanar Flaws. Fatigue test data for butt welds containing embedded porosity or slag inclusions have been interpreted on the basis of the set of quality category *S-N* curves. Hence, the tolerable flaw levels to achieve particular quality categories have been identified, and are summarized in Table 5-4. Limits for slag inclusions are given in terms of length of an inclusion measured in a direction normal to the applied principal stress, taking into account interaction with adjacent inclusions, as shown in Fig. 5-4. The benefit of postweld heat treatment on the slag inclusion limits derives from the release of hydrogen and not to the relief of residual stress.

With regard to porosity, the radiograph reference area is bounded by the maximum width of the weld and the length of weld affected by the porosity. Apart from strength considerations, the limits specified diminish the possibility of the porosity masking more harmful defects. The limits given in Table 5-4 apply to butt or fillet welded joints in both steels and aluminum alloys, noting that the allowable stresses for aluminum alloys are lower (Fig. 5-1 and Table 5-5). Surface-breaking flaws are excluded; they should be treated as planar flaws. Similarly, multiple inclusions may have to be regarded as planar flaws (Fig. 5-4). If there is any doubt about the type of an embedded flaw, it should be treated as planar.

It would be necessary to assess any flaw as planar that was required to determine the progress of the fatigue crack which propagated from it. Such a situation would arise if the growing fatigue crack could initiate brittle fracture, or it was required to establish inspection periods. The acceptance limits are based on fatigue lives which effectively

Table 5-2 Acceptance levels for misalignment

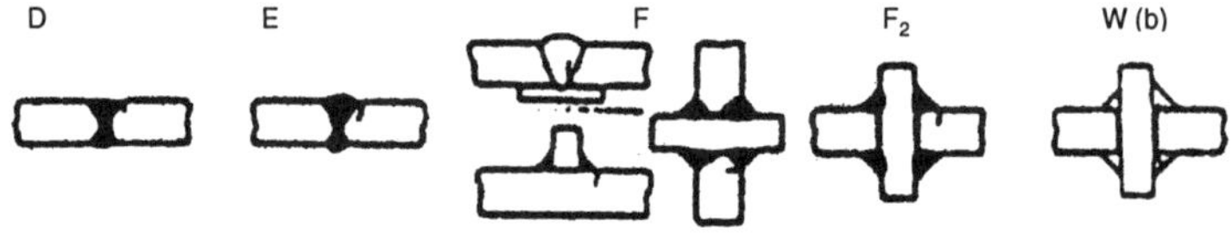

Quality category of misaligned joint	Allowable K_m (a) D	E	F	F_2	W (b)
Q1	1.0	...	...	...	...
Q2	1.14	1.0	...	...	...
Q3	1.34	1.18	1.0	...	...
Q4	1.52	1.34	1.13	1.0	...
Q5	1.84	1.61	1.37	1.21	...
Q6	2.16	1.88	1.61	1.42	1.0
Q7	2.48	2.18	1.85	1.63	1.15
Q8	2.92	2.56	2.18	1.92	1.35
Q9	3.40	2.99	2.53	2.23	1.58
Q10	4.00	3.52	2.98	2.63	1.85

(a) K_m = magnification factor due to any type of misalignment including combinations of more than one type. (b) Assessment of class W joints refers to possible fatigue cracking in weld and is based on stress range on weld throat

correspond to the attainment of through-section fatigue cracking, and they are therefore not directly applicable for other failure criteria or for judging the fatigue crack size at various stages during the fatigue life.

Planar Flaws. The treatment of embedded nonplanar flaws is simplified by the fact that flaw severity can be characterized using one dimension—inclusion length and projected area of gas pores—with the result that their fatigue strengths be readily represented by a few *S-N* curves based on test data. However, the treatment of planar flaws is more complex, because the severity of the flaw depends not only upon size but also upon shape and location, all of which can vary from one weld to the next. These factors can be taken into account if the fatigue strength of the flaw is calculated using fracture mechanics as described in Chapter 3. A simplified treatment of planar flaws is presented in BSI PD 6493 which enables the limits on planar flaws to be readily related to the quality categories.

Using the procedure presented, it is possible to relate quality category (i.e., a particular *S-N* curve) to planar flaw size for embedded flaws of any shape in butt welds and surface flaws of any shape in fillet welds and flush-ground butt welds. It is also possible to specify or estimate the critical crack size. Condensed presentations of the recommendations in BSI PD 6493 are given in the form of nomograms in Fig. 5-5 and 5-6 (Ref 2). Figure 5-5 refers to surface flaws in plain material or flush-ground butt welds and embedded flaws in butt welds. The recommendations for embedded flaws refer specifically to flaws at mid-thickness. However, other embedded flaws can still be assessed on the basis that they are in the middle of a thinner plate of effective thickness, $B' = 2a + 2p$, where p is the distance from the edge of the flaw to the nearest plate surface. Figure 5-6 (Ref 2) refers to surface flaws at butt or fillet weld toes; these are cases where the flaw is situated in the field of stress concentration due to the welded joint geometry. Both nomograms can be applied to steel or aluminum alloy joints. Failure is defined as the attainment of through-thickness cracking, which was essentially the definition of failure in the fatigue tests used to determine the data upon which the main fatigue rules are based. To achieve simplicity, it has been necessary to introduce approximations into the nomograms, with the result that the allowable sizes of some flaws, particularly embedded flaws in butt welds and small surface flaws, can be underestimated compared to assessments based directly on BSI PD 6493. However, since embedded and

Table 5-3 Acceptance levels for weld toe undercut

Quality category	Depth of undercut, plate thickness: Butt welds	Depth of undercut, plate thickness: Fillet welds
Q1	0.025	...
Q2	0.05	...
Q3	0.075	0.025
Q4	0.10	0.05
Q5	0.10	0.075
Q6–Q10	0.10	0.10

Maximum depth of undercut in any thickness = 1 mm.

small surface flaws are particularly difficult to size, this extra conservatism is not unreasonable.

As an example of the application of the nomogram for butt welds (Fig. 5-5), consider its use, together with the information in Table 5-4, for the assessment of quality requirements for embedded flaws in an as-welded butt weld in the 35 mm thick flange of a steel girder, which experiences the same fatigue loading as a stiffener fillet welded to the flange. The welds attaching the stiffener have a lower fatigue strength than the highest fatigue strength that could be obtained from the butt weld (i.e., corresponding to Class E), so they determine the highest quality category required. This corresponds to Class F and is therefore category Q3. However, because the stiffener is attached to a plate thicker than 22 mm, the effective quality category is lower. Referring to Table 5-5 and applying the thickness correction of $(22/B) \times 0.25$ to the stress range corresponding to Q3 reduces it to 165 N/mm^2. Therefore, the effective quality category is Q4 or Class F2. From Table 5-4, the butt weld could contain a 35 mm long slag inclusion and ≤5% porosity. To use the nomogram, Fig. 5-5, enter the diagram at the plate thickness (broken line) and move on a vertical line to meet the required quality category. Movement horizontally to the embedded flaw axis enables the allowable depth of embedded flaw for different shapes to be read. In this example, the acceptable depths of embedded flaws at mid-thickness are between 4.2 and 20 mm depending on the flaw shape. Thus, for a shape $a/2c = 0.1$ $\times 2a/B = 0.29$, which corresponds to a flaw depth of $2a = 10$ mm. If the butt weld was flush ground, Fig. 5-5 could also be used to determine acceptable surface breaking flaws. Figure 5-6 would be used to assess surface flaws at the toe of the as-deposited butt weld. The procedure is to move vertically downward from the diagonal line and to read off surface flaw depths for the required flaw shape in the same way as for embedded flaws. In the present example, acceptable flaws range from 1.2 to 8 mm in depth (e.g., $a/B =$ 0.11, or $a = 3.8$ mm, for $a/2c = 0.1$). The nomogram can be used in the reverse direction to determine the fatigue strength of a particular flaw or service crack.

In the assessment of flaws at fillet or butt weld toes, it is necessary to take into account the attachment size (L) as well as plate thickness (B), since these influence the correction factor M_k as discussed in Chapter 3. Consider a 25 mm thick member with a

Table 5-4 Limits for nonplanar flaws in steel and aluminum alloy joints

	Allowable length of slag inclusion (a), mm		Limits of porosity	
Quality category	As-welded	Postweld heat treated	% of area on radiograph	Individual pore diameter
Q1	2.5	19	3	6 mm or ¼ in. plate thickness if less
Q2	4	58	3	
Q3	10	Continuous	5	
Q4	35	Continuous	5	
Q5	Continuous		5	
Q6–Q10	Continuous		5	

(a) Tungsten inclusions in aluminum alloy welds need not be considered; their presence does not affect fatigue strength.

12 mm thick stiffener attached by 9 mm leg length fillet welds (i.e., $L/B = 1.2$) which is found to contain a fatigue crack estimated to be 2.5 mm deep by 25 mm long (i.e., $a = 2.5$ mm, $a/2c = 0.1$). The member is required to have a fatigue strength corresponding to the Q6 *S-N* curve for fatigue service. Should the crack be left unrepaired? The routes taken on the nomogram (Fig. 5-6) are shown as a broken line. The horizontal and vertical lines meet at a point which falls between Q5 and Q6, indicating that the crack will not propagate through the plate thickness during the required life.

The nomograms illustrate the situations in which an increase in plate thickness will lead to an increase or decrease in fatigue strength. In surface cracking in a butt weld without overfill, there is no surface stress concentration, so fatigue strength increases with plate thickness. In contrast, the curves for weld toe cracks show how fatigue strength can decrease with increase in plate thickness.

Methods for Improving the Fatigue Life of Welded Joints

Three factors influence fatigue of welded joints (Ref 3):

- Stress concentration due to joint and weld geometry
- Stress concentration due to localized defects
- Welding residual stresses

Improvement in fatigue life can be obtained by reducing the effects of one or more of these parameters. This is true for cracks starting from the weld toe, which is by far the most common failure site. In load-carrying fillet welds, however, cracks can start at the weld root and propagate through the weld throat until failure occurs. In this case, additional weld metal to increase the weld throat dimension will result in reduction in shear stress and, hence, a corresponding increase in fatigue strength.

Table 5-5 Details of quality category *S-N* curves for as-welded joints in steel

	Constants in equation $S^mN = A$		Stress range, N/mm^2			
Quality category	m	A	10^5 cycles	10^6 cycles	10^7 cycles	S'_0, N/mm^2
Q1	3	1.52×10^{12}	248	115	53	42
Q2	3	1.04×10^{12}	218	101	47	37
Q3	3	6.32×10^{11}	185	86	40	32
Q4	3	4.31×10^{11}	163	76	35	28
Q5	3	2.48×10^{11}	135	63	29	23
Q6	3	1.52×10^{11}	115	53	25	20
Q7	3	1.00×10^{11}	100	46	22	17
Q8	3	6.14×10^{10}	85	39	18	15
Q9	3	3.89×10^{10}	73	34	16	12.5
Q10	3	2.38×10^{10}	62	29	13	11

For aluminum alloys, divide A by 27 and stress range by 3 and assume $S'_0 = 0$.

Because the most common failure site is the weld toe, many postweld treatments for this region have been developed to improve the fatigue life. These techniques rely largely on removing the detrimental intrusions at the weld toe, reducing the joint stress concentration, or modifying the residual stress distribution. There are primarily two broad categories of postweld techniques: modification of the weld geometry and modification of the residual stress distribution (Ref 4, 5). A complete list is given in Fig. 5-7 (Ref 3). Methods that reduce the severity of the stress concentration or removal of weld toe intrusions include grinding, machining, or remelting. These techniques essentially focus on altering the local weld geometry by removing the intrusions and at the same

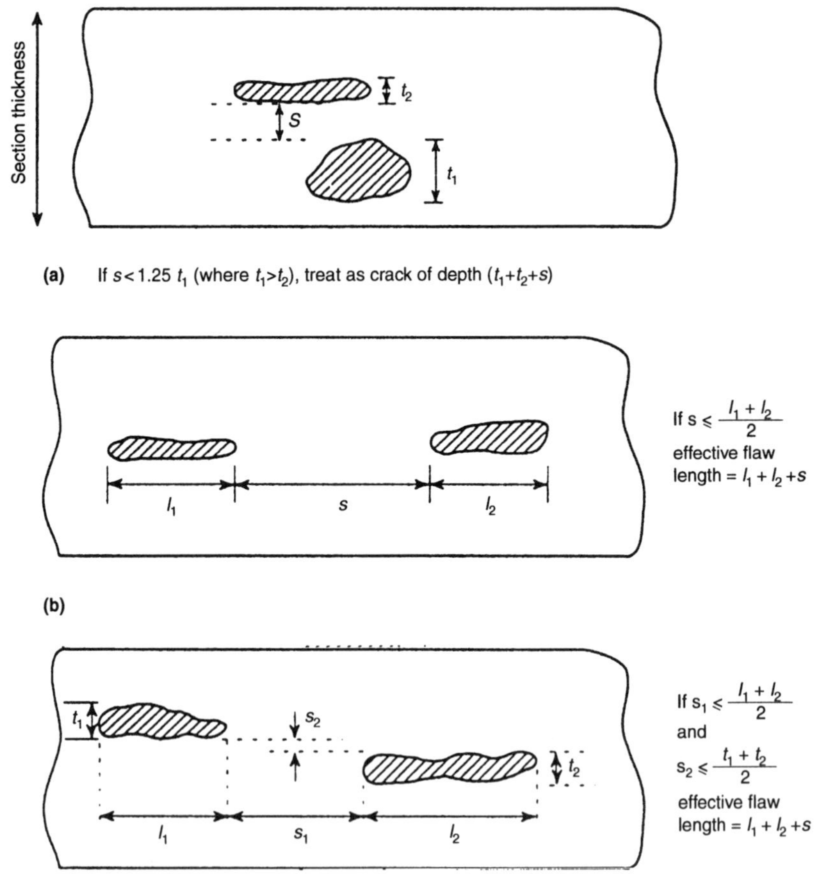

Fig. 5-4 Interaction of slag inclusions. (a) Coplanar embedded slag inclusions. (b) Effective length of multiple slag inclusions in same or different plane(s)

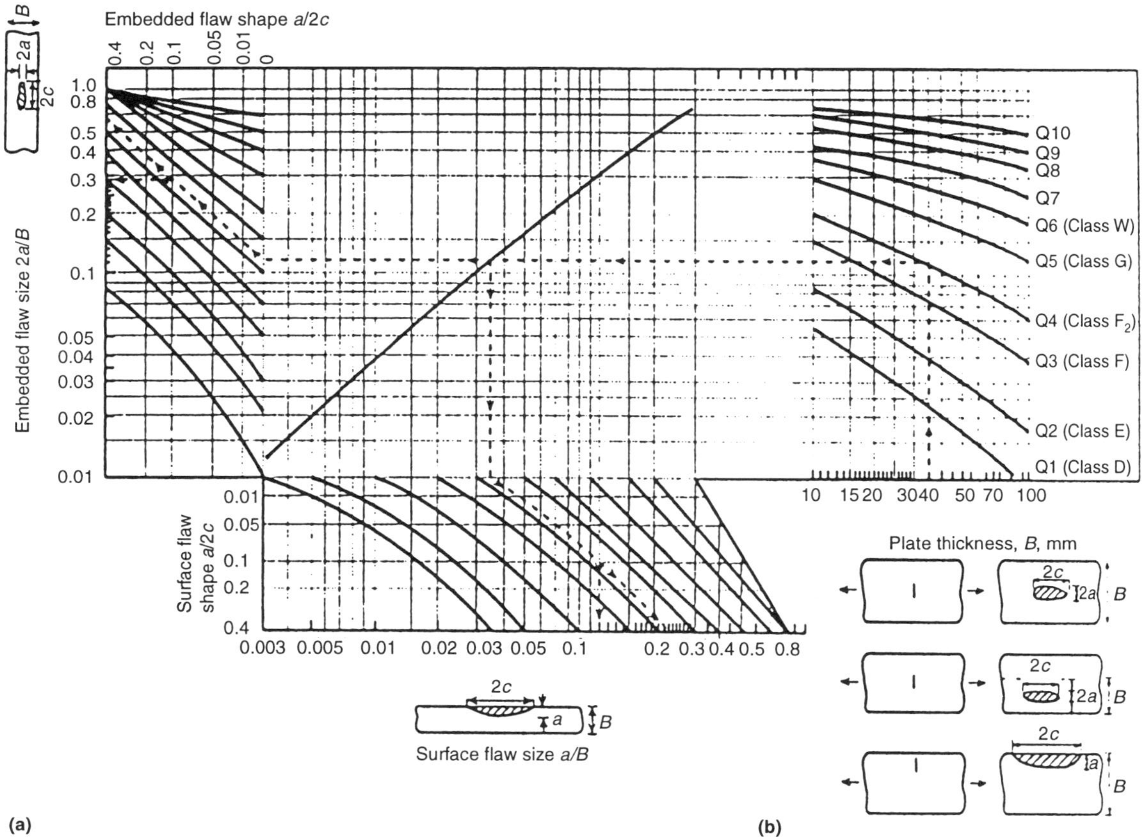

Fig. 5-5 (a) Nomogram to determine allowable embedded or surface planar flaw sizes. (b) Flaws that can be assessed using (a)

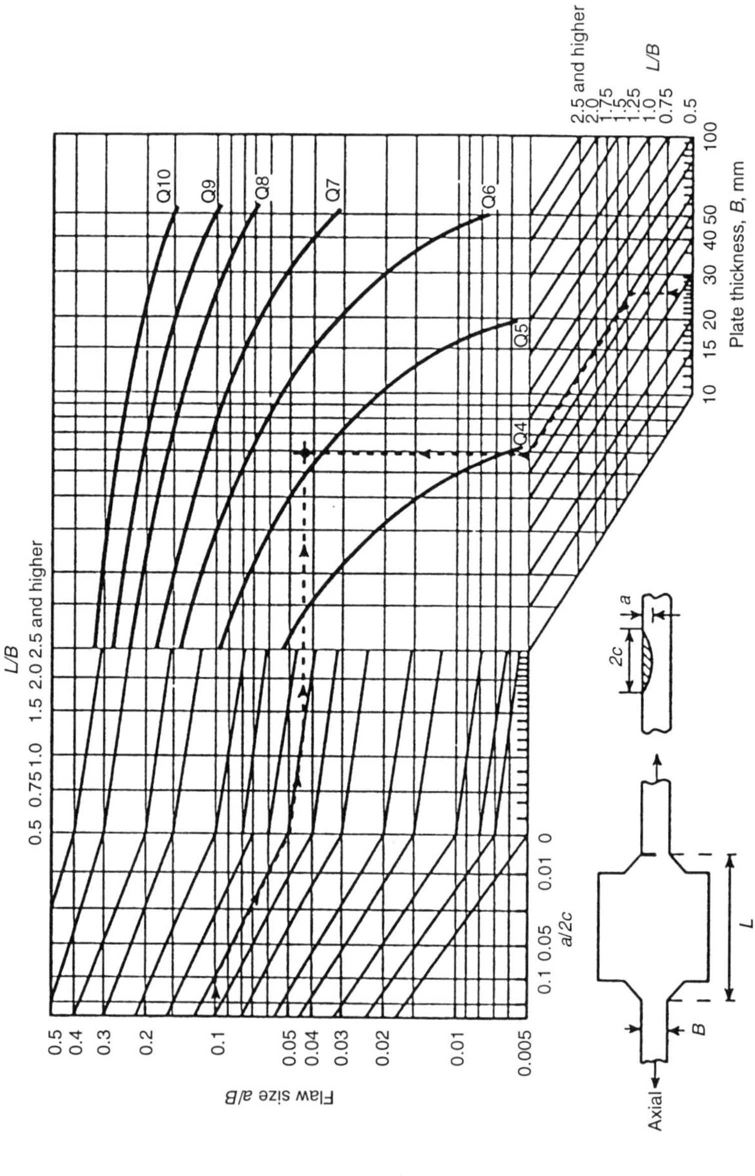

Fig. 5-6 Nomogram to determine sizes of planar flaws at butt or fillet weld toes

time on achieving a smooth transition between the weld and the plate. Typical profiles of burr ground and tungsten inert gas (TIG) dressed welds are given in Fig. 5-8. Because weld toe intrusions can be up to 0.4 mm (0.016 in.) in depth, the general guideline is that the grinding or machining operation must penetrate at least 0.5 mm (0.02 in.) into the parent plate. In these procedures, the depth, the diameter of the groove, and the direction of the grinding marks become important issues. In TIG and plasma dressing, the weld toe is remelted in order to improve the local profile and also to "burn" away or move the intrusions. With these techniques, the position of the arc with respect to the weld toe and the depth of the remelt zone are two critical variables. Controlling the shape of the overall weld profile has also been recognized as a method of improving the fatigue life. The American Welding Society (AWS) weld profile procedure using the dime test reduces the weld geometry K_t (see Chapter 2) and hence increases the fatigue life (Ref 6, 7).

Methods that modify the residual stress field include heat treatment, hammer and shot peening, and overloading. Postweld heat treatment is known to reduce tensile residual stresses but does not eliminate them completely. The benefits of postweld heat treatment can be realized only if the applied loads are either partially or fully compres-

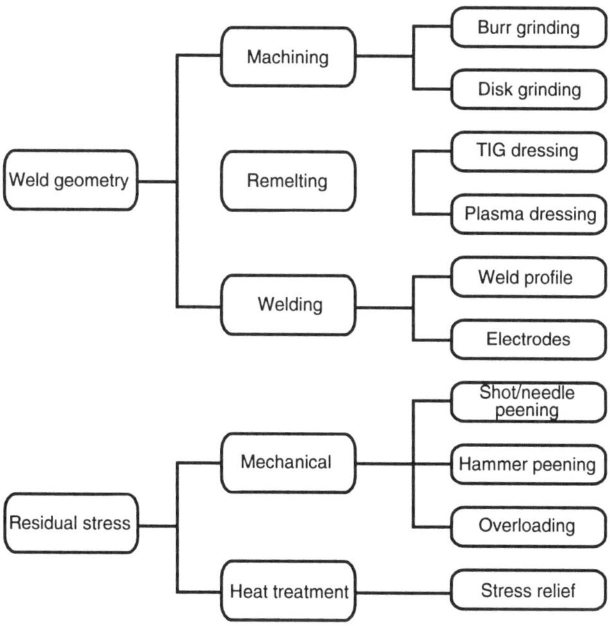

Fig. 5-7 Range of postweld fatigue life improvement techniques

sive. Overloading techniques rely on reducing the tensile residual stress field and/or introducing compressive stresses at the weld toe. Exact loading conditions for a complex structure are difficult to establish, so this technique is rarely used. To obtain significant improvement in fatigue strength, it is necessary to introduce compressive stresses in the local area in a consistent and repeatable manner. The three peening techniques (shot, needle, and hammer peening) aim to achieve this by cold working the surface of the weld toe. As with the grinding methods, it is necessary to penetrate the parent plate and deform to a depth of at least 0.5 mm (0.02 in.). Hammer peening is perhaps the best technique for this, and it has the advantage of removing the weld toe intrusions by cold working them out. However, peening techniques require special equipment and can present safety challenges for noise control. In addition, they are very difficult to automate and control in production.

A comparison of the improvement in fatigue strength obtained by some of these techniques is given in Fig. 5-9. It can be seen that hammer peening is perhaps the best

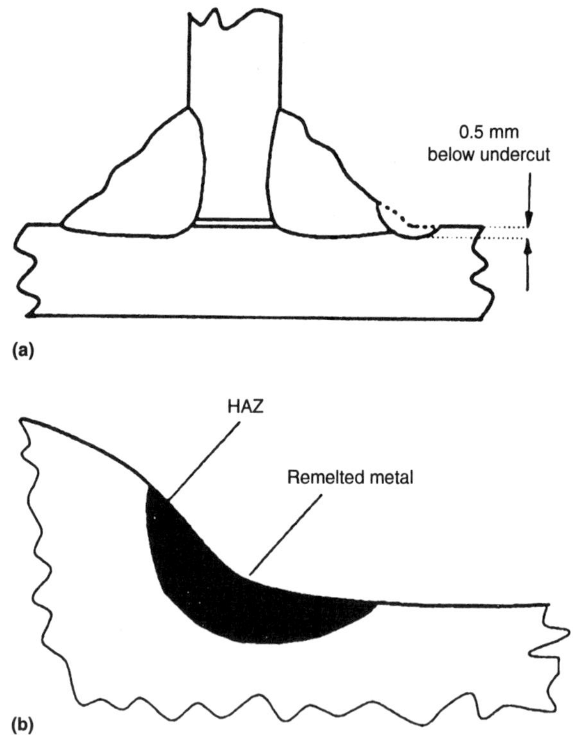

Fig. 5-8 Typical profiles for (a) burr grinding and (b) tungsten inert gas dressing of weld toe

technique, because it reduces or eliminates intrusions as well as introducing compressive stresses. Burr grinding, which is easier to implement, also produces significant improvement in fatigue life. Disc grinding and burr grinding are probably the most frequently used methods for improving the fatigue life of welded joints.

Types of Weld Discontinuities

Discontinuities can be classified into two separate categories: welder-induced discontinuities and hydrogen stress cracking (HSC). When subjected to a sustained load, a steel containing hydrogen can fail at a stress level that is much below its tensile strength as measured in a short time test. This behavior is variously termed HSC, delayed failure, or delayed low-stress brittle failure. It is a common type of failure encountered in service applications of steels containing hydrogen or exposed to hydrogen environments (Ref 8).

Solutions to Welder-Induced Discontinuities

Welder-induced discontinuities include slag inclusions, porosity, undercut, lack of fusion, and others and can be controlled by good welder technique and by following a qualified welding procedure.

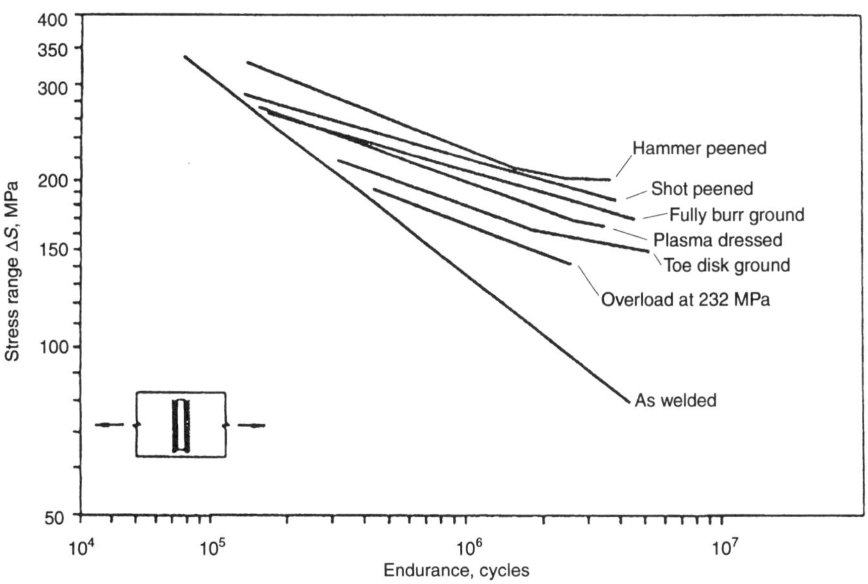

Fig. 5-9 Comparison of different improvement techniques

A qualification is used to demonstrate that a procedure is capable of producing sound welds. Welder qualification is used to show that a particular welder is capable of executing the qualified procedure. Proper use of welding procedure and welder qualification, combined with close monitoring in the field to ensure that the welding procedure is being followed, should minimize the occurrence of welder-induced discontinuities.

Hydrogen Cracking

HSC will only occur when the conditions are right and generally cannot be controlled by the skill of the welder. HSC is also the most significant discontinuity with respect to integrity. The formation of heat affected zone (HAZ) HSC in welds made onto in-service pipelines, for example, requires three primary, independent conditions to be fulfilled. The requirements are as follows (Ref 9).

High Hydrogen Level. All arc welding processes introduce hydrogen into the weld to some extent. Hydrogen can originate from moisture that exists in electrode coatings, in the atmosphere (humidity), or on the pipe surface (condensation). Hydrogen can also originate from hydrocarbons, grease, rust, or other organic contaminants on the pipe or on the welding wire.

Susceptible Microstructure. In general, only hard HAZ microstructures are susceptible to hydrogen cracking. Such microstructures are promoted by steel that has a high carbon-equivalent value and by rapid welding cooling rates. The carbon equivalent is expressed as:

$$CE = \mathrm{C} + \frac{\mathrm{Mn}}{6} + \frac{\mathrm{Cu} + \mathrm{Ni}}{15} + \frac{\mathrm{Cr} + \mathrm{Mo} + \mathrm{V}}{5}$$

where CE = carbon equivalent.

Weld cooling rates are determined by welding heat input and pipeline operating conditions. Operating conditions that influence weld cooling rates include ambient temperatures, pipeline flow rates, and pipe wall thickness.

Tensile Stresses Acting on the Weld. Tensile stresses can be either applied or residual. Applied stresses can result from movement of the pipeline due to soil settlement. Residual stresses arise from the restraint of the welded connection and strains imposed by the contraction of the weld on cooling.

The relationship of the conditions necessary for hydrogen cracking to occur is shown schematically in Fig. 5-10. (Ref 8).

Hydrogen Control

It is possible to severely limit the amount of hydrogen entering the weld by the use of suitably prepared low-hydrogen consumables and by rigorous cleanliness. It should be stressed that the term *low-hydrogen* when applied to consumables (EXX18- or EXX16-type) does not mean "no hydrogen." Even low-hydrogen consumables can lead to hydrogen cracking under adverse conditions, because hydrogen can be introduced through moisture that can be absorbed by the electrode coating. Low-hydrogen electrodes should be stored at an appropriate temperature in portable field ovens or used from freshly opened, airtight containers.

The moisture content of low-hydrogen electrode coatings can be reduced following exposure to the atmosphere by drying and storing them according to manufacturer's recommendations. This typically involves drying at temperatures between 250 °C and 450 °C and storing at temperatures above 100 °C. Ovens should not be overloaded, and drying should last long enough to ensure that all the contents receive the minimum baking time as specified by the manufacturer.

Moisture removal from cellulosic-coated electrodes (EXX10-type) is undesirable because these electrode types require a reducing atmosphere (as provided by hydrogen) in the arc to produce the deep penetration for which they are characterized. In the past, many welds onto in-service pipelines have been made using cellulosic-coated electrodes. Research (Ref 10, 11) has shown that the risk of hydrogen cracking is significantly greater using these electrodes than for low-hydrogen electrodes. Therefore, cellulosic

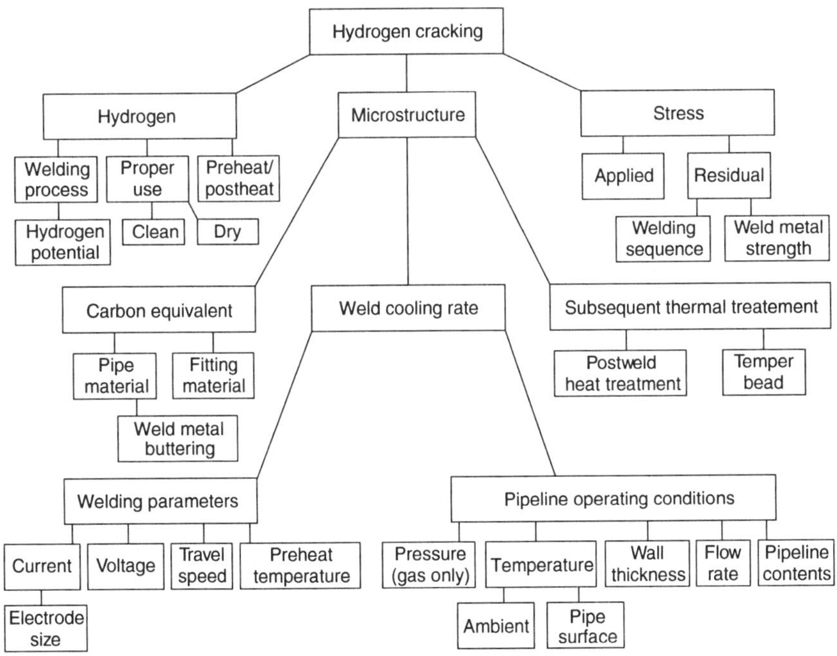

Fig. 5-10 Conditions required for HSC

electrodes generally are not recommended for welding onto in-service transmission pipelines.

The input of hydrogen to the welds can also be reduced by the use of low-hydrogen processes, in particular gas-shielded processes such as gas metal arc welding (GMAW) or gas tungsten arc welding (GTAW). Under certain conditions, GMAW has the added benefit of higher deposition rates and, therefore, greater productivity. However, the GMAW process (short circuit transfer mode) has been known to be susceptible to producing lack-of-fusion discontinuities, particularly for welding onto in-service pipelines as the result of heat sink effects.

Cleanliness of weld preparation, welding wire, and welding apparatus are also important. Paint, rust, grease, and degreasing agents can all be hydrogen sources. Lubricants from the wire-drawing operation are another potential source. Hydrogen levels can also be reduced by controlling the humidity level in the immediate vicinity of the welding operation by removing standing water, particularly if the pipe surface temperature is below the dew point.

An additional approach to hydrogen control is to allow hydrogen removal by diffusion, so the levels are reduced to acceptable values by the time the weld has cooled. This is achieved by increasing the weld thermal cycle, usually by applying a preheat, which reduces the weld cooling rate, thereby allowing more time for the hydrogen to diffuse. This is difficult, however, for in-service pipeline applications, because the flowing contents can quickly remove heat from the pipe wall. The use of preheating is also beneficial because moisture and contamination are burned off prior to welding. Using higher welding heat inputs is also beneficial because this also increases the weld thermal cycle time. However, using high heat inputs is often impractical for welding in some positions and can increase the risk of burn-through in thinner wall pipe. The thermal cycle from a "hot pass" (i.e., a second pass applied within a few minutes of the first pass), which usually requires the use of more than one welder on larger-diameter pipe, also allows hydrogen removal by diffusion. In severe cases where it is not possible to apply preheating or dry low-hydrogen electrodes satisfactorily, austenitic (i.e., stainless steel) weld metal can be used. This technique is particularly valuable when welding onto high-carbon-equivalent pipe steel. Austenitic weld metal has a comparatively high solubility for hydrogen, coupled with a low diffusivity. The hydrogen, therefore, tends to remain in the austenitic weld metal where it is rendered harmless.

Microstructural Control

The risk of hydrogen cracking in the HAZ increases with hardness. For a given weld cooling rate, HAZ hardness is a function of the carbon equivalent of the pipe and fitting steels (hardness increasing with carbon equivalent).

Carbon Equivalent

Records that contain chemical composition information for older, existing pipelines are usually difficult or impossible to locate. In these cases, estimates of chemical composition are often made based on the maximum allowable limits of the specification to which the steels were produced. This usually results in an overestimation of the tendency for unacceptably high hardness levels to result and is therefore restrictively overconservative. The chemical composition of an in-service pipeline can be determined by removal of samples for laboratory analysis, provided that care is taken in removing and analyzing the samples. The chemical composition of fitting or sleeve steels can be controlled by proper use of purchase specification requirements, although low-carbon-equivalent steel is not always available. The buttering technique can be used to deposit a layer of weld metal, which has an inherently low carbon equivalent, onto the weld preparation of higher carbon-equivalent fitting steels using preheating prior to installation. The buttering technique can also be used to deposit a layer of weld metal onto the pipe steel under low restraint conditions. The welds attaching the branch or fitting are then deposited onto the low-carbon weld metal. This procedure, although cumbersome, results in a decreased risk of hydrogen cracking when welding onto high-carbon-equivalent pipe steels.

Weld Cooling Rate

The HAZ hardness in pipeline steels will increase as the cooling rate through the transformation temperature range increases, due to the formation of lower-temperature transformation products, such as martensite and lower bainite.

The cooling rate of welds made onto in-service pipelines is a function of the welding parameters and the pipeline operating conditions. Heat input can be increased by increasing the welding current or by decreasing the travel speed. Increasing the heat input results in a decrease in the weld cooling rate. Preheating is used in conventional welding applications to reduce weld cooling rates. Preheating using conventional techniques (e.g. flame methods) is difficult due to the ability of the in-service pipeline to remove heat from the pipe wall, particularly with thinner wall pipe. Induction heating coils with large power sources have been used to reduce the cooling rates of welds made onto in-service pipelines, however. The development of procedures for welding onto in-service pipelines has focused primarily on the use of high-heat input levels.

Pipeline operating conditions that affect welding cooling rates include the pipeline contents, flow rate, pressure (gas only), pipe surface, ambient temperature, and pipe wall thickness. Unlike in other welding applications, as pipe wall thickness decreases, the cooling rate of welds made onto in-service pipelines increases. Welds made onto thin-wall pipe are made in close proximity to the flowing pipeline contents, which removes heat from the pipe wall. This effect becomes less prominent as pipe wall thickness exceeds 0.500 in. and reverts back to a conventional relationship due to the thermal mass at greater wall thicknesses. Preheating of thicker wall pipe is often practical as the result of this less prominent effect. Also, the gap between the pipe and the fitting often allows effective use of preheating of the fitting material and can reduce bulk weld cooling rates.

Subsequent Thermal Treatment

Post-weld heat treatment (PWHT) is used in conventional welding applications to reduce hardness level by tempering. As with the application of preheating, PWHT is difficult due to the ability of the in-service pipeline to remove heat from the pipe wall, particularly with thinner wall pipe. However, induction heating has been used for this purpose. The use of PWHT also allows for hydrogen removal by diffusion.

Weld bead deposition sequence can influence the resulting HAZ hardness. The stringer bead technique, with the last pass deposited on a previous pass (i.e., the last pass is not allowed to impinge on the pipe steel), has been shown to result in a lower HAZ hardness level than the stacked wave bead technique (Ref 12). The use of a suitably developed multi-pass or temper bead procedure can be used to reheat the hard HAZ constituents of one pass by a subsequent pass. This can result in retransformation to form softer microstructures, or tempering of the hard structures without subsequent transformation. The use of buttering layers as part of a temper bead procedure is also beneficial in that weld cooling rates in the fitting side of the joint are reduced by increasing the local thickness of the run pipe.

Stresses Acting on the Weld

The tensile stress required for HSC can be applied externally or can develop as a residual stress. Examples of external loads that can occur during or soon after welding include loads imposed by pipeline movement or by the back-filling operation. Applied stresses should be minimized by restricting pipeline movement and preventing soil settlement. Residual stresses develop as the result of the restraint imposed by the welded assembly and the thermal contraction of the weld as it cools. The restraint of a weld is often difficult to control, although a certain degree of control can be exercised by altering the design of the welded connection. For example, using unnecessarily thick fitting materials and/or over-matching strength electrodes results in a higher level of weld stress than a properly designed connection. Lower-strength electrodes allow strains to be accumulated in the less-crack-susceptible weld metal (as opposed to the HAZ). In production, the stress imposed on two identical joints may vary considerably. For instance, a circumferential weld made on the first end of a full-encirclement fitting will be lightly restrained, but the identical weld on the opposite end will be highly restrained. Opposite end tack welds are of particular concern, because a higher level of stress can occur as the result of their relatively small cross-sectional area. Stresses can also be reduced by minimizing the gap between the fitting and the pipe.

REFERENCES

1. British Standards Institute, Public Document 6493, 1991
2. S.J. Maddox, *Fatigue Strength of Welded Structures,* Abington, 1994
3. T. Jutla, Fatigue and Fracture of Weldments, *ASM Handbook,* Vol 19, *Fatigue and Fracture,* ASM International, 1996

4. G.S. Booth, Ed., *Improving the Fatigue Performance of Welded Joints,* TWI, 1983
5. *Fatigue Handbook—Offshore Steel Structures,* TAPIR, 1985
6. P.W. Marshall, Fatigue Design Rules in the USA *Proc. 9th Int. Conf. on Offshore Mechanics and Arctic Engineering,* Vol III, Part A, ASME, 1994, p 175–184
7. "Structural Welding Code-Steels," AWS D1.1, American Welding Society, 1994
8. P.F. Timmins, *Solutions to Hydrogen Attack in Steels,* ASM International, 1997
9. P.H. Threadgill, Avoiding HAZ Defects in Welded Structures, *Met. Mater.,* May 1985, p 422–429
10. J.F. Kiefner et al., "Further Studies of Two Methods For Repairing Defects in Pipelines", NG-18 Report #112, Battelle Columbus Laboratories
11. P.L. Threadgill and M.J. Cola, "Criteria for Hot Tap Welding," Edson Welding Institute, May 1989
12. J.F. Kiefner et al., "Development of Guidelines for Repair and Hot-Tap Welding on Pressurized Pipelines," Battelle Columbus Laboratories, Sept 1981

CHAPTER 6

Solutions to Equipment Failures in Hydrocarbon Processing

Solutions to equipment failures in hydrocarbon processing, including piping and process systems, encompass the inspection (monitoring), maintenance, repair, life assessment, and life extension of the various engineering components used. For the various types of equipment in process systems, the significance of equipment failure can be ranked as a function of failure rate and the cost of failure. The types of process equipment most important to controlling failure rate or loss (listed in decreasing order of importance) are piping, reactors, tanks, and process towers (Fig. 6-1). This ranking suggests the priority areas for inspection, maintenance, repair, life assessment, and life extension, as described in Chapter 2.

Maintenance can have a dramatic impact on profitability (Fig. 6-2). Corrective maintenance or unplanned maintenance is the most costly. In Fig. 6-2, this form of maintenance is reflected as unity on the unit cost basis shown. Preventive maintenance (planned maintenance on a fixed time scale) is ~60% of the cost of corrective maintenance. Predictive maintenance (maintenance on a sliding time scale) is ~40% of the cost of corrective maintenance. Clearly, predictive maintenance is most desirable. If inspection and maintenance are organized and performed in the predictive mode, then cost will be reduced, equipment failure rate will decrease, and the profitability of the hydrocarbon processing operation will consequently increase.

This chapter focuses on predictive maintenance and inspection, with particular emphasis on the control and prediction of equipment damage for life extension and failure rate reduction.

Predictive Maintenance and Inspection

Predictive maintenance and inspection differ from traditional preventive maintenance in that key equipment parameters are monitored as preliminary indications of failure. Vibration monitoring and signature analysis of rotating or reciprocating equipment is one industrial example of a predictive maintenance and inspection system.

Another example is a leak detection system with acoustic emission sensors. The objectives of this approach are to monitor the performance and failure rates of critical components or subcomponents and to reduce age-related failures in the traditional "bathtub" failure rate curve (Fig. 6-3).

Implementation of a predictive maintenance and inspection program requires the identification of critical components or weak links that could justify continuous monitoring or detailed time-based examination. For example, a risk matrix and an FMEA are ways of identifying equipment priorities for failure reduction. Each major piece of equipment is then critically examined in terms of weak links and their failure mechanisms.

Many failures involve more than one factor; that is, one process may initiate cracking and other processes may eventually lead to failure. For example, fatigue failures can start at corrosion pits, as shown in Chapters 1 and 2. Failure mechanisms for a given component or process should be understood for effective implementation of predictive maintenance.

Evaluation of defects and the type of analysis applied often depend on the failure mechanism or mode of operation that led to the defect (Ref 1). For instance, a manufacturing defect that is found after some years of service requires a remaining-life analysis based on fracture mechanics (see Chapters 3 and 5). A defect produced by misuse of equipment may require a different kind of analysis. Because it is very important to

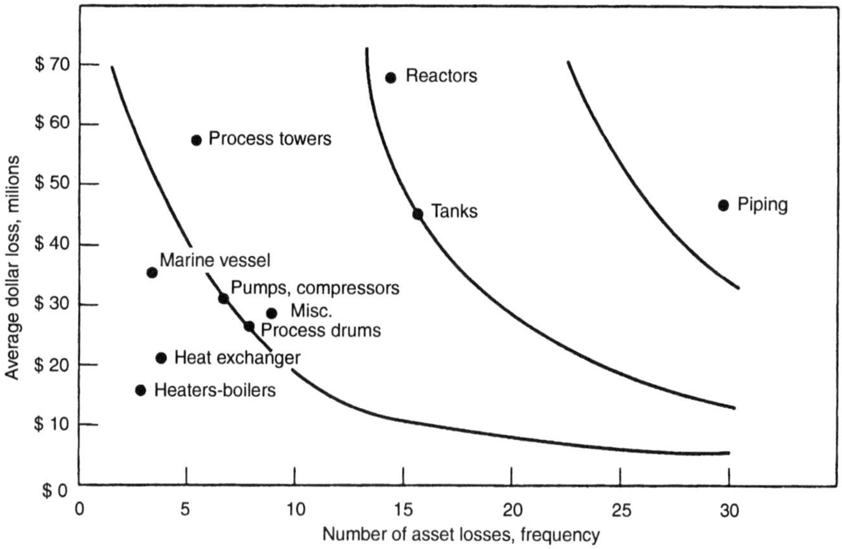

Fig. 6-1 Asset loss risk as a function of equipment type. Source: W.G. Garrison, Loss Prevention, *Hydrocarbon Processing,* Sept 1988

determine the root cause of a defect in order to apply the appropriate analysis, an understanding of the applicable failure categories is necessary, as outlined in Chapters 1 through 5.

Nondestructive testing and inspection techniques are being improved and used more often to reduce the observed, increasing failure rates and to improve safety of chemical and refinery plans, for example. Early meaningful detection of deterioration can provide the basis for repairs or changes in operating conditions that may extend equipment life.

Useful nondestructive methods (Table 6-1) include not only conventional methods (e.g., ultrasound, radiography, magnetic particles, and dye penetrants) but also eddy current testing, capacitive strain gage testing, replication metallography, and borescope examination of internal piping.

Over the years, experience and advancements in nondestructive examination (NDE) techniques and equipment have led to an improvement in the ability to detect defects in process equipment. This improvement has led to the detection of original manufacturing defects in existing process equipment that may not have been detected at the time of manufacture. This does not necessarily indicate a lack of quality in older process equipment, but it can indicate that many manufacturing defects have no effect on the integrity or service life of equipment and that a high level of confidence can be placed in the current NDE technology.

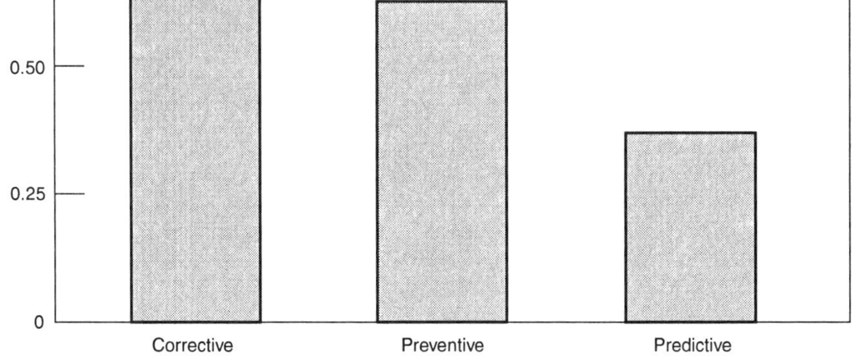

Fig. 6-2 Relative cost of maintenance approaches. Source: Power Plant Diagnostics Go On-Line, *Mechanical Engineering*, Dec 1989

Once a piece of equipment is in service, it cannot be assumed that it will not require further evaluation. It is imperative to obtain and maintain all the design documents for process equipment and never to accept shipment of new equipment or changes without those documents. It is far easier to verify design drawings and engineering changes (see Chapter 1) than to destructively examine equipment in order to resolve an issue. A typical example is weld backing rings left in place on the inside of a vessel. These backing rings assist the manufacturer in making the closing weld on process equipment. If the backing ring is discovered during an in-service radiographic examination, the analyst should determine whether the material is compatible with the process fluid and whether the ring is integral or protected from contact with the process fluid. This can be resolved when thorough documentation is available; otherwise, it may be impossible to determine without destructive examination. Other improper designs can make use of the incorrect material for the service or process that the equipment will support. Poor construction details can lead to in-service failure. For example, rough edges or joints can collect corrosive material that leads to pitting and accelerated degradation.

During examination, indications should be characterized and recorded as to type and size. Each indication should then be evaluated. The first step in classifying indications is to compare them to the acceptance criteria of the applicable consensus code (e.g., ASME Section VI). All indications that meet the code acceptance criteria (Ref 2, 3) are considered adequate, and no further evaluation is required. Indications that exceed the

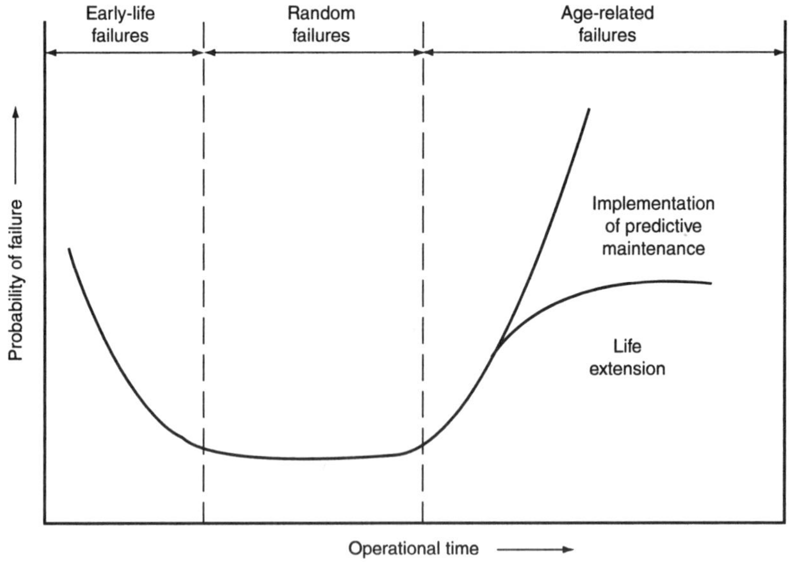

Fig. 6-3 Typical reliability curve illustrating the conceptual objective of life extension

code acceptance criteria are normally considered defects and require additional evaluation. Code acceptance criteria are established based on quality control levels that are arbitrary yet conservative. As such, indications that do not exceed the code acceptance criteria are accepted without further consideration. Indications that exceed the code acceptance criteria are rejectable based on that code, but they are not necessarily unacceptable. Further analysis as to their fitness for continued service is required.

Corrosion Monitoring

Corrosion monitoring has become an important aspect of the design and operation of modern industrial plants, because it alerts inspection and maintenance personnel to damage caused by corrosion and the rate of the deterioration. A variety of techniques are available for corrosion monitoring in plant corrosion tests. The most widely used and simplest method involves the exposure and evaluation of the corrosion in actual test coupons (specimens). The ASTM standard G4 was designed to provide guidance for this type of testing.

Table 6-1 Inspection and measurement techniques for corrosion control

Technique	Description
Acoustic emission	Measures the location, initiation, and propagation of cracks and defects under stress in metals
Dye penetrant	Simple procedure for locating surface cracks. Requires shutdown for internals
Magnetic particle	Surface and subsurface crack and defect location, seams, and inclusions
Radiography	X-ray and gamma-ray (Co^{60}, Ir^{192}) wall-thickness measurement and crack and defect location
Thermography	Identifies lack of bond, hot spots, local thinning, and temperature changes due to poor/wet lagging
Ultrasonics	Indicates internal defects, porosity, lack of fusion, crack location, and wall thickness
Visual exam	Identifies localized corrosion, erosion pitting, deposit scaling and fouling problems, and staining and corrosion leakage due to cracking
Sentinel holes	Small telltale holes drilled from outside of carbon steel pipework to depth of corrosion allowance. Weepage indicates replacement is required
Weight loss coupons	Traditional method of limited sensitivity but used in all environments. Type of corrosion observed is an important indicator
Electrical resistance	Measures change in probe resistance. Widely used for carbon steel fabrications in gas and liquid phases. Automated readings
Corrosion potential	Identifies corrosion conditions (i.e., active, passive, pitting, stress-corrosion cracking regimes) in ionically conductive media
Linear polarization resistance	Al electrochemical dc method used for measurement of uniform corrosion. Standard electrochemical technique. Typically requires a conductive electrolyte, but some new probes have a conductive separator between the metal probe elements
Zero resistance ammetry	Established method for assessing galvanic corrosion between dissimilar metals, but can be used with nominally "identical" electrodes in some applications
Hydrogen probes	Measures rate of diffusion of hydrogen through steels, either by means of a pressure gage or by electrochemical techniques
Thin layer activation	Measures change of radioactivity as a local irradiated area corrodes
Electrochemical impedance	An ac method used for general corrosion measurements similar to linear polarization resistance. More versatile and accurate than dc measurements
Electrochemical noise	A more recent technique used for assessing general corrosion and potential fluctuations associated with localized corrosion

When selecting a corrosion monitoring method, a variety of factors should be considered. First, the purpose of the test should be understood by everyone concerned with the corrosion monitoring program. The cost and applicability of the methods under consideration should be known, and it is also important to consider the reliability of the method selected. In many cases, it will be desirable to include more than one method in order to provide more confidence in the information generated.

Monitoring Locations

A principal part of any corrosion monitoring program is deciding where to locate the monitoring devices. Because corrosion will probably not occur uniformly throughout the plant, it is desirable to find sites at which the highest corrosion rates will be experienced.

The problems involved in developing corrosion monitoring programs for a plant are exemplified by a distillation column in a refinery crude unit. The most logical points for corrosion monitoring in a distillation column are the feed point, the overhead product receiver, and the reboiler or bottoms product line. These points are the locations at which the highest and lowest temperatures are encountered, as well as the points at which the most and least volatile products are concentrated. A monitoring program can be restricted to the most corrosive location within the column once this area has been identified. The problem reduces to identifying the principal structural element (see Chapter 2) for that vicinity.

Another problem with distillation columns is that the liquid on trays tends to be frothy, which creates difficulties for electrochemical methods. One solution is to install bypass loops that remove liquid from the column, pass it over the corrosion monitoring probes, and reinject it at a lower point in the column. This practice avoids the problem of foam and froth and provides a more controlled flow rate over the corrosion monitoring equipment. Use of a bypass loop also allows removal of liquid samples at times of high corrosion rates.

Redundancy is also important in designing corrosion monitoring programs. The use of at least two different types of monitoring devices at any location is often desirable. For example, an electrical resistance probe with a polarization resistance probe allows measurement of both instantaneous corrosion rates and an average corrosion rate. The data thus obtained can be correlated, and this is very helpful in identifying spurious or inaccurate readings.

In another approach, the polarization resistance probe is weighed before and after the test in order to correlate mass loss with the average corrosion rate that the probe suffered. There is reason to expect that the electrochemical value is in error if the average corrosion rate and the mass loss of the probe do not agree. Also, to obtain time-averaged corrosion rate values, independent coupons can be installed together with polarization resistance probes.

A variety of corrosion monitoring approaches must be used when designing pilot or demonstration plants. Coupon tests can be very helpful in selecting optimum materials for processes based on pilot plant experience. Polarization resistance monitoring is very useful for determining whether certain processing conditions cause corrosive situations to develop. Because the corrosion mechanisms may not be understood, and because the result of erroneous information can be serious overdesign or exposure to unexpected hazards, redundancy in the design of such monitoring systems is important in pilot plants.

The process stream can be sampled in locations other than distillation columns. A sample tap can be helpful in conjunction with polarization resistance devices. The alarm on polarization resistance monitoring equipment can be used to signal the need to remove samples. This is particularly desirable in the case of pilot plant operation, in which wide variations in processing conditions are encountered. It is also helpful in plants that produce different products in the same equipment.

High-velocity gas streams in pipes can cause problems with conventional monitoring systems. In this case, the presence of an aqueous phase is usually restricted to a thin layer on the surface of the pipe. A probe that protrudes into the pipe may miss the liquid layer that is present only close to the pipe wall. A flush-mounted surface probe can be used in such cases. This probe permits the measurement of polarization resistance in order to estimate the corrosion rate of the pipe wall.

Probe location is also critical in storage tanks containing nonaqueous liquids. The most corrosive location in these tanks can be at the liquid level if the liquid in the tank has a density exceeding that of water. In this case, the corrosion monitoring probe should be mounted on a floating platform so it can detect the presence of a corrosive aqueous phase. However, when the liquid stored in the tank is less dense than water, the probe should probably be positioned at the bottom of the tank.

Interpretation and Reporting

In-service monitoring, more than any other type of corrosion testing, requires the utmost skill in interpretation and reporting. Important economic decisions are often based on the test results. A number of standards provide guidelines for certain procedures, but none is comprehensive. To plan an appropriate test program, an inspector should know or have good advice on both the chemistry and the mechanics of the processes involved. That is, an inspector should understand the entire corrosion system and be adequately trained as outlined by the Flixborough Court of Inquiry (see Chapter 1). There should be a strong emphasis on the strategy of the program and a searching analysis of the test results. It is important to prepare detailed records of what is done as part of the inspection practice, and it is also important to document any unplanned changes that occur in the process stream or the equipment during inspection. Without a valid interpretation and effective (timely) reporting, the price of the work can be significantly greater than the cost of time and materials. The consequences of corrosion failures go beyond additional costs. Personal safety risks and liability, hazard potentials, product quality, and pollution problems are involved.

Corrosion Monitoring Methods

The primary purpose of corrosion monitoring is to estimate the condition of equipment so that planned turnarounds can be scheduled by maintenance personnel and replacement of equipment can be anticipated. Frequent measuring also helps minimize the duration of corrosive upset conditions.

Corrosion Coupons

If properly located, the coupon is a reasonably accurate tool. In addition to producing corrosion data in measurable terms (mpy or mm/yr), the coupon is helpful in identifying the type of corrosion activity present at the point where the coupon is located in the system. Pitting on the coupon surface indicates activity from a strong acid or acid salt, such as hydrochloric acid or ammonium chloride. General thinning indicates the activity of hydrogen sulfide on the coupon.

Coupons are widely used to monitor inhibitor programs, for example, in water treatment or refinery overhead streams. With retractable coupon holders, the coupons can be extracted from the process without having to shut down in order to determine the corrosion rate. Coupons can be designed to detect such phenomena as crevice corrosion, pitting, and dealloying corrosion. For example, some pulp mill bleach plant washer drums are electrochemically protected to mitigate crevice corrosion. Specifically designed crevice corrosion test coupons are used to monitor the effectiveness of the electrochemical protection program. These coupons are periodically removed from the equipment and examined for evidence of crevice corrosion.

Coupon testing does have important limitations. First, coupon testing cannot be used to detect rapid changes in the corrosivity of a process. Second, localized corrosion cannot be guaranteed to initiate before the coupons are removed, even with extended test durations. Coupons only show corrosion that has already taken place, and a single coupon will not show whether corrosion was uniform or occurred all at once. Generally, a retractable coupon should be the in the system for 20 to 30 days before reliable information can be obtained. The coupon will not record upset conditions rapidly or measure the true corrosion activity in the system if it is improperly placed.

Third, the calculated or measured corrosion rate of the coupon may not translate directly to that of the equipment. Despite every effort to achieve equivalence, differences in mass and coupon area/solution volume ratio are usually sufficient to render direct comparison meaningless. For example, the metal surface temperature of the coupon is governed by the process stream that surrounds it. Where a cooling medium is present on the other side of the metal surface, the skin temperature of the tube is lower than the temperature of the surrounding hydrocarbon. The cool tube skin could cause surface water condensation at temperatures above the anticipated water dew point in the process fluid. Coupons would not indicate corrosion activity of this nature. Useful correlations can be established by monitoring the corrosion rate of the equipment with ultrasonic thickness monitoring and by comparing this corrosion rate with the calculated rate for equivalent coupons.

Lastly, certain forms of corrosion cannot be detected with coupons. The principal limitation is the simulation of erosion-corrosion and heat transfer effects. Careful placement of the coupons in the process equipment can slightly offset these weaknesses.

Erosion-corrosion is related to process turbulence, and process turbulence is often a function of equipment design. Because coupons tend to shield one another from the effects of process turbulence, field coupon testing is not reliable as a method of simulating erosion-corrosion.

For heat transfer effects, specially designed coupons are required that simulate effects such as those found in heating elements or condenser tubes. Coupons range in design from thermowell-shaped devices to sample tubes in a test heat exchanger. Thermowell-shaped devices are heated or cooled on the inside and project into the process stream. Heat transfer tests can also be conducted in the laboratory. In this environment, the coupon forms part of the wall of the test vessel and can therefore be heated or cooled from one side. Because of the cost involved, heat transfer coupon tests are usually carried out on only one (or perhaps two) alloys that have been selected from a larger group.

Preparing, Installing, and Interpreting Coupons

Corrosion coupons can be flat or cylindrical and can be installed in any accessible location. It should be remembered that coupons measure corrosion only where they are placed. Coupons show corrosion that has already taken place, and a single coupon will not show whether the corrosion was uniform or occurred all at once.

Different types of coupon holders or chucks are used, depending on the system, the pressure, the location, and other factors. Most coupons are run in a 25 or 50 mm (1 or 2 in.) threaded plug. Flat coupon holders hold two coupons, while cylindrical coupon chucks can contain eight or more. The multichuck coupons allow a coupon to be pulled at intervals to see if the corrosion rate is uniform or not.

High-pressure systems require a special coupon check and insertion device. The insertion tool fits into a special attachment on the pipe or vessel that has a high-pressure chamber with a valve on each end. The inner valve is closed, the retrieval tool inserted, and the inner valve opened. The tool is then run in and left. The procedure is reversed to remove the coupon.

The industry guide for preparing, installing, and interpreting coupons is NACE standard RP-07-75. The primary consideration is that all coupons should be treated exactly alike. A method of preparation that does not alter the metallurgy of the coupon is required. Grinding and sanding of coupons should be controlled to avoid metallurgical changes and to provide a consistent and reproducible surface finish.

Coupons should be handled carefully and stored under noncorrosive envelopes until they are installed. Rust spots caused by improper handling (e.g., fingerprints) can initiate a pit that is not representative of the system being evaluated. Prior to installation, the weight, serial number, date installed, name of system, location, and orientation of the coupon and holder should be recorded. A coupon is left in the system for a predetermined number of days and then removed.

When a coupon is removed, the serial number, date removed, observations of any erosion or mechanical damage, and appearance should be recorded. A photograph of the coupon can be valuable in some cases. The coupon should then be placed in a moisture-proof envelope impregnated with a vapor phase inhibitor and taken immediately to the laboratory for cleaning and weighing. A coupon can be blotted (not wiped) dry prior to being placed in the envelope. The laboratory receives the coupon and inspects, cleans, and weighs it. A report is issued describing any thickness loss, pitting observations, and other observations of interest.

Electrical resistance probes are specially designed corrosion coupons. Their corrosion rate is calculated from measurement of electrical resistance rather than mass loss. These measurements are made by installing a wire or other device fabricated from the material in question in such a way that its electrical resistance can be conveniently measured. Corrosion reduces the cross section of the exposed element; therefore, its electrical resistance will increase with exposure time if corrosion is taking place. A temperature-compensating element should be incorporated in such a probe because the resistance of the probe is also influenced by the temperature.

Electrical resistance probes measure the remaining average metal thickness. To obtain the corrosion rate, measurements are made over a period of time, and the results are plotted as a function of exposure time. The corrosion rate can be determined from the slope of the resulting plot. There are several advantages to this approach. Because probes are relatively small, they can be installed easily. For determination of the metal remaining, the probe can be wired directly to a control room location or to a portable resistance bridge at the probe location.

There are also some disadvantages to this approach. Vessel and piping walls must be penetrated in order to install the probes; such penetration results in potential leaks. It is expensive to direct-wire the probe to a control room location, and such work must be carried out with care to avoid spurious signals and errors. On the other hand, it is time consuming and sometimes impossible to take measurements at the probe site with a portable bridge. The temperature compensation device reacts slowly, and it can be a source of error if the temperature varies when the measurement is taken. Corrosion rate measurements obtained in short periods of time can be inaccurate because the method measures only the remaining metal, not the rate of attack; this increases the signal-to-noise ratio in short exposures. This method provides no information on localized attack.

Ultrasonic thickness measurements can be used to monitor corrosion rates in situ. Ultrasonic thickness measurements involve placement of a transducer against the exterior of the vessel in question. The transducer produces an ultrasonic signal. This signal passes through the vessel wall, bounces off the interior surface, and returns to the transducer. The thickness is calculated by using the time that elapses between emission of the signal and its subsequent reception, along with the velocity of sound in the material. To obtain a corrosion rate, a series of measurements should be made over a time interval, and the metal loss per unit time should be determined.

Measurement errors can occur when vessel walls are at high or low temperatures. Serious problems can exist in equipment that has a metallurgically bonded internal lining, because it is not obvious from which surface the returning signal will originate. Despite these drawbacks, the ultrasonic thickness approach is widely practiced where it is necessary to evaluate vessel life and suitability for further service. It should be kept in mind, however, that depending on the type of transducer used, the ultrasonic thickness method can overestimate metal thicknesses when the remaining thickness is less than ~1.3 mm (0.05 in.).

Polarization Resistance Measurement

Unlike the previously discussed methods, which provide information on remaining thickness, the technique of polarization resistance provides an estimate of the corrosion rate. The theory behind the technique is that the corrosion rate of a probe is inversely proportional to its polarization resistance—that is, the slope of the potential-current response curve near the corrosion potential. It is necessary in a plant situation to use a probe that enters the vessel in the area where the corrosion rate is desired. The electrodes of the probe are fabricated from the material in question. An electronic power supply polarizes the specimen ~10 mV from the corrosion potential. The resulting current is recorded as a measure of the corrosion rate. Polarization resistance yields an instantaneous estimate of corrosion rate.

There are several limitations to this approach. The corroding environment should be an electrolyte with reasonably low resistivity. High resistivity electrolytes produce erroneously low corrosion rates. The vessel wall must be penetrated, and this involves concerns regarding leaks, personnel safety, and other problems. The ability to use direct wiring from the probe location to a remote control room is desirable, but the installation of these wiring systems is costly. In addition, these systems do not provide information on localized corrosion, such as pitting and stress-corrosion cracking. Also, the corrosion rate values are approximate at best, and the method is best suited for use during periods when substantial corrosion rate changes occur.

Measurement of Corrosion Potentials

The use of corrosion potential measurements for in-service corrosion monitoring is not as widespread as the use of polarization resistance. However, this approach can be valuable in some cases, particularly where an alloy could show both active and passive corrosion behavior in a given process stream. For example, stainless steels can provide excellent service as long as they remain passive. However, if an upset occurs that would introduce either chlorides or reducing agents into the process stream, stainless alloys can become active and exhibit excessive corrosion rates. Corrosion potential measurements indicate the development of active corrosion and can be coupled with polarization resistance measurements as additional confirmation of high corrosion rates.

The success of corrosion potential measurements depends on the long-term stable performance of a standard reference electrode. Such electrodes have been developed for continuous pH monitoring of process streams, and their application for measuring corrosion potentials is straightforward. However, temperature, pressure, electrolyte composition, pH, and other possible variables can limit the applications of these electrodes for corrosion monitoring service.

Hydrogen Probes

The concept of the hydrogen probe is based on hydrogen being one of the cathodic reaction products in non-oxidizing acidic systems. The hydrogen atoms thus generated diffuse through the thickness of the vessel and are liberated at the exterior surface. Hydrogen probe analysis measures the corrosion rate, unlike ultrasonic thickness measurements and other techniques that measure remaining wall thickness. However, hydrogen probe analysis is limited to systems in which the temperature is close to ambient and the diffusion rate of hydrogen is high. Gas pipeline service is the most common application. In this case, corrosion can occur when hydrogen sulfide, water, and sometimes carbon dioxide are present. This approach has a variation that consists of simply attaching a chamber to the exterior of the pipe and monitoring hydrogen liberation through increasing pressure.

Commercial devices are available for corrosion monitoring by this technique, although it is questionable whether such devices could be positioned and allowed to operate unattended for extended periods of time. In addition, these units have all of the problems associated with any electrochemical measuring device—namely, the need for complex electronic equipment and wiring, as well as operators and installers with a sensitivity to the requirements of such equipment. Also, this method is in practice limited to steel, which has a high hydrogen diffusivity and low solubility of hydrogen. Exterior hydrogen monitoring does not supply a quantitative measurement of hydrogen damage.

Analysis of Process Streams

Another useful in-service corrosion monitoring technique is analysis of the process streams for the presence of corrosion products. This straightforward approach usually does not require the installation of specialized equipment. For example, process streams from the bottom of the distillation column can be routinely sampled and atomic absorption analysis techniques can be used to determine such heavy metals as iron, nickel, and chromium at very low levels. The concentration of such impurities is then directly proportional to the corrosion rate multiplied by the area of metal corroding, if the only source of metal ions is corrosion and if the corrosion products are not precipitating. One problem is that the corroding area may not be known with certainty; if not, the results are relative. However, they do help to determine whether conditions have improved.

Predictive Corrosion Control

Predictive corrosion control follows the predictive maintenance approach by reducing age-related failures (Fig. 6-3). The general approach involves the collection and systematic analysis of operational corrosion behavior and data on key plant equipment critical for safety or reduced failure rates. A general example of a predictive control flow chart is shown in Fig. 6-4.

A key step in developing a predictive corrosion control program is to identify critical components in a given process unit, based on a risk assessment such as that in Fig. 6-1 (or some other prioritization criterion, as described in Chapters 1 and 2). The first step is to subdivide each process unit into functional sections (e.g., preheat train, heater, reactor, overheads, pumparounds, etc.). For a given unit, it is helpful to sketch the system under consideration and draw in areas of concern (Fig. 6-5). Key components can be identified with a HAZOP, FMEA (Chapter 1), and a risk analysis. For example Fig. 6-1 ranks piping high, followed by reactors. The piping with the greatest risk in a

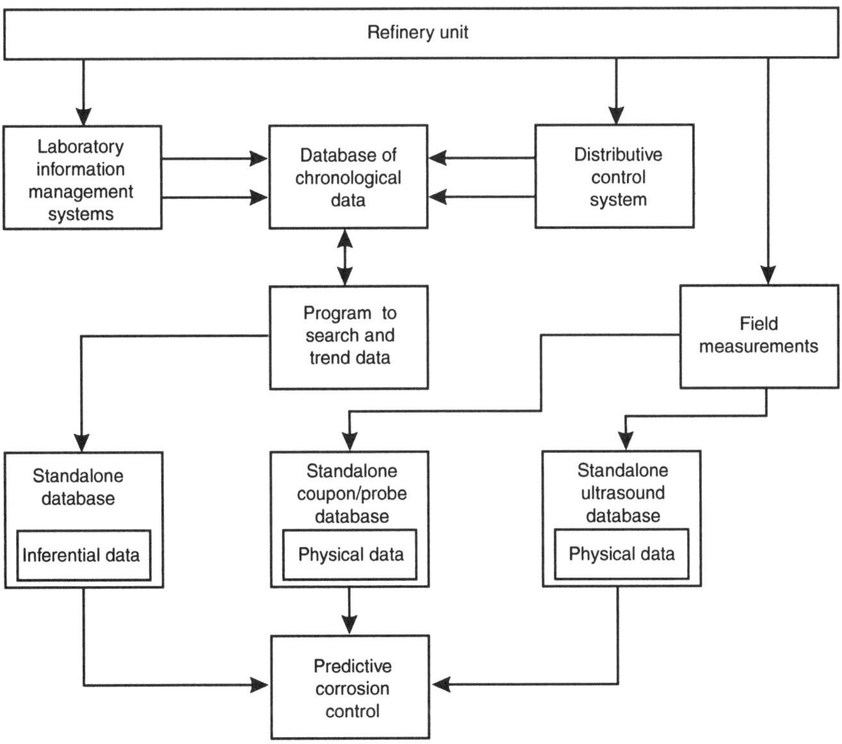

Fig. 6-4 Flowchart of a typical predictive corrosion control system

given unit or section is determined from a knowledge of all the degradation mechanisms taking place.

The next step is to analyze the corrosion mechanisms and failures and determine critical areas for monitoring. Sometimes sections are divided according to the mechanisms of aqueous corrosion or high temperature. For example, aqueous corrosion is predominant in overheads and pumparounds, whereas high-temperature corrosion is the main concern in a preheat train, reactor, or bottoms. Aqueous phase corrosion is generally controlled by process control and chemical inhibition, because the mechanisms are typically wet hydrogen sulfide, ammonium bisulfide, carbonates, hydrogen

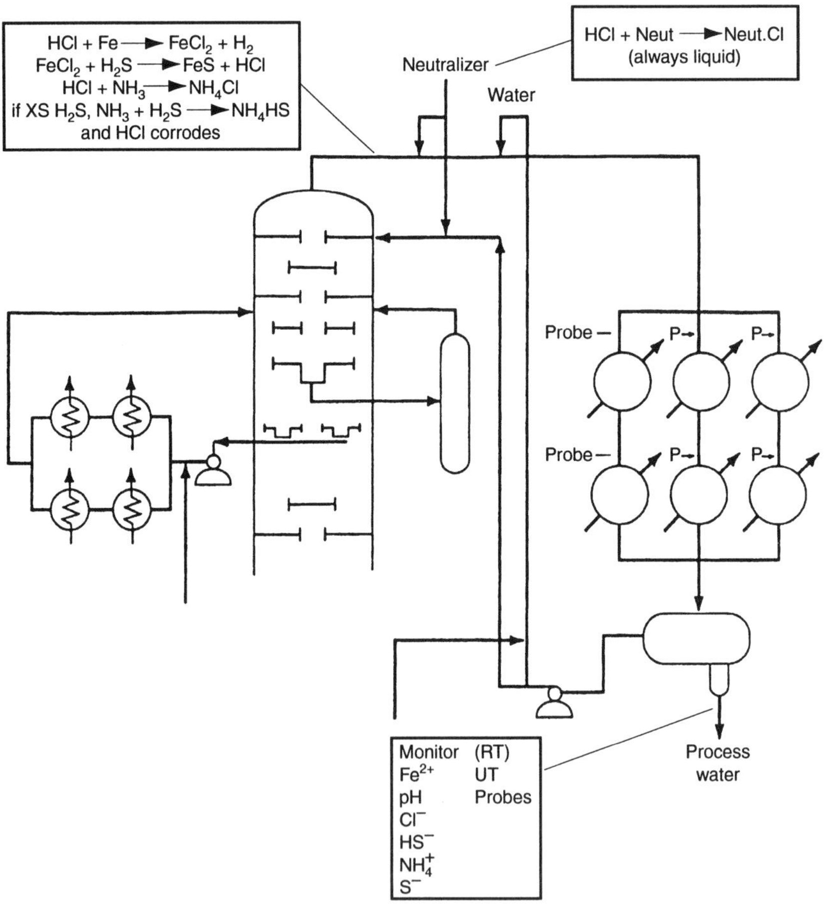

Fig. 6-5 Example of system schematic used to define critical points for corrosion control

chloride, and ammonium chloride. High-temperature corrosion is generally controlled by metallurgy, because the mechanisms are typically hot sulfur attack and hot hydrogen attack.

Complete understanding of the corrosion mechanisms and the system is essential for further development of a predictive corrosion control program with a given unit or section. To illustrate the complexity of such a program, corrosion factors for a crude unit overhead are summarized in the following section.

Corrosion Mechanisms in a Crude Unit Overhead

A typical crude unit overhead system is representative of the aqueous corrosion experienced in many refinery systems. In a crude unit, the most aggressive corrosion is typically from concentrated hydrochloric acid (HCl). A substantial portion of the acid comes from ammonium chloride near the dew point of water in a system. Few metallurgies will withstand the combination of corrosives that are found near the dew point of water in a sour crude unit overhead system. Consequently, mitigation and corrosion control of carbon steel handling ammonium chloride (NH_4Cl) is generally by filming inhibitors and sustained wash water rates.

Oxygen and Oxidizing Agents

Oxygen and other materials present in the feedstock (e.g., ferric chloride and cupric chloride) act as oxidizing agents and can play a role in the corrosion process that occurs on any unit. In an environment such as the crude unit, free oxygen is generally not a source of corrosion attack, because oxygen readily reacts with hydrogen sulfide (H_2S) to form elemental sulfur:

$$O_2 + 2H_2S \rightarrow 2H_2O + 2S \downarrow$$

The presence of free sulfur in an overhead system confirms that oxygen is getting into the system. Elemental sulfur cannot distill up a crude tower. Sulfur has a boiling point of 444 °C (831 °F) and would be in the bottoms from the fractionator if it were present in the crude.

It is widely recognized that the presence of oxygen or oxidizing agents rapidly accelerates corrosion in a process unit. In one laboratory study, the following relationship between oxygen and H_2S was observed:

Corrodent, 6.0 pH	Corrosion rate of metal loss, mm/yr (mils/yr)
H_2S only	0.76 (30)
O_2, 1 ppm	0.35 (14)
O_2, 1 ppm, plus H_2S	3.0 (120)

As can be seen from these data, there is a synergistic influence on corrosion when the corrosives are present together. In this case, the rate of metal loss was increased by four times over the highest corrosion shown by either corrosive. In a second study, corrosion activity increased from 0.7 to 2.4 mm/yr (27 to 95 mils/yr) when oxygen was added to H_2S. A similar increase in corrosion activity would be present with other oxidizing agents, such as ferric and cupric chloride.

Oxygen contamination sources include water washes. In one instance, corrosion rates more than tripled when accumulator washwater rates were recycled ten times rather than three. Oxygen and other oxidizing agents can be introduced with crude feedstock, slop oils, and desalter washwater. Oxygen levels in water must be controlled to be < 50 parts per billion (wt).

Hydrochloric Acid

Hydrochloric acid is a major cause of corrosion in most crude units. The acid is formed in the crude preheat exchanger and in the crude atmospheric furnace by hydrolyzing magnesium chloride and calcium chloride present in the crude to form hydrogen chloride. The reactions are of the type:

$$MgC_2 + 2H_2O \rightarrow Mg(OH)_2 + 2HCl$$

In the overhead, tower top, or top pumparound near the water dew point, a low pH condition is produced by the corrosion mechanism:

$$2HCl \rightarrow 2H^+ + 2Cl^-$$

$$2HCl + Fe \rightarrow FeCl_2 + H_2 \uparrow$$

At lower temperatures, or at an elevated pH in the presence of H_2S, the corrosion mechanism is:

$$FeCl_2 + H_2S \rightarrow FeS + 2HCl$$

After hydrogen chloride dissolves in the water near the point of initial condensation, the strong acid attacks the metal surface, producing a pitted appearance. The corrosion products are metal chloride salts. Further, when the pH of the water in this region is below 5.0, HCl also dissolves any metal sulfide corrosion products that are present. These corrosion products tend to be carried away from the site of the corrosion attack. Hydrogen sulfide, a weak acid that is present in most systems, dissolves in the water where it reacts with the metal chloride— that is, ferrous chloride— producing iron sulfide. Iron sulfide (FeS) is insoluble in water and precipitates from the aqueous phase, causing fouling on the surface of the heat transfer equipment. Because FeS tends to precipitate from the water phase, HCl is then free to react a second or third time with the metal surface.

It is important to understand that the system is extremely dynamic, such that there is a tendency for small pockets to form where each corrosion reaction is occurring. All corrosion reactions tend to occur simultaneously in these systems. Mitigation and corrosion control of carbon steels handling HCl is by neutralization and/or inhibition by chemical treatments, in addition to sustained washwater rates.

Ammonium Chloride

Reactions involving ammonia and hydrogen chloride result in a vapor phase reaction between ammonia and hydrogen chloride gas:

$$NH_3 + HCl \leftrightarrow NH_4Cl_{solid}$$

In the presence of heat and a small amount of water:

$$NH_4Cl \rightarrow NH_4^+ + Cl^-$$

$$NH_4^+ + Cl^- \rightarrow NH_3 + HCl$$

This HCl is free to attack the metal surface:

$$HCl + Fe \rightarrow FeCl_2 + H_2 \uparrow$$

and in the presence of H_2S:

$$FeCl_2 + H_2S \rightarrow FeS \downarrow + 2HCl$$

Ammonium chloride (NH_4Cl) is formed between a strong acid and a moderately weak base. For this reason, the salt is an acid salt. Ammonium chloride has the ability to absorb water from steam at temperatures above the dew point of water. This can produce a saturated solution of NH_4Cl when at the boiling point of water, and the salt can easily have a pH as low as 3.3. Also, slightly moistened NH_4CI has good adhesive qualities. Because of this, the salt has a tendency to stick to wetted metal surfaces.

When the hygroscopic NH_4Cl contacts water, it ionizes into ammonium ions and chloride ions. Once the chloride ion enters the water it is very stable and stays in the aqueous phase. The ammonium is not nearly as stable and decomposes to a more stable hydrogen ion and ammonia gas (NH_3). Ammonia has a very high vapor pressure and tends to escape from the aqueous phase back into the vapor.

As ammonia flashes into the vapor phase, it leaves a concentrated hydrochloric acid solution behind. Hydrochloric acid attacks the metal surface, producing a metal chloride corrosion product. The strong acid will dissolve any metal sulfide or metal oxides that can be present on the metal surfaces of the equipment. This type of secondary attack on sulfide-oxide metal films is particularly aggressive at a pH below 5.0.

Other Corrosives

Some types of crude release small quantities of sulfur trioxide when they are heated. Sulfur trioxide is very hygroscopic and readily forms sulfuric acid. Sulfuric acid is a strong acid and can rapidly accelerate corrosion. In a crude unit, this acid is not usually found in high concentrations in the atmospheric tower, but high concentrations have been found on occasion in the vacuum tower overhead water. This is probably because the vacuum furnace operates at 70 to 100 °C (130 to 185 °F) higher than the atmospheric furnace. Control of corrosion from this acid is handled in a manner very similar to control over corrosion from HCl and ammonium chloride.

In some oilfield operations, carbon dioxide is injected into the wells to increase crude oil recovery and production. When carbon dioxide dissolves in water, it becomes carbonic acid. This acid is weak and can drive the pH of a system down to only ~4.5 at the temperatures present in the system. However, the addition of H_2S increases the aggressive nature of the weak acid. Neutralization and inhibition with a filming inhibitor provide acceptable control over the acid.

Ammonia can also be classified as a corrosive in systems that have been alloyed with copper-base metallurgies. This type of corrosion attack accelerates rapidly with increasing pH and is especially aggressive at pH ranges > 8.5.

Corrosion Control in a Crude Unit Overhead

Crude Oil Desalting

The first step to failure control on a crude unit is optimizing the desalter unit. The major corrosive coming from the desalting unit is magnesium chloride, which is hydrolyzed to form HCl in the desalter, preheat exchangers, and furnace. The desalter can also be a source of ammonia and other corrosives that can influence overhead corrosion control.

Chloride-containing salts in crude are found in three forms: sodium chloride (NaCl), calcium chloride ($CaCl_2$), and magnesium chloride ($MgCl_2$). The chloride distribution may be different for each crude type, but in general the distribution is NaCl, 75%; $CaCl_2$, 15%; and $MgCl_2$, 10%.

Magnesium chloride starts to hydrolyze at temperatures > 120 °C (250 °F). Calcium chloride similarly hydrolyzes to form hydrogen chloride. Suppose the desalted crude from a unit contains sufficient salt to generate a concentration of 115 ppm of chlorides in the condensed water in the overhead. If the system is sour and ammonia is used for pH control, the level of HCl found in the water that initially condenses could be as high as 1150 ppm. This level of hydrochloric acid could easily drive the pH of the water at dew point down to 1.5. No metallurgy could withstand this level of acidity at the temperatures present at dew point. Thus, HCl concentrations in the overhead water require the optimization of desalting.

A less apparent reason to optimize the desalting operation is the cost associated with poor crude dehydration or water separation from the crude. Poor water separation increases the quantity of water that must be condensed in the overhead system, and this costs the refinery extra money to vaporize and then cool the water. In addition, steam can occupy 7 to 10 times more volume than crude in the preheat. An increase in steam could increase pressures in the atmospheric tower, which could reduce the quantity of light ends that are separated from the flashed crude.

From the standpoint of corrosion control, increased quantities of water in the desalted crude increase the quantity of steam in the overhead. This raises the dewpoint temperature of water in the overhead. Consider a dewpoint of 107 °C (224 °F) in the overhead. Water carryover in the desalter can easily raise the dewpoint of the water to 125 °C (260 °F) in the overhead. If the tower top temperature is 116 °C (242 °F), a relatively dry, noncorrosive tower top can become a very wet tower top, which can result in an aggressive level of corrosion attack on the top tray in the tower. Corrosion in the tower top is aggressive dewpoint corrosion.

Ammonia contamination in the desalter stabilizes crude emulsions and can upset overhead corrosion control. Some refiners have injected sulfuric acid into the desalter with the washwater. This procedure consumes ammonia because ammonium sulfate is formed. Ammonium sulfate is extremely stable and will not sublime or dissociate at the temperatures present in the preheat. The problem with sulfuric acid is that the material can be carried over, causing sulfuric acid corrosion activity in the overhead. This type of attack can be extremely aggressive. The best pH for good desalting is between 5.0 and 6.5. If bases are introduced, emulsion stability in the desalter is increased.

The primary source of ammonia in the desalter is from the sour water strippers that process sour water from delayed cokers, thermal crackers, and fluid cracking units. These units produce large quantities of ammonia and H_2S. Phenolic compounds produced in the cracking operation are partially soluble in high-pH waters. Phenols are pollutants that are closely monitored in discharge water from the refinery. Because of the overload that phenols place on water treatment facilities, many plants use the desalter as a waste disposal unit for phenols. If the sour water stripper is operated to minimize both the H_2S content and the ammonia content in the stripped water, there is no problem with using this water in the desalter. Unfortunately, few refineries operate their strippers to optimize ammonia removal. When this situation is allowed to exist, corrosion control becomes increasingly difficult.

Caustic Injection

Good corrosion control in a crude unit overhead system is relatively easy to obtain if overhead water chloride concentrations are maintained in the range of 40 to 50 ppm, measured as NaCl. When chlorides exceed this level, control becomes increasingly difficult. At a chloride level > 100 ppm in the overhead water, corrosion control is very difficult. If the desalting unit has been optimized and overhead chlorides are above the target of 40 to 50 ppm, the use of caustic soda should be considered to reduce overhead chloride concentrations.

Sodium chloride is a stable salt that does not hydrolyze in appreciable quantities until a temperature of ~525 °C (980 °F) is reached. Sodium hydroxide converts the salts that hydrolyze easily into more stable NaCl:

$$MgCl_2 + 2NaOH \rightarrow Mg(OH)_2 + 2NaCl$$

$$CaCl_2 + 2NaOH \rightarrow Ca(OH)_2 + 2NaCl$$

$$HCl + 2NaOH \rightarrow H_2O + NaCl$$

$$FeCl_3 + 3NaOH \rightarrow Fe(OH)_3 + 3NaCl$$

Note that caustic neutralizes any acid in the crude and, being a strong base, also displaces weaker bases in the salts found in crude. It is almost impossible to stoichiometrically calculate the amount of caustic needed to reduce overhead concentrations to acceptable levels.

Disadvantages of sodium hydroxide injection into desalted crude include the following:

- The reaction products, hydroxides, are insoluble in crude and contribute to exchanger fowling and furnace coking in atmospheric and vacuum furnaces.
- Unreacted caustic can increase fouling and coking tendencies in the preheat system.
- On occasion, unreacted caustic has caused furnace tube stress cracking and embrittlement.
- Caustic can influence the foaming and emulsification characteristics of the crude.
- The strong base and the reaction products can cause foaming in the atmospheric and vacuum tower flash zones.
- Sodium contents in residual fuels can be increased.
- Sodium content in coke produced at the delayed coker increases.
- Sodium is a poison on equilibrium catalysts.
- Caustic can reduce the activity of many antifoulants.

An extremely conservative approach should be taken to injecting caustic into any preheat system.

Filming Inhibitors

Adequate control over the corrosive process is not possible without adequate neutralization of the acids in the system and proper use of a filming inhibitor. Inhibition and neutralization are important, because without the proper application of both, the corrosion process will continue. Corrosion is inhibited if the metal is separated from its environment by an inert and impervious barrier. The barrier, or film, should satisfy a number of requirements to function as a suitable inhibitor. For example, it should adhere firmly to metal surfaces; be continuous, dense, and nonporous; and be relatively non reactive.

Two types of inhibitor are used in process units, oil soluble and water soluble. The filming mechanism is slightly different for each type, although both inhibitors are polar compounds. The oil-soluble inhibitor uses the hydrocarbon in the process stream to provide the protective barrier. The inhibitor molecule is polar. One end of the molecule is called an oil-soluble tail. Its function is to dissolve in the hydrocarbon that is present in the stream. The chemical composition of this portion of the inhibitor determines in what hydrocarbon streams the inhibitor will be soluble. Many inhibitors are soluble in butane and heavier hydrocarbon. Extremely hydrocarbon-soluble inhibitors generally provide better filming inhibition.

At the opposite end of the filming inhibitor is the polar head, the portion of the inhibitor that is attracted to the metal surface. The exact mechanism of this attractive force is not completely understood. What probably occurs, however, is that the polar head forms a coordinate bond with the metal surface. The attraction is called absorption. The bond is pH-dependent, and high-pH conditions break the bond between the metal surface and the polar head. When such a condition exists, the polar head portion of the inhibitor can be preferentially attracted toward water, as opposed to the metal surface.

Measurement

Corrosion Coupons

If properly located, the coupon is a reasonably accurate tool. The primary purpose of measuring the corrosion activity on a unit is to provide an estimate of the condition of the equipment so that planned turnarounds can be scheduled and replacement of equipment anticipated. Frequent measuring also helps minimize the duration of corrosive upset conditions. It is well documented that short upset conditions can cause very serious corrosion activity in an overhead system.

In addition to producing corrosion data in measurable terms (mpy or mm/yr), the coupon is helpful in identifying the type of corrosion activity present at the point where the coupon is located in the system. Pitting on the coupon surface would indicate activity from a strong acid or acid salt, such as hydrochloric acid or ammonium chloride. General thinning indicates the activity of hydrogen sulfide on the coupon.

However, where a cooling medium is present on the other side of the metal surface, the skin temperature of the tube is lower than the temperature of the surrounding hydrocarbon. The cool tube skin could cause surface water condensation at temperatures above the anticipated water dew point in the hydrocarbon. Coupons would not indicate corrosion activity of this nature. This is important in units using cold crude as a heat exchange medium for hot overhead vapors.

Corrosometer Probe

If properly placed, the corrosometer probe is a fast, accurate method for determining relative corrosion activity in a system. The probe produces data in meaningful terms (mpy or mm/yr) for periods as short as one week. Upset conditions can be observed in even shorter periods. The major limitation of the probe lies in the placement. Also, the probe is not sensitive to cool tube skin condensation conditions; and it is not a reliable corrosion measuring tool in systems that are extremely sour, because the probe element measures corrosion product laydown as metal thickness. If sulfiding corrosion activity accelerates, the corrosometer probe tends to show increasing metal thickness rather than metal loss.

Sample Analysis

One of the more useful measuring methods available is sample analysis. Unlike the probe or coupon, the sample indicates corrosion activity through the entire overhead system at one point in time, or continuously, depending on the type of sampling (i.e., remote or online). Sample analysis is used to identify the level of corrosives in the system and the products produced.

The level of metal corrosion product found in overhead water from a process unit is a function of corrosion activity, filmer inhibitor emulsification tendencies, the pH of the water, and the H_2S concentration. Inhibitors that emulsify hydrocarbon badly tend to increase the level of metal corrosion products found in hydrocarbon streams. In most units, the principal corrosion product is a metal sulfide. Inhibitors that cause excessive emulsification tend to hold these corrosion products up in the hydrocarbon stream. Also, as the pH rises above 6.0, FeS can preferentially stay with the hydrocarbon phase, or it can concentrate at the interface between the water and hydrocarbon in the accumulator water leg. When this condition exists, sample analyses of overhead water alone are not very significant.

Samples of overhead hydrocarbon product should be run to supplement data generated from analyses of water samples. Typical sample analyses from an overhead system should include:

Sample	Analyses
Overhead water	pH, chlorides, total iron, ammonia, hydrogen sulfide
Overhead hydrocarbon	Total iron, inhibitor residual, water content

Sample analysis data are difficult to interpret and can be misleading, so sample data should be subjected to trend analysis. The major value of the sample is that it provides the ability to identify and correct upset conditions rapidly on the unit. This capability is extremely important and has ensured the success of many corrosion control programs.

Condition and Life Assessment

Condition and life assessment involve defining a critical crack size for a given component and measuring existing crack sizes by NDE or other inspection techniques. Life assessment requires an understanding of the factors affecting critical crack size as described in Chapter 3 and the methods for crack measurement. Under certain circumstances, any defect or flaw observable in a component by visual or other NDE methods constitutes grounds for retirement. Under other circumstances, NDE observations are combined with crack growth analysis to determine remaining life. The conservative approach is to replace components based on crack initiation. However, with increasing awareness of fracture mechanics considerations, it is possible to consider the crack tolerance of components (Chapters 2, 3, and 5).

Techniques that use crack initiation as a failure criterion include calculations based on history, extrapolations of failure statistics, strain measurements, accelerated mechanical testing, microstructural evaluations, oxide scale growth, hardness measurements, and advanced NDE techniques. For crack growth–based analysis, the NDE information, results from stress analysis, and crack growth data are integrated and evaluated with reference to a failure criterion.

Remaining-life determination is generally confined to equipment that is degraded by fatigue, creep, hot hydrogen attack, or combination of these mechanisms. Condition assessment is required, because it is fundamental to the continued safe and reliable operation of refinery or hydrocarbon processing equipment. For the most part, however, life assessment is based on estimates from nondestructive, deterministic, or phenomenological approaches. A recent review by Scasso (Ref 4) of the Italian Welding Institute focuses on the regulatory approach adopted by Europe and the United States. The findings are summarized in Tables 6-2 to 6-5.

The Finnish standard SFS 3280:E provides a credible deterministic approach (Ref 5). The scope of SFS 3280:E is to provide guidance in the monitoring of creep and the assessment of the degree of creep in tubes, pipe systems, and headers, that have been designed on the basis of creep strength, and in comparable steel constructions. The objectives of assessing the degree of creep are to establish the intervals between inspections and to estimate the residual service life of the constructions.

Critical Crack Sizes

The critical crack size (a_C) can be defined in a number of ways based on fracture toughness, ligament size, crack-growth-rate transitions, or other considerations as appropriate. A common definition for many heavy-section components is based on the fracture toughness of the material. Several alternative criteria can be used to establish a value for the critical crack size: a flow-stress-governed failure (PFM) criterion, a linear elastic fracture mechanics (LEFM) or elastic-plastic fracture mechanics (EPFM) controlled failure criterion, and a limiting creep-crack growth rate criterion.

The first two criteria are employed in a scenario in which rupture occurs during or immediately following a startup transient in the absence of creep. The third criterion is used in a scenario where failure occurs by creep-crack growth under operating conditions. The lowest value of critical crack size determined by use of these criteria is then used for remaining life analysis. The use of combined failure criteria to define the safe operating pressure for a pipe, for example, is illustrated in Fig. 6-6.

Stress level is not the only factor that affects the definition of a_C. Often a_C decreases with time due to embrittlement or aging. For example, severely embrittled bolts, vessels, and rotors can have a_C below the detection limits of conventional testing techniques. In other instances, a_C can be large but the rate of crack growth is so high that once a crack initiates, it reaches critical size rapidly. Many environmentally induced failures in highly stressed components exhibit this behavior. For instance, in generator retaining rings and steam turbine blades where crack growth under corrosive conditions is encountered, the presence of a pit or pit-like defect is cause for retirement.

With respect to material behavior, the major problem is the unavailability of data pertaining to crack growth and toughness in the service-degraded condition specific to the component, although this problem is addressed to a degree in Chapter 4. Although considerable data are available on materials in the virgin condition, the data bank on service-exposed materials is very small. Nondestructive methods are needed to determine those properties with specific reference to a given component.

In welded components, the problem is further compounded by the fact that a weldment contains a complex microstructure of many zones with varying material properties. Failure can occur through any of these zones or at the interfaces between them. In cases where crack growth rates might be rapid, conventional NDE techniques are often inadequate to detect the initial crack. The uncertainties in interpretation of NDE results can sometimes be overwhelming. Difficulties in distinguishing between innocuous and harmful flaws and identifying their orientations can lead to uncertainties in life assessment. If there are numerous indications closely spaced, the manner of treating them in terms of a linkup analysis could be crucial. Geometric discontinuities, such as fillets, section transitions, and weld backing rings, interfere with NDE signals and mask flaws. Guidance on the assessment of cracks and flaws is found in Chapters 3 and 5. Information on determining appropriate inspection intervals is given in Chapter 2.

Table 6-2 Regulations and standards on life assessment

Country	Regulation or standard
Denmark	Instructions for manufacturing and operation of high-pressure piping issued by the Danish Power Companies
Finland	SFS 3280:E (1984-10-22). Inspection of pressure vessels. Assessment of degree of creep
Germany	TRD 508 (1978-10). Additional tests on components calculated with time-dependent design strength values—inspection and testing. TRD 508, Annex 1. Method for the calculation of components having time-dependent design strength values. VdTuV MB 451 (1986-3). Investigation of the surface structure of building components subjected to creep rupture according to TRD 508
Italy	IEPESL (1992). Components of steam generators and components of vessels under steam or gas pressure operating in creep condition of materials—calculations and testing
The Netherlands	TO 1o2V (1985-2), Appendix 1. Rules for pressure vessels—periodic inspection: parts in the creep range
Portugal	ISQ-AVR-002. General procedure for the remanent life assessment of static components that have undergone high temperature
United Kingdom	BSI-PD 6510: 1982. A review of the present state of the art of assessing remanent life of pressure vessels and pressurized systems designed from high-temperature service
United States	API Recommended Practice 530, Appendix E (1988-89). Calculation of heater tube thickness in petroleum refineries—estimation of remaining tube life. ASME Code Case N-47-26, Appendix T (1986-2-23)(a). Class 1 components in elevated-temperature service (Section III, Division 1)—rules for strain, deformation, and fatigue limits at elevated temperatures

(a) The ASME Code Case N-47-26 is a design document and therefore is not aimed at assessing the remanent life; however, the creep fatigue damage can be calculated according to its rules.

Creep Life Assessment

Failure due to creep can be classified as resulting from either widespread bulk damage or localized damage. The structural components that are vulnerable to bulk damage (e.g. boiler tubes) are subjected to uniform loading and temperature distribution during service. If a sample of material is examined, it will truly represent the state of damage in the material surrounding it. The life of such a component is related to the creep-rupture properties. On the other hand, components subjected to stress (strain) and temperature gradients (typical of thick-section components) may not fail by bulk creep rupture. It is likely that at the end of the predicted creep-rupture life, a crack will develop at the critical location and propagate to cause failure. A similar situation exists where failure originates at a stress concentration or at pre-existing defects in the component. In this case, most of the life of the component is spent in crack propagation, and creep-rupture based criteria are of little value. Creep life assessment is discussed from the perspective of creep-rupture properties and creep crack growth. Practical methods based on replication and parametric approaches are also described.

Creep Rupture Life. Although it is relatively easy to quantify damage in laboratory creep tests conducted at constant temperature and stress (load), components in service hardly ever operate under constant conditions. Start-stop cycles, reduced power operation, thermal dents, and other factors result in variations in stresses and temperatures. Procedures are needed that will permit estimation of the cumulative damage under changing exposure conditions.

Damage Rules. The most common approach to calculation of cumulative creep damage is to compute the amount of life expended by using time or strain fractions as measures of damage. When the fractional damages add up to unity, then failure is postulated to occur. The most prominent rules are:

Table 6-3 Nondestructive evaluation requirements of regulations and standards given in Table 6-2

Country	Visual exam (a)	Dimensional check (b)	Magnetoscopic test	Penetrants	Replica exams	Ultrasound	Radiography	Magnetic permeability (c)
Denmark		X	X	X	(d)	X	X	
Finland	X	X	X	X	X	X		
Germany	X	X	X	X	X	X		
Italy	X	X	X	X	X	X	X	
The Netherlands (e)	X	X	X	X	X			
Portugal	X	X	X	X	X	X		X
United Kingdom	X	X	X	X	X	X		

Note: Not all the mentioned inspections are always demanded; on the contrary, other examinations different from those indicated, can be required according to the results of the previous one. (a) Supported by appropriate devices such as endoscopes. (b) Thickness measurements included. (c) In chemical plants, for tubes subject to carburization. (d) Metallurgical examination on parts cut out of the piping. (e) The minimum extent of nondestructive examinations is indicated in the document.

1. Life-fraction rule (LFR) (Ref 6):

$$\Sigma \frac{t_i}{t_{ri}} = 1$$

2. Strain-fraction rule (Ref 7):

$$\Sigma \frac{\varepsilon_I}{\varepsilon_{ri}} = 1$$

3. Mixed rule (Ref 8):

$$\Sigma \left(\frac{t_i}{t_{ri}}\right)^{1/2} \left(\frac{\varepsilon_I}{\varepsilon_{ri}}\right)^{1/2} = 1$$

4. Mixed rule (Ref 9):

$$k\Sigma \left(\frac{t_i}{t_{ri}}\right) + (1-k)\,\Sigma \left(\frac{\varepsilon_I}{\varepsilon_{ri}}\right) = 1$$

where k is a constant; t_i and ε_I are the time spent and strain accrued at condition i; and t_{ri} and ε_{ri} are the rupture life and rupture strain under the same conditions.

Table 6-4 Inspection frequencies of regulations and standards given in Table 6-2

Country	Timing of inspection: First check (a)	Timing of inspection: Subsequent checks
Denmark	Approx. 1 yr after commissioning of the plant	Every third year, counting from first measurement until registration of 1% creep. Then every year until registration of 2% creep. After that the pipe system must be scrapped or the Inspection Authority must be approached.
Finland	TLC <60% TTL ST <80% DTB PS expected >1%	According to inspection results but at least every 4 yr
Germany	TLC < 60% TTL TLC by fatigue >50% TTL (at least at the first periodic inspection after the mentioned limits have been attained)	According to inspection results; after the 100% TTL or 1% permanent set has been attained, the test intervals can be reduced.
Italy	TLC <60% TTL (b) ST <100% DTB (c) TLC ≥60% TTL (d)	According to inspection results, but if ST <100% DTB, at least every 50,000 h or 20% TTL, whichever is smaller and if ST >100% DTB, at least every 25,000 h
The Netherlands	TLC ≥60% TTL (e)	Planned schedule (if there is no reason not to use it)
Portugal	ST <80% DTB	According to inspection checks
United Kingdom	Not specified	Not specified

TLC, total life consumption; TTL, total theoretical life; ST, service time; DTB, design time base; PS, permanent strain. (a) The results should be compared with the results of the fabrication checks, if available. (b) Only for dimensional checks. (c) Only for visual inspections, dimensional checks, penetrant tests, magnetic tests, ultrasound, or radiographic tests. (d) Only for replica examinations. (e) At least at the first periodic inspection after 60% TTL has been attained

Example: LFR Calculation

The purpose of this example is to illustrate the use of the LFR. A piping system made of 1¼ Cr - ½ Mo steel designed for a hoop stress of 7 ksi was operated at 540 °C (1000 °F) for 42,500 h and at 550 °C (1025 °F) for the next 42,500 h. From the minimum curve of Larson-Miller parameter for the steel, it is found that, at σ = 48 MPa (7 ksi):

$$t_r \text{ at } 1000\ °\text{F} = 220{,}000 \text{ h}$$

$$t_r \text{ at } 1025\ °\text{F} = 82{,}380 \text{ h}$$

Life fraction expended, t/t_r, at 1000 ° = 42,500/220,000 = 0.19. Life fraction expended, t/t_r at 1025 °F = 42,500/82,380 = 0.516. The total life fraction expended is 0.71.

Validity of Damage Rules

Goldhoff and Woodford (Ref 10) studied the Robinson life-fraction rule and determined that for a Cr-Mo-V rotor steel, it worked well for small changes in stress and temperature. Goldhoff (Ref 11) assessed strain-hardening, life-fraction, and strain-fraction rules under unsteady conditions for this steel. While all studies gave similar results, the strain-fraction rule was found to be the most accurate.

From careful and critical examination of the available results, the following overall observations can be stated (Ref 12):

- Although several damage rules have been proposed, none has been demonstrated to have a clear-cut superiority over any of the others. The LFR is therefore the rule most commonly used.

Table 6-5 Rejection criteria of regulations and standards given in Table 6-2

Country	Rejection criteria
Denmark	Attainment of a permanent set of 2%
Finland	Not specified (according to the calculation and testing results)
Germany	Presence of nonrepairable cracks Consumption of total theoretical creep life and / or total theoretical fatigue life, unless safe, continued operation can be demonstrated Attainment of a permanent set of 2%, if reference measurement results were available from the beginning of service Attainment of a permanent set of 1%, if reference measurements were not available after 60% of the total theoretical life had been reached
Italy	According to the calculation and testing results Consumption of total theoretical life Attainment of a permanent set of 2%, if reference measurement results were available from the beginning of service Attainment of a permanent set of 1%, if reference measurement results were not available after 60% of the total theoretical life had been reached
The Netherlands	According to the calculation and testing results, particularly in the presence of nonre pairable cracks
Portugal	Not specified (according to the calculation and testing results)
United States	API recommended practice 530, Appendix E: not specified (according to the calculation and testing results) ASME Code Case N-47-26, Appendix T: not applicable

- The LFR is clearly not valid for stress-change experiments. Under service conditions where stress may be steadily increasing due to corrosion-related wastage (e.g. boiler tubes), application of the LFR will yield nonconservative life estimates; that is, the actual life will be less than the predicted life. On the other hand, residual-life predictions using postexposure tests at high stresses will yield unduly pessimistic and conservative results.
- The LFR is generally valid for variable temperature conditions as long as changing creep mechanisms and environmental interactions do not interfere with test results.

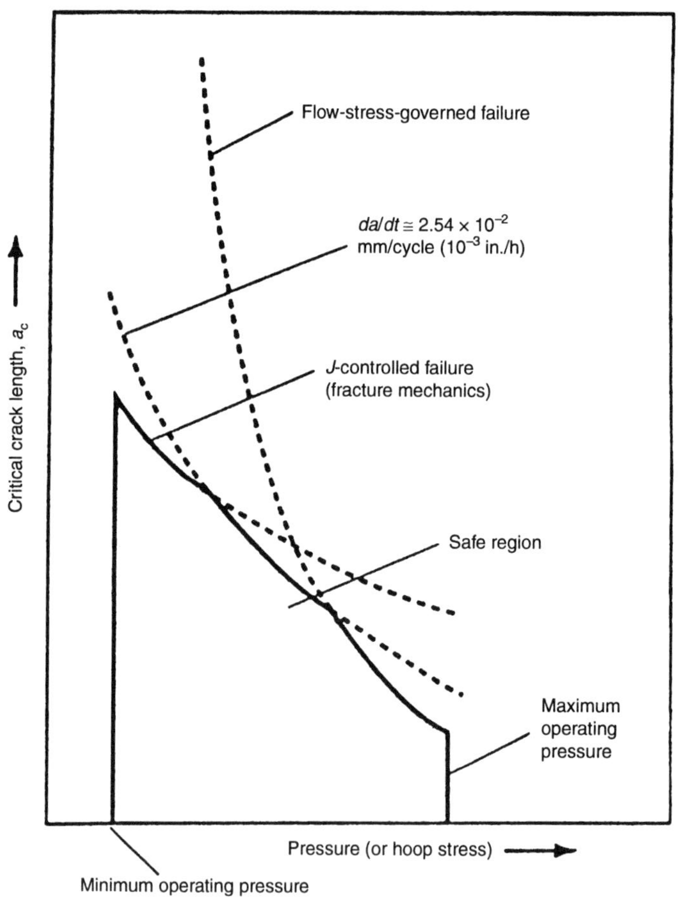

Fig. 6-6 Combined failure criteria for a pipe under internal pressure. Source: Ref 12

Hence, service life under fluctuating temperatures and residual life based on accelerated-temperature tests can be predicted with reasonable accuracy by use of the LFR.

- The possible effects of material ductility (if any) on the applicability of the LFR need to be investigated. A major limitation in applying the LFR is that the properties of the virgin material must be known or assumed. Postexposure tests using multiple specimens often can obviate the need for assuming any damage rule.

Creep Crack Growth

Gross and uniform creep deformation of components is usually the exception rather than the rule. Localized defects and stress concentrations often play decisive roles in failure. Under these circumstances, the growth of cracks and defects is governed by the creep ductility of the material. Extensive creep crack growth data pertaining to Cr-Mo piping steels have been collected, analyzed, and consolidated (Ref 13, 14). It has been observed that a crack-tip driving force parameter, C_t, that takes time dependent creep deformation into account correlates much better with crack growth rate, da/dt, than the traditionally used elastic stress intensity factor, K. The relationship between da/dt and C_t can be expressed as:

$$da/dt = bC^m$$

To perform a remaining-life assessment of a component under creep crack growth conditions, two principal ingredients are needed: an appropriate expression for relating the driving force C_t to the nominal stress, crack size, material constants, and geometry of the component being analyzed; and a correlation between this driving force and the crack growth rate in the material, which has been established on the basis of prior data or by laboratory testing of samples from the component. Once these two ingredients are available, they can be combined to derive the crack size as a function of time. The general methodology for setting inspection intervals using this approach has been described elsewhere (Ref 12, 15).

A number of variables have been identified as affecting the crack growth rate by modifying b, C_t, or m (Ref 16):

- In-service degradation increases da/dt by increasing C_t in the case of ductile materials and by increasing m and/or b for brittle material.
- Crack growth rates in welds, fusion lines, and HAZ materials are at least 5 times higher than those in base metal.
- The presence of localized chains of inclusion assisted by segregation of impurities to interfaces such as grain boundaries and fusion lines causes significant increases in creep-crack growth rates.
- The presence of large amounts of impurities in the steel accelerates crack growth by increasing m.
- All material and experimental variables that reduce creep ductility result in higher crack growth rates.

- Temperature can have mixed effects on crack growth. In cases where the effect of temperature is merely to increase creep rate, the *da/dt* increases with increase in temperature due to increase in C_t. On the other hand, if a transition from a brittle to a ductile condition is involved, increase in temperature can actually decrease the crack growth rates.
- Crack tip constraint has a pronounced effect on crack growth. Assumptions regarding plane-stress or plane-strain conditions can have a pronounced effect on *da/dt*.
- Inclusion of primary creep in addition to the secondary creep in calculating C_t results in a larger value of *da/dt* and reduced remaining life.

Practical Life Assessment in Creep Regime

Replication Methods

Surface replication is a well-known sample preparation technique that can be used to assess the condition of high-temperature power plant and petrochemical components from creep damage. The usual method of metallographic investigation involves cutting large pieces from components, which thus renders the component unfit for service. In contrast, surface replication allows examination of microstructural damage without cutting sections from the component.

Replication techniques are sufficiently sophisticated to allow classifications of microstructural damage (Table 6-6) that can be directly correlated to life fractions (Fig. 6-7). A distinct correlation exists for these data, such that a minimum and maximum remaining-life fraction can be specified as in Table 6-7, for example (Ref 17). For assessed consumed-life fraction, X, after exposure time, T_{exp}, the remaining life, T_{rem}, is:

$$T_{rem} = T_{exp}\,(1/X - 1)$$

The qualitative-quantitative relation is advantageous because data from surface replication are predictive in terms of generating a conservative minimum and a maximum life estimate. The maximum life is useful in a predictive maintenance and inspection environment, because it would dictate the planning of future repairs or replacements.

Table 6-6 Neubauer classification of creep damage

Damage level (Neubauer)	Description	Recommended action
1	Undamaged	No creep damage detected
2	Isolated	Observed
3	Oriented	Observe, fix inspection intervals
4	Microcracked	Limited service until repair
5	Microcracked	Immediate repair

Source: Ref 17

Parametric Methods

Extrapolation to service stress and temperature are performed using the appropriate extrapolation rule of the general form,

$$P(\sigma) = C(T, T_{rup})$$

where the stress (σ), temperature (T), and rupture life (T_{rup}) are related by various functional forms, P and C. This is similar to the Larson-Miller parametric relationship (Ref 18):

$$P = (T_d + 460)\,(C + \log L_d)\ \times 10^{-3}$$

where C is a constant, P is the Larson-Miller parameter, T_d is the design steel temperature, and L_d is the design life. By adopting the preceding approaches, either singularly or combined (depending on the availability of data), useful practical assessments of the remaining life of refinery or hydrocarbon processing equipment in the creep regime can be achieved.

Hydrogen Cracking in Wet H_2S

Cracking resistance of steels is a major concern in refining and petrochemical industries where aqueous H_2S is present. The generally accepted theory of the mecha-

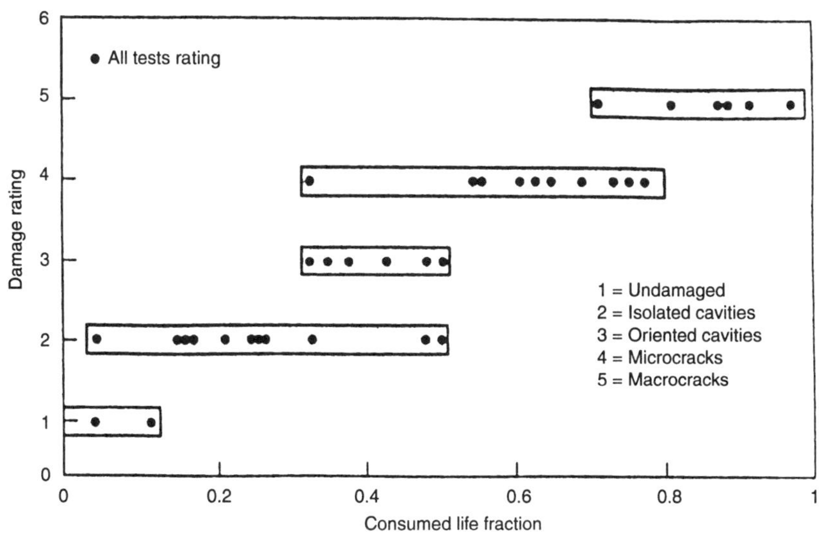

Fig. 6-7 Relation between Neubauer damage rating (Table 6-6) and consumed life fraction. Source: Ref 17

nism for hydrogen damage in wet H_2S environments is that monatomic hydrogen is charged into steel as a result of sulfide corrosion reactions that take place on the material surface. The primary source of atomic hydrogen available at internal surfaces of pipeline and vessel steels is generally the oxygen-accelerated dissociation of the H_2S gas molecule in the presence of water. The basic reaction is:

$$Fe + H_2S \rightarrow FeS + 2H$$

The FeS formed on the surface of the steel is readily permeated by atomic hydrogen, which diffuses further into the steel. This diffusion of atomic hydrogen into steel is associated with three distinct forms of cracking:

- Hydrogen-induced cracking (HIC)
- Stress-oriented hydrogen-induced cracking (SOHIC)
- Sulfide stress cracking or sulfide stress-corrosion cracking (SSC)

HIC and SOHIC are both caused by the formation of hydrogen gas (H_2) blisters in steel. HIC, also called stepwise cracking or blister cracking, is found primarily in lower-strength steels, typically with tensile strengths <550 MPa (80 ksi). It is found primarily in line pipe steels.

In contrast, SSC does not involve blister formation, but it does involve cracking from the simultaneous presence of high stress and hydrogen embrittlement of the steel. SSC occurs in higher-strength steels or at localized hard spots associated with welds or steel treatment. As a general rule, SSC can be expected to occur in process streams containing in excess of 50 ppm H_2S (although cracking has been found to occur at lower concentrations).

The basic factors of these cracking modes include temperature, pH, pressure, chemical species and their concentration, steel composition and condition, and welding or the condition of the weld HAZ. For a detailed review of HIC, SOHIC, SSC, and other forms of hydrogen attack in line pipe and pressure vessel steels in aqueous H_2S environments, the reader is referred to Ref 19.

Table 6-7 Correlation of damage level and life fraction consumed

Damage level	Consumed life fraction Range X	Remaining life factor ($1/X-1$) Minimum	Maximum
1	0.00–0.12	7.33	Unknown
2	0.04–0.46	1.17	24.00
3	0.3–0.5	1.0	2.33
4	0.3–0.84	0.19	2.33
5	0.72–1.00	0 (failed)	0.39

Source: Ref 17

REFERENCES

1. D.H. Stoneburg et al., Ed., Practical Considerations for Engineering Evaluation and Analysis Applied to Process Equipment, *Conf. Proc. Improving Reliability in Petroleum Refineries and Chemical and Natural Gas Plants,* I. Chem. E., 1993
2. "National Board Inspection Code," ANSI NB-23, The National Board of Boiler and Pressure Vessel Inspectors
3. "Pressure Vessel Inspection Code," API 510 *Maintenance Inspection, Rating, Repair and Alteration,* American Petroleum Institute, 1995
4. M. Scasso, "Comparison Among the Present Regulations Concerning the Remanent Life Assessment of Components Designed for High Temperature Service," Italian Institute of Welding, 1993
5. "Inspection of Pressure Vessels: Assessment of Degree of Creep," SFS 3280:E, Finnish Pressure Vessel Commission, 1986
6. E.L. Robinson, Effect of Temperature Variation on the Creep Strength of Steels, *Trans. ASME,* Vol 160, 1938, p 253–259
7. Y. Lieberman, Relaxation, Tensile Strength and Failure of E1 512 and Khl F-L Steels, *Metalloved. Term. Obrab. Met.,* Vol 4, 1962, p 6–13
8. H.R. Voorhees and F.W. Freeman, "Notch Sensitivity of Aircraft Structural and Engine Alloys," Wright Air Development Center Technical Report, Part II, Jan 1959, p 23
9. M.M. Abo El Ata and I. Finnie, "A Study of Creep Damage Rules," ASME Paper 71-WA/Met-l, ASME, Dec 1971
10. R.M. Goldhoffand, D.A. Woodford, "The Evaluation of Creep Damage in a Cr-Mo-V Steel," STP 515, American Society for Testing and Materials, 1982, p 89
11. R.M. Goldhoff, Stress Concentration and Size Effects in a Cr-Mo-V Steel at Elevated Temperatures, *Joint Int. Conf. on Creep,* Institute of Mechanical Engineers, London, 1963
12. R. Viswanathan, *Damage Mechanisms and Life Assessment of High Temperature Components,* ASM International, 1989, p 73, 87–103
13. A. Saxena, J. Han, and K. Banergi, "Creep Crack Growth in Boiler and Steam Pipe Steels," Report CS-5583, Electric Power Research Institute, Jan 1988
14. A. Saxena, "Creep Crack Growth in Cr-Mo-V Rotor Steels," Report RP2481-5, Electric Power Research Institute
15. P.K. Liaw and A. Saxena, "Remaining Life Estimation of Boiler Pressure Parts—Crack Growth Studies", Report CS4688, Electric Power Research Institute, July 1986
16. R. Viswanathan and S. Gehl, Creep Life Assessment Techniques for Piping, *Mechanical Behavior of Materials,* VI, Vol 2, ASME 1983, p 117–122
17. B. Neubauer, *Creep and Fracture of Engineering Materials and Structures,* Pineridge, U.K., 1981, p 617
18. F.R. Larson and J.Miller, *Trans. ASME,* Vol 74, 1952, p 765
19. P.F.Timmins, *Solutions to Hydrogen Attack in Steels,* ASM International, 1997

CHAPTER 7

Summation and Direction

It has been established that in the aviation sector, failure rates are decreasing, whereas in hydrocarbon processing, failure rates are increasing. In both sectors, massive capital investment equipment failures can and have resulted in the loss of human life as well as substantial dollar loss. In terms of specific mechanisms involved in aviation failures, fatigue (~60%) and corrosion dominate the failure mechanisms. With causes of failure, it appears in one survey that poor maintenance (~44%), design conditions (~26%), and manufacturing defects (~24%) dominate. In hydrocarbon processing failures, there appears to be less dominance of fatigue (~40%), with corrosion (~40%) and static fracture (~20%) having more dominant effects. Causes of failure are apparently attributable to design (~60%), manufacture (~20%), and maintenance and inspection (~20%).

Predicting possible causes of failure at the design stage or at a proposed change of design has been addressed by the introduction of a team approach to identify, evaluate, and recommend ways to control hazards associated with the design, operation, and maintenance of process facilities. In the two catastrophic situations considered, it was illustrated that relatively simple approaches to the identification of hazards could possibly have averted the catastrophes. These tools have the common appeal of assessing what can go wrong and estimating some level of probability of occurrence. Their usefulness can be extended if some attempt is also made at assessing risk.

The aftermath of the Flixborough catastrophe led the Court of Inquiry to indicate that in making a change to a part of the plant that required careful consideration, on-site personnel were being asked to do work outside those areas in which they were trained and competent. This led the Court of Inquiry to further recommendations.

- The training of engineers should be broadly based. Although it may well be that the occasion to use such knowledge will not arise in an acute form until an engineer has to take executive responsibilities, it is impossible at the training stage to know who will achieve such a position. All engineers should, therefore, learn at least the elements of branches of engineering other than their own in both academic and practical training.

- It is also essential that:
  - Persons given certain responsibilities are competent to carry out those responsibilities.
  - Top management has a clear knowledge and understanding of individuals and the magnitude and type of demand made upon them.
  - Top management has a clear knowledge and understanding of the total workload placed on each individual in relation to his capacity. Even good and competent individuals have increased potential for errors of judgment when overworked. Also, in times of crisis and extreme demand, it is easy to overwork the "willing horses," some of whom may not know their own limitations.

Examinations of those factors which contributed to the observed decrease in failure rate in aviation led to the development of a methodology for solving equipment failure problems that can be applied to other sectors of industry. The methodology included the following factors:

- The adoption of the fracture mechanics–based defect tolerant design
- The concept of principal structural elements (PSE)
- The development of alloys for corrosion and fatigue resistance
- The assumptions made in the retardation of multiple site damage (MSD)
- The control of manufacturing procedures
- The need for careful maintenance and procedures
- The requirement of a corrosion prevention and control plan
- The need for detailed inspection procedures and precise inspection Intervals

The derived methodology outlined would have prevented a number of the aviation and hydrocarbon processing service failures presented in this work as typical examples. It has been demonstrated that fracture mechanics can be used to:

- Determine the maximum safe operating loads that an engineering structure can sustain for the sizes and locations of existing flaws. Such cracks might be those actually found during an inspection, whereupon the continued safe operation of the structure is in question.
- Determine for given loads the largest crack size that can exist without fracture, in order to provide specifications to be set in advance of an inspection.

Scenarios in corrosion design that can act as guidelines to facilitate predicting possible causes of failure at the design stage or at a proposed change of design have been addressed by the introduction of a team approach to identify, evaluate, and recommend ways to control hazards associated with the design or design change.

Furthermore, in recent years there has been an additional requirement in certain sectors of industry that the corrosion lifetime of various components be predicted and that the technical (and economic) benefits of life extension actions be quantitatively defined. Such requirements for life extension are seen in the aviation, nuclear power, gas processing, and refinery sectors of industry. To meet these particular design criteria, verifiable life prediction methods have been developed. One such approach has been described which attempts to unify life prediction of SCC, SIC, and CF.

Almost all welded joints contain stress concentrators other than welding flaws, such that depending on their severity, the incidence of these flaws may or may not be of consequence. In practice, the majority of cases of fatigue cracking in service are associated with the inherent stress concentration features of the basic welded joint geometry and only a relatively small number with welding flaws.

A rational basis for the assessment of imperfections is engineering critical assessment (ECA) based on fitness for purpose, whereby a flaw could be accepted, provided it could be shown that its presence would not reduce the fatigue strength of the welded joint below that required. The avoidance of unnecessary repair of welding flaws on the basis of ECA and consequent savings in fabrication costs and delays to projects has saved millions of dollars in the offshore and power generation industries. The ECA procedures used in the British Standards Institute Public Document 6493 have been reviewed to provide solutions to equipment failures at welded joints.

In addition, methods of improving resistance to fatigue in welded joints have been considered, together with a commentary on the nature of flaws and the ease by which weldments in steels can become embrittled by hydrogen. Solutions to equipment failures in hydrocarbon processing, including piping and process systems, encompass the inspection (monitoring), maintenance, repair, life assessment, and life extension of the various engineering components used.

The significance of equipment failure can be ranked as a function of failure rate and the cost of failure. The types of process equipment most important in controlling failure rate or loss, in decreasing order of importance, appear to be piping, reactors, tanks, and process towers. This ranking has suggested the priority areas for inspection, maintenance, repair, life assessment, and life extension. Furthermore, when inspection and maintenance are organized and performed in a predictive mode, then costs are reduced and equipment failure rates decrease. The profitability of the hydrocarbon processing operation therefore increases.

This focus on predictive maintenance and inspection, with particular emphasis on the control and prediction of equipment damage for life extension and failure rate reduction, has been demonstrated to be effective in a California refinery, which draws from the principles outlined, to run lean and profitable. This proven methodology, derived from the aviation sector, can be applied to gas processing facilities (both on-shore and offshore), power generating facilities, pulp and paper production, and the petrochemical and chemical sectors.

Bibliography

Books

- D. Broek, *Fracture Control for the Process Industries,* MTI Manual No. 8, 1983
- D. Broek, *Elementary Engineering Fracture Mechanics,* Kluwer Academic Publishers, Dordrecht, 1991
- L. Cartz, *Nondestructive Testing,* ASM International, 1995
- G.P. Cherepanov, *Mechanics of Brittle Fracture,* McGraw-Hill, 1979
- K.A. Eskalul, Ed., *Handbook of Case Histories in Failure Analysis,* Vol 1, ASM International, 1992
- K.A. Eskalul, Ed., *Handbook of Case Histories in Failure Analysis,* Vol 2, ASM International, 1993
- N.E. Frost, J.J. Marsh, and L.P. Pook, *Metal Fatigue,* Oxford University Press, 1974
- R.W. Hertzberg, *Deformation and Fracture of Engineering Structures,* Wiley, 1976
- M.F. Kanninen and C.H. Popelar, *Advanced Fracture Mechanics,* Oxford University Press, 1985
- J.F. Knott, *Fundamentals of Fracture Mechanics,* Butterworths, 1973
- L.J. Korb and D.L. Olson, Ed., *Corrosion,* Vol 13, *ASM Handbook,* ASM International, 1987
- S. Lampman, Ed., *Fatigue and Fracture,* Vol 19, *ASM Handbook,* ASM International, 1996
- H.K. Leibowitz, Ed., *Fracture—An Advanced Treatise,* Academic, 1968
- G.E. Linnert, *Welding Metallurgy: Carbon and Alloy Steels,* 4th ed., Vol 1 *Fundamentals,* American Welding Society, 1994
- S.T. Rolfe, and J.M. Barsom, *Fracture and Fatigue Control in Structures,* Prentice-Hall, 1977
- H. Tada, P.C. Paris, and G.R. Irwin, *The Stress Analysis of Cracks Handbook,* Del Research Corporation, 1973
- P.F. Timmins, *Solutions to Hydrogen Attack in Steels,* ASM International, 1997

CD-ROM

- *Heat Treating Library on CD-ROM,* ASM International, 1995

Standards and Specifications

- American Petroleum Institute, API RP 941, current edition, British Standards Institute, Public Document 6493, 1997, NACE International, NCE Recommended Practices

Index

Note: Page numbers followed by f indicate figures; page numbers followed by t indicate tables.

D

G

H

T

U

W